C0-AUE-872

The Evolution of Social Wasps

JAMES H. HUNT

THE EVOLUTION OF SOCIAL WASPS

UNIVERSITY PRESS

2007

Oxford University Press, Inc., publishes works that further
Oxford University's objective of excellence
in research, scholarship, and education.

Oxford New York
Auckland Cape Town Dar es Salaam Hong Kong Karachi
Kuala Lumpur Madrid Melbourne Mexico City Nairobi
New Delhi Shanghai Taipei Toronto

With offices in
Argentina Austria Brazil Chile Czech Republic France Greece
Guatemala Hungary Italy Japan Poland Portugal Singapore
South Korea Switzerland Thailand Turkey Ukraine Vietnam

Published by Oxford University Press, Inc.
198 Madison Avenue, New York, New York 10016

www.oup.com

Library of Congress Cataloging-in-Publication Data
Hunt, James H.
The evolution of social wasps / James H. Hunt.
p. cm.
Includes bibliographical references and index.
ISBN: 978-0-19-530785-6
ISBN: 978-0-19-530797-9 (pbk.)
1. Vespidae—Behavior. 2. Social behavior in animals. I. Title.
QL568.V5H775 2007
595.79'8156—dc22 2006015474

9 8 7 6 5 4 3 2 1

Printed in the United States of America
on acid-free paper

For Noah, Tyler, and Jesse

Foreword

It is not surprising that insect societies have long intrigued and fascinated people. Their colonial life with workers who help to rear the offspring of queens, caste system, division of labor, and communication—not to mention cooperation, altruism, and suicide—have unmistakable parallels with the achievements and predicaments of our own social lives. But evolutionary biologists have an important additional reason to be intrigued and fascinated by insect societies. The origin and persistence of the sterile worker caste is a major evolutionary puzzle, in many ways more puzzling than the evolution of the honey bee queen, capable of laying thousands of eggs per day, or the honey bee worker, capable of building perfectly hexagonal cells in a comb of wax during the first half of her life and estimating and communicating the distance and direction to newly discovered food sources by means of a symbolic dance language during the second half of her life. It is also more puzzling than the invention of agriculture through the cultivation of fungus gardens by leaf cutter ants 50 million years ago or the colonial organization of African army ants with more than 20 million workers weighing a total of more than 20 kg per colony. The reason for this is that sterility and self-sacrifice are the last things we expect natural selection to promote. Not surprisingly, the evolution of a worker caste is the quintessential problem of the evolution of sociality.

This problem can be approached in many different ways. At one extreme it can be approached as an abstract problem of how non-reproducing or more slowly reproducing units (e.g., cells, animals, or even robots) can increase in population relative to fertile or faster-reproducing entities. This approach was made possible due to W. D. Hamilton's inclusive fitness theory and requires mathematical skills and a deep understanding of population genetic principles.

It does not necessarily require familiarity with the biology of any particular group of social insects. (That W. D. Hamilton was a naturalist *par excellence* is another matter, however!) At the other extreme it can be approached by inquiring into the specific circumstances and historical sequence of events in the evolution of the worker caste in a particular evolutionary lineage. Although researchers have attempted to occupy all possible niches along this continuum, niches relatively close to the former extreme are better populated than those closer to the latter extreme. The closest anyone has gotten to the latter extreme is the author of the present book. Hunt's chosen group is the lineage of social wasps belonging to the family Vespidae, and what a wonderful choice it is.

If paleontologists trace evolutionary trajectories by holding a magnifying glass to rocks through real time (albeit in millions of years), Hunt's brand of evolutionary biology entails a microscopic examination of every aspect of the biology of all salient extant taxa. This is much harder work for at least two reasons. First, one has to cut-across disciplinary and taxonomic boundaries and understand and synthesize vast amounts of information about every taxon. Second, there is no help from carbon dating, so one has to do considerable detective work to sequence and date evolutionary events. Our knowledge of social wasps, let alone of other social insects, is extraordinarily sparse relative to what is needed for a satisfactory practice of this enterprise. Where information is inadequate, Hunt speculates, and where he feels that the current wisdom is at apparent loggerheads with known facts, he proposes radically new interpretations.

Perhaps Hunt's two most controversial points concern identification of the extant wasp taxon that can be considered to most closely resemble the solitary ancestor of social wasps and the number of times sociality has arisen independently in the family Vespidae. Here and elsewhere, Hunt's speculations and interpretations are bound to provoke, even anger, custodians of current wisdom, but because he provides powerful arguments for why current knowledge is inadequate to settle these questions, they are also sure to lead to more investigations and better understanding of wasp biology and phylogeny. But the single most important point of Hunt's synthesis is to argue that the life cycle of a social wasp such as *Polistes* is based on the underlying reproductive ground plan of a partially bivoltine solitary wasp. This argument, if true, suggests the need for altogether new lines of investigation at the morphological, physiological, and genomic levels and thus brings the hitherto esoteric field of social evolution in wasps to the doorsteps of diverse disciplines of modern biology. This is no mean achievement indeed.

Hunt offers us much more than a new perspective on and a new synthesis of our current understanding of the evolution of social wasps. He ends the book with a scathing attack on kin selection; inclusive fitness theory; sex ratio theory; behavioral ecology; the use of potentially loaded terms such as selfishness, altruism, worker policing, and even eusociality; and he questions the very validity of asking "why" questions in evolutionary biology. I expect that while, on the one hand, this will attract much attention among historians and philosophers of science, it will also make most members of the criticized camps at first see red, then get defensive, and finally do much soul searching. Having taught behavioral ecology for more than 20 years, made inclusive fitness theory the center-stage of my life's research, and written one book on "why" questions in animal behavior and another book on the evolution of eusociality, I cannot now deny that I belong to the camp that Hunt so penetratingly strikes. All I can say after reading the book is that I could not have asked for a more competent "adversary!"

Raghavendra Gadagkar
Centre for Ecological Sciences,
Indian Institute of Science, Bangalore
National Academy of Sciences, USA

Preface

When a desert is ablaze with springtime flowers, snow-capped mountains are tinted with alpenglow, or multicolored birds punctuate the greenness of a tropical rainforest, one's attention is not normally drawn to the ants at one's feet nor the wasp alighting on a nearby twig—unless, that is, one happens to be an entomologist, and it is these objects of study that have brought one to these vistas. Such has been my good fortune. Memories of many such vistas linger, some of the earliest preserved on grainy Ekatchrome, but even memories so preserved recede into shadows cast by light of the quest that fostered these travels. For 30 years I have sought the origin of insect sociality, using paper wasps as my model system. The book now before you is not a travelogue, but if you read it to the end I hope you will reflect on it as such. Meanderings both geographic and disciplinary have brought me to a destination envisioned only vaguely at the outset. I knew that a destination was out there, and I had imaginings of what it would be, but details of the journey and its outcome have often surprised and delighted me.

Many have embarked on this journey before me, and I have benefited from the trails—no, superhighways—that they constructed. Social insects are truly one of the wonders of the natural world, so it is no surprise that the origins of their sociality have stimulated deep inquiry by many and lifelong investigations by some. Although no listing can be comprehensive, notables among the students of insect social evolution include Darwin, Wheeler, Roubaud, Evans, Michener, Wilson, Hamilton, West-Eberhard, Eickwort, Crozier, Gadagkar,—and the list could go on for some time. None of these, however, has reached quite the same destination I describe in this book. If, therefore, you read the book to its end and contemplate it as travelogue, you may decide whether the destination I have reached is a shining citadel or a vermin-infested backwater. From colleagues, I already have sampled both opinions.

Charles Darwin famously raised issues of both origin and elaboration of insect sociality in a passage, which is often misinterpreted, describing "one special difficulty, which at first appeared to me insuperable, and actually fatal to my whole theory [of natural selection]" (Darwin 1859, p. 236). In the decades after Darwin, however, those who pursued the evolution of sociality did so as naturalists rather than as natural selection theorists. Some work was scholarly and insightful, with contributions by Emile Roubaud, whose photograph hangs above my desk, exemplifying the best work of the times. The first Darwinian century closed with the Modern Synthesis (Huxley 1964), and soon thereafter the real fun began.

The dividing line between initial and recent approaches to the problem of insect social evolution was 1964, when William D. Hamilton published his now-famous formalization of inclusive fitness (Hamilton 1964a) and its application in understanding a range of situations in the biology of social taxa, including the evolution of social insects (Hamilton 1964b). It took about a decade for Hamilton's ideas to rise to prominence. Contests of ideas such as mutualism (Lin and Michener 1972) and parental manipulation (Alexander 1974), as well as kin selection (West-Eberhard 1975), intermingled with debate on topics such as subsocial versus semisocial routes to sociality. Studies of sex ratio (Trivers and Hare 1976) and relatedness (see Gadagkar 1991a) established research domains that remain active today. By the late 1970s kin selection had become the prime motivator of the burgeoning field of behavioral ecology, while Hamilton's three-quarter relatedness hypothesis (Hamilton 1964b), often called the haplodiploidy hypothesis (West-Eberhard 1975; Freeman and Herron 2003), stimulated an explosion of research on the evolution of insect sociality.

An explosion of a different sort was occasioned by "The spandrels of San Marco" (Gould and Lewontin 1979), which exposed (others might say fostered) a rift among evolutionary biologists. Once the dust had settled a bit, Antonovics (1987) clarified the existence and character of the dichotomy between those evolutionary biologists who infer past events and those who study present-day processes. West-Eberhard (1988, p. 123) brought social insects into the discussion by noting that the preceding two decades of research had been marked by confusion between "two different groups of evolutionary biologists . . . observers of social insects (naturalists) and makers of mathematical models (geneticists)." She argued that each group had a different interpretation of kin selection: naturalists think in terms of facultatively expressed phenotypes, whereas geneticists think about a phenotype underlain by a particular allele. Although these camps of social insect biologists may not map precisely onto Antonovics' dichotomy, the concordance is high. West-Eberhard's "naturalists" are the intellectual descendents of the post-Darwin, pre-Synthesis students of social evolution, whereas the "geneticists" have primarily been inspired by W. D. Hamilton. In a similar vein, Reeve and Sherman (1993) distinguished questions of evolutionary history, as pursued by paleobiologists and systematists, from questions of phenotype existence, as pursued by behavioral and evolutionary ecologists.

Although the partitioning of natural historical and theoretical approaches may be the norm in much evolutionary research and publication, it was the intimate interweaving of these that made Darwin's expositions so compelling. Can the evolution of sociality be a platform for integration across realms of current inquiry? I believe it can if one delineates one's question with care. A multi-taxon question that seeks answers equally applicable to ants and aphids, bees and birds, wasps and whales, by its very nature excludes historical analysis from inclusion in the answer. A single-taxon question, on the other hand, risks lack of general applicability, leading to a narrow readership of only taxonomic specialists. But if synthesis of historical and mechanistic approaches is a worthwhile goal (Autumn et al. 2002), then a single-taxon approach seems necessary. The major pitfall to be avoided in a single-taxon approach is not narrow taxonomic focus but narrow-minded focus on one type of investigation or one line of reasoning. An open mind, although not an uncritical one, is essential to synthesis.

In this volume I ask, how did social behavior evolve and how is it maintained in wasps of the family Vespidae? This question combines taxonomic focus with freedom from conceptual constraint. The question as framed does presume a temporal sequence—that history preceded maintenance. The question does not, however, presuppose primacy for either a historical or procedural approach in understanding social insect evolution. Instead, it necessitates a melding of both.

Before beginning, I must address the question, why wasps? A first-order answer is that these are the insects I know. I began to study them the first weekend of my assistant professorship, when, less than a mile from where Lewis and Clark had passed on the first day of their epic journey of exploration, I cautiously (but not cautiously enough) approached my first nest of *Polistes metricus*—and was promptly stung. I had envisioned a route to knowing insect sociality only 4 months before, and that was the first step in my 30-year journey. Over time I acquired an inordinate fondness for wasps. They fascinate me, and I wish that many more curious naturalists were similarly entranced.

A second-order answer to the question, why wasps?, is that research on wasps—some of my own and much more by others—has yielded insights into components of insect sociality that can be applied to other taxa and that address basic topics of evolutionary biology. Review of the relevant literature can synthesize this information for specialists, and it can paint a rather complete picture of the organisms and our understanding of them for nonspecialists. Wasps, notably *Polistes*, constitute a model system (Dugatkin 2001) that has been, and can continue to be, used to enlighten fundamental issues in evolutionary biology that have broad interest and general applicability.

I travel now with sunset hues starting to tinge the horizon. Some will read this volume in the light of early dawn. If you are one of these and are embarking on your own journey, I commend to you the study of social insects. I cannot say what your journey will be like, but I can assure you that you will enjoy it—especially if you study wasps.

Acknowledgments

Writing a first book when one's beard has turned gray instills thoughts that this might be the best opportunity to look back and acknowledge some people who played key roles—sometimes briefly but nonetheless importantly—in the early course of one's career. My parents, Nile and Lucile Hunt, were perplexed by their son's enchantment with everything that wriggled, flew, swam, or crawled, but they also encouraged and supported my eccentric pursuits. Thomas L. Quay showed me that there could be a career in academic zoology, and he gave me a job, an office, and a key to the building when I was still a freshman. James F. Parnell exposed me to the sheer pleasure of doing field research, as did William M. Palmer, who also instilled in me a lifelong deportment of respect for the animals, whatever they may be, that are the objects of one's study. Alistair M. Stuart taught my most rigorous undergraduate course, in animal behavior, in an exemplary way. Bernard S. Martof encouraged me, and assured my parents, that my pursuit of zoological knowledge would not be misspent. George H. Lowery, Jr., sent me to the tropics. Gordon H. Orians, Daniel H. Janzen, and my fellow students in an Organization for Tropical Studies course lastingly shaped my attitude toward scholarship. Frank A. Pitelka admitted me to Berkeley, and Robert K. Colwell saw me through. Roy R. Snelling mentored me in formicology. Harold A. Mooney sent me to Chile. Otto T. Solbrig brought me to Harvard, where Edward O. Wilson gave me a home and, at a moment when the course of my research interests changed forever, some key advice. Martin Sage ensured that the University of Missouri-St. Louis retained me at a critical early rung on the academic ladder. All of these helped shape my career, and I am grateful. Mary Penney taught me English composition, and perhaps readers of this volume will share a small fraction of the gratitude I have for her.

Donald E. Grogan was an early collaborator at the University of Missouri-St. Louis. Collaborators and principal colleagues at other institutions have included Robert L. Jeanne, John W. Wenzel, Christine A. Nalepa, Sean O'Donnell, James M. Carpenter, Diana E. Wheeler, Malcolm G. Keeping, Christopher K. Starr, Stefano Turillazzi, Mary Jane West-Eberhard, and Gro V. Amdam. Master of Science students who completed theses on wasps in my lab and thereby contributed to my knowledge and understanding are Barry M. Kayes, Pamela S. Mitchell, Bernice B. DeMarco, Carol S. Ferguson, David A. Landes, Anthony M. Rossi, Margaret A. Dove, and Jon N. Seal. Particularly important undergraduate research assistants were Margaret McCarthy, Karen Sago, and Margaret Williams. There have been many more colleagues, students, and friends, and I thank them all.

This book was written during a fellowship year at the Wissenschaftskolleg zu Berlin. Raghavendra Gadagkar was responsible for my good fortune in becoming a fellow, and I am deeply grateful. The University of Missouri-St. Louis continued to support me while away. The staff of the Wissenschaftskolleg enabled me to accomplish the task I had set for myself. The Kolleg's library services, in particular, are nonpareil. Among the fellows, John Rieser was a stalwart of camaraderie; Christof Rapp answered my questions about classical languages; Stefan Litwin inspired me to try to be as good at my craft as he is at his; Cosima Rughiniş helped me see the rest of my life in changed perspective; and David Poeppel bought cheap schnapps. All of the Kolleg's staff and fellows were marvelous good company.

Chapter 4 was discussed during a seminar held at the Wissenschaftskolleg zu Berlin. I thank the Otto und Martha Fischbeck Stiftung for funding the seminar, and I thank the seminar participants for their comments. Conversations with Robert E. Page, Jr., and especially with Gro V. Amdam, profoundly affected chapter 8. Comments and questions from colleagues during a colloquium I presented at the Wissenschaftskolleg helped shape chapters 9 and 10. Sandra D. Mitchell issued challenges that changed the objectives of chapter 11.

Selected portions of the early manuscript were reviewed by Jeremy Field, Sandra D. Mitchell, Stefano Turillazzi, and James B. Whitfield. Longer portions were reviewed by Gro V. Amdam and Sara Shettleworth. Raghavendra Gadagkar reviewed the full first draft. Peter J. Prescott of Oxford University Press oversaw review and acceptance of the submitted manuscript. I thank three anonymous reviewers for their opinions that led to its acceptance and for suggestions that led to its improvement. Joan M. Herbers then graciously agreed to read the full manuscript, and the text was further improved by her insights and advice. By virtue of keen eyes and thoughtful comments, members of a graduate seminar in entomology at the University of Illinois at Urbana-Champaign left their mark on the penultimate draft. Admonitions from Sydney A. Cameron and especially from Raghavendra Gadagkar and Peter J. Prescott gave impetus to the final substantive revisions, which were executed during a leave granted by the University of Missouri-St. Louis. Gene E. Robinson generously provided an academic home away from home where those revisions could be executed, and Amy L. Toth and Moushumi Sen Sarma insightfully and constructively reviewed them. I thank all of these learned colleagues for their assistance, comments, suggestions, and admonitions, and I absolve them of any taint by association.

Noah, Tyler, and Jesse Hunt, my now-grown children who accommodate my unabated eccentricities with boundless good humor, are my touchstones for what is important in life.

Contents

The Evolution of Social Wasps

Introduction

Order Hymenoptera is a grand and captivating taxon. Within it are ants and bees, sawflies and horntails, a staggering array of species called parasitoids because of their mode of larval feeding, and a splendid profusion of wasps (Gauld and Bolton 1988). Imbedded in this diversity is an astonishing phenomenon: social behavior has evolved more times independently in Hymenoptera than in any other insect order (Wheeler 1928; Wilson 1971). Many social hymenopterans rank among Earth's dominant terrestrial macroinvertebrates and can be of considerable ecological importance (Chapman and Bourke 2001; Hunt 2003). One social species, the honey bee *Apis mellifera*, is the insect most beneficial to humankind. Edward O. Wilson (1971, 1975) has called social hymenopterans one of the crowning achievements of evolution. And in our human culture where success attracts attention, social hymenopterans are no exception. Considerable effort has been devoted to learning about them. Evolution of hymenopteran sociality is the topic of this book, and to bring that topic into focus I look at ecology, behavior, and development, as well as other topics. The picture I want to paint is not a minimalist abstraction of simple elements. (Picasso's "Femme" leaps to mind.) Abstraction can be aesthetically pleasing, but inquiring minds seek intimate knowledge. Instead, I want to paint a grand panoramic landscape with perspective, vibrant color, and rich detail. Anything less would be a disservice to my model.

In this book I use the terms "social" and "sociality" rather than more specific labels. By these terms I mean a life history in which some offspring of a reproductive female remain at their natal nest together with their mother, and these offspring generally do not produce offspring of their own, but instead they engage in alloparental behaviors including provisioning immatures, constructing the nest, and defending the colony to the benefit of their mother and the nestmates they care for. Among the hymenopterans that independently evolved such sociality, we can learn more about social evolution from the wasp family Vespidae than from any other taxon. Among the reasons is that Vespidae is the only insect family in which diverse genera and species span a full spectrum of levels of organization, including solitary life, various presocial arrangements, simple sociality, and more than one form of complex sociality. Such diversity and richness invite phylogenetic interpretation of social evolution (Hunt 1999), which is the starting point for the present book. Other lineages are now yielding secrets of their social evolution through phylogenetic analysis (halictine bees: Danforth 2002; Danforth et al. 2003; allodapine bees: Schwarz et al. 2003; thrips: Morris et al. 2002), but no other taxon encompasses the range of social levels found in Vespidae. No other taxon, therefore, offers the same richness of investigative and informative possibilities.

A second reason to focus on Vespidae is that vespids at the threshold of sociality live in simple, accessible, easily observed nests—a circumstance that has invited the attention of countless naturalists. In particular, the cosmopolitan and abundant paper wasp genus *Polistes* is critically situated just at the sociality threshold and is a key taxon for understanding the evolution of wasp sociality (Evans 1958). We know a great deal about *Polistes*, and much of what we know impinges on our understanding of social evolution. Therefore much of this book focuses on *Polistes*.

Order Hymenoptera is one of the four megadiverse orders of insects. Each of these—Coleoptera (beetles), Lepidoptera (moths and butterflies), and Diptera (true flies), as well as Hymenoptera—have as many or more species as the remaining 26 or so insect orders combined. Clearly, something in the biology of these four orders has enabled extraordinary adaptive radiations. At first blush, complete metamorphosis seems a likely candidate. These insects have larvae and adults as distinctly different in appearance, behavior, and ecology as—well, as different as a caterpillar and a butterfly. Orders such as Orthoptera (grasshoppers and kin) have gradual metamorphosis in which juveniles and adults are similar in most respects. There are more orders with gradual than with complete metamorphosis, yet none is among the big four. Metamorphosis alone can't be the key to success, however, because there are orders such as Mecoptera (scorpionflies) that have complete metamorphosis yet come nowhere close to the big four in their adaptive radiation. Metamorphosis must have been coupled with additional traits, and very likely different ones in each of the big four orders, to have fostered their predominance (Kristensen 1999). This line of reasoning sets the stage for inquiry into the success, measured by number of occurrences, of social evolution within the Hymenoptera. Traits of the order may be essential, and they may have been crucially involved at key junctures in the evolutionary history of sociality, but just as adaptive radiation is not randomly distributed among insect orders, sociality is not randomly distributed among families of Hymenoptera. Thus, in addition to traits of the order, other traits must be involved. Family Vespidae can best reveal those traits, their mode of action, and their consequences.

This book is divided into three parts. Parts I and II each have three data-based chapters followed by a synthesis chapter. Part I adopts a historical approach to the evolution of sociality in Vespidae. Chapters 1 through 3 will seem to many readers to be little more than a meandering trip through the wilderness of biodiversity. They are much more. Knowledge of the order that contains the wasps at the focus of this book has been essential to me in developing the perspective and insight without which the syntheses presented in chapters 4 and 8 would have been impossible. Part II addresses current dynamics of sociality in Vespidae through a focus on factors that are integral to the individual development and daily lives of social vespids. Chapters 5 through 7 summarize much of my own research and embed it in a larger literature. The first three chapters of parts I and II conclude with a short list of empirical questions that I wish I could explore.

Part III turns to the paradigmatic framework in which the study of social evolution is pursued. In chapter 9 I look at inclusive fitness, which is the conceptual paradigm that has dominated the field for the past three decades, and I find it wanting as a framework to understand how social wasps evolved. In chapter 10 I address some current practices in behavioral ecology that have more often impeded than enhanced our knowledge of social evolution. In chapter 11 I propose that natural selection, acting at the level of the colony and unrestricted by narrow focus on inclusive fitness, is an adequate and appropriate paradigm within which to pursue the study of social insect evolution, and I offer both a conceptual framework and action plan for future research. I conclude with an encapsulation of the evolution of social vespid wasps as I currently understand it.

PART I

History

Every social organism had solitary ancestors at some point in its past. To understand the evolutionary origin of social behavior in any taxon, therefore, I can see no way around a need to know the biology of its solitary ancestors. Thus, to set out to learn the history of social evolution in Vespidae is not an excursion apart from other ways of knowing vespid sociality; it is an essential undertaking to shape and inform the way we perceive and understand vespid sociality in the present tense. The history of social evolution is integral, fundamental, and central to the full evolutionary story.

Phylogenetic systematics has become the methodology of choice to construct historical hypotheses when fossil evidence is inadequate. Certainly this is the case with wasps. The first phylogenetic revolution, in the 1980s, was based on morphological traits and strict parsimony. The second phylogenetic revolution, in the 1990s, was based on peptide or nucleotide sequences, and it accommodated analytical methods more suited to molecules than to mandibles. Continuing researches in these realms are yielding phylogenies of increasing resolution and with ever-stronger support. These phylogenies provide the framework within which the search for the historical origins of sociality in Vespidae can be informatively and rigorously pursued.

A fully resolved phylogeny for Insecta remains a work in progress, and even if it were available, it might not add much to the search for social origins. A fully resolved phylogeny for Hymenoptera also remains a work in progress, although the resolution is becoming increasingly clear and the support increasingly solid. The same can be said for Aculeata, the clade of Hymenoptera that contains all social forms. Vespidae, the family that is the focus of this book, was first treated phylogenetically in 1982. That phylogeny has been challenged, and it will be challenged again in these pages, but these challenges do not diminish the power that the initial phylogenetic analysis brought to my search for the origins of vespid sociality. That analysis shaped my thinking about the origin of vespid sociality for more than 20 years, and so it has been an invaluable tool in my search for knowledge. Thus, phylogenies of Hymenoptera, Aculeata, and Vespidae are the foundation and starting point in the effort to understand the evolution of vespid sociality. The scientists who have produced these phylogenies have my thanks and respect.

Insects treated in chapters 1 and 2 are less well known than fully social forms, and they are less well studied by students of insect sociality. Therefore I have tried to give literature citations that can be gateways to broader knowledge of these insects. I cite original work in preference to reviews and also try to acknowledge pioneering work when possible. Some citations are exemplary rather than comprehensive, but I chose not to use "e.g." to facilitate smoother reading. Chapter 3 treats wasps for which there is an enormous literature, which if cited comprehensively would cause the text to disappear between long parenthetic enclosures. In some cases I have tried to cite pioneering work, but many of the citations are exemplary. I apologize to the many researchers whose work has not been cited. Assertions unsupported by citation are general knowledge. These can be checked by turning to basic sources. Readers wanting to learn more about Polistes *should begin with the monograph by West-Eberhard (West Eberhard 1969). Corresponding monographs on* Mischocyttarus *are those by Jeanne (1972) and Litte (1977, 1979, 1981). A single species of* Ropalidia *is treated in depth by Gadagkar (2001). Spradbery (1973) and Edwards (1980) present basic information on such aspects as anatomy, and both feature European Vespinae. Matsuura and Yamane (1990) coauthored a treatise on Vespinae with emphasis on Oriental species. The edited volume by Ross and Matthews (1991) is a broadly comprehensive treatment of the social biology of wasps. The volume edited by Turillazzi and West-Eberhard (1996) presents topics on paper wasps. All of these publications can be gateways to information, ideas, and to the literature at large.*

Plant Feeders and Parasitoids

1

Some Hymenoptera are the size of unicellular protistans; some drag tarantulas across the ground. Some live under water; others feed on leaves or wood. Many pass their larval development inside another insect; some build nests in which their larvae live. All these hymenopterans develop in distinct life stages: egg, larva, pupa, adult. Adults have biting mandibles, two pairs of membranous wings, a sclerotized (hardened) ovipositor in females, and prominent antennae. Females are normal diploids, but males develop from unfertilized ova and so are haploid. Among these traits only the wing type is unique to the order, hence the ordinal appellation: *hymen* (membrane) + *ptera* (wing). The forewing is larger than the hindwing, and the two are typically joined into a single airfoil by a set of small hooks called hamuli, which are on the leading edge of the hindwing and ensnare the trailing edge of the forewing. A few years ago I enthusiastically favored the suggestion—and I lament that I don't know who inventively made it—to name a new journal "Hamuli." The clear but prosaic *Journal of Hymenoptera Research* was chosen instead.

Symphytans

Each of the features mentioned above plays a role in hymenopteran sociality. To place these features into a context of greater rigor than a just-so story calls for specific interpretive methods. Character mapping is one such method (Brooks et al. 1995). Cladograms can be used to mark the first occurrence of key traits along the route to sociality and place them in context vis-à-vis one another (Hunt 1999). Figure 1.1 shows a cladogram for the basal taxa of Hymenoptera, collectively often called symphytans, plus the more derived Apocrita. All the traits mentioned above fall at the base of the clade, on the branch numbered 1. Sociality occurs only in some Apocrita, indicated by the pointer. Clearly, traits of Apocrita must contribute to the origin of sociality. Some traits that arose after the origin of Hymenoptera but antecedent to Apocrita also play a role.

In addition to having the traits mentioned above, adults of all taxa in figure 1.1, except Apocrita, share a trait in common with most nonhymenopteran insects: the thorax and abdomen are broadly joined, and the abdomen is unconstricted (figure 1.2). That is, symphytans lack the familiar "wasp waist." Some symphytans do not feed as adults; others feed on nectar, honeydew, sap, and other liquid foods; some ingest pollen; some capture and eat other insects and so are insectivorous (Jervis and Vilhelmsen 2000). Some lay eggs in batches; others lay eggs singly. Larvae of symphytans have sclerotized head capsules and mandibles, and some are remarkably caterpillarlike (figure 1.3). Larvae of some species are gregarious and have evolved specific behaviors and modes of communication to maintain group cohesion (Costa and Louque 2001; Flowers and Costa 2003), which strongly suggests adaptive advantage (Boeve 1991) and reflects the placement of eggs in batches (Codella and Raffa 1995). Such gregariousness among larvae, but not adults, is the closest approach to sociality among symphytans. Some females guard newly eclosed larvae (Schiff 2004), which is the closest approach among symphytans to parental care.

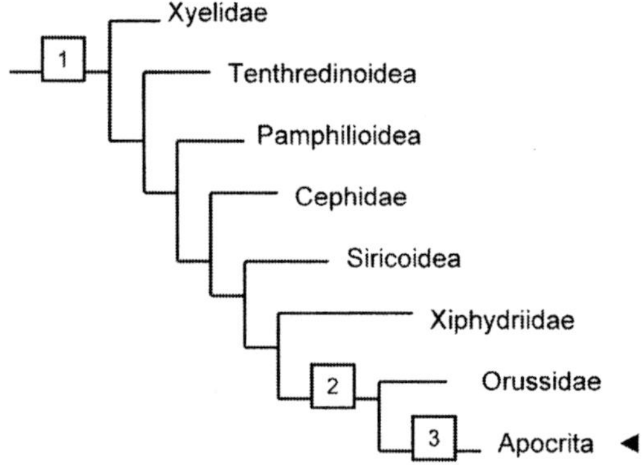

Figure 1.1
Cladogram of Hymenoptera showing basal taxa plus Apocrita. Numbered branches correspond to placement of traits described in the text. Larvae of all except Orussidae and Apocrita feed on plants (leaves, stems, galls, fungus-infected wood), and so the name "Symphyta" (plant-loving) has long been used to refer to them. All social hymenopterans are members of the Apocrita, indicated by the pointer. Traits are: (1) mandibles, sclerotized ovipositor, membranous wings with hamuli, haplodiploidy; (2) larvae carnivorous, larvae legless; and (3) closed larval hindgut, petiolate abdomen. Redrawn from Schulmeister (2003).

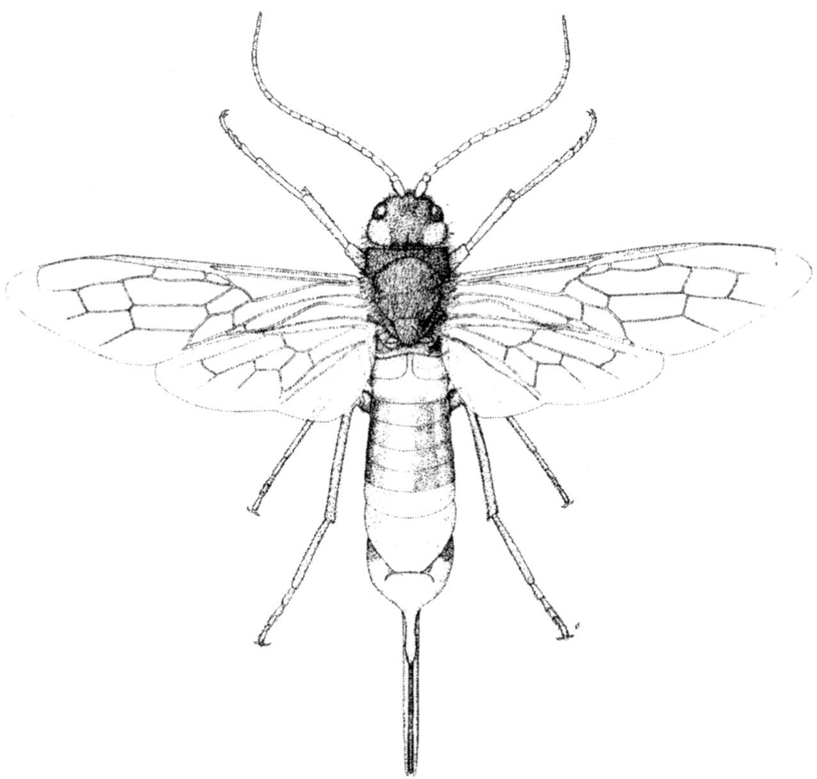

Figure 1.2
A female symphytan, the horntail *Urocerus* (Siricidae). The wings, coupled by hamuli, are typical for the order Hymenoptera. The thorax and abdomen are broadly joined, and the abdomen is not constricted. The rigid ovipositor of horntails serves only for oviposition and is not a stinger. From Gauld and Bolton (1988). © Natural History Museum, London. Reprinted with permission of the Natural History Museum, London.

Apocrita

If symphytans were the only hymenopterans, they would attract about as much attention as do scorpionflies. The great adaptive radiation of Hymenoptera is not among the thick-waisted plant feeders; it is among the Apocrita. One key adaptation leading to the Apocrita is marked on branch number 2 in figure 1.1. Whereas larvae of most symphytans feed on plants, larvae of Orussidae feed on the larvae of other insects. Excepting some secondary reversions to specialized modes of plant feeding and of scavenging by some social forms, the larvae of almost all Apocrita also feed on other insects. A scenario for this transition is not hard to envision. Larvae of Siricidae and Xiphydriidae feed on wood that has

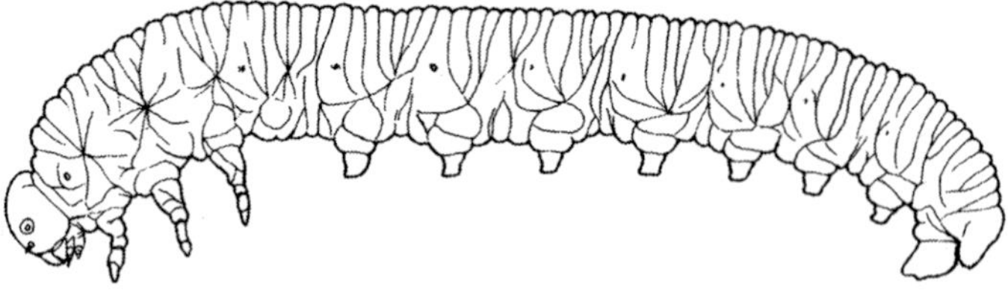

Figure 1.3
Larva of a plant-feeding symphytan, the sawfly *Tenthredo* (Tenthredinidae). Such larvae feed on leaves of herbaceous plants and so are similar to caterpillars in more than appearance. From Gauld and Bolton (1988). © Natural History Museum, London. Reprinted with permission of the Natural History Museum, London.

been softened and made more nutritious by specific fungi (Francke-Grossman 1967). Larvae of Orussidae are generalist feeders on larvae of wood-feeding beetles (Powell and Turner 1975) or Hymenoptera (Rawlings 1957). The proto-orussid, tunneling through wood, encountered whole protein and found it good. Selection favored traits of oviposition behavior that fostered such encounters (Handlirsch 1907) as well as the dietary shift itself (Eggleton and Belshaw 1992).

A second trait of significance to sociality also falls on branch 2 in figure 1.1 (Vilhelmsen 2001, 2003): larvae of Orussidae and Apocrita are legless. Leglessness and larval insectivory have been described as coadapted traits (Vilhelmsen 2003), although leglessness was preceded by leg reduction in larvae of the stem-boring Cephidae and the wood-feeding Siricoidea and Xiphydriidae (Vilhelmsen 2001, 2003). Leglessness in larvae whose food habits do not require foraging could reflect either adaptive developmental economy or selective neutrality accompanied by degenerative mutation fixed by genetic drift (Fong et al. 1995).

Sometime following the origin of larval insectivory but before the radiation of Apocrita (branch 3 in figure 1.1), two traits of importance to sociality appeared. Both concern aspects of development. Insect larvae normally have complete digestive tracts and void feces as they feed and grow. Apocritan larvae, however, have a hindgut that remains closed until the end of larval growth and onset of metamorphosis, excepting reversions in some bee larvae (Michener 1953, 1974). Evolution of this trait was perhaps enabled by the transition to a largely liquid diet, namely the hemolymph of the beetle or hymenopteran larvae on which the proto-apocritan larva fed. The few undigestible solids that accumulate in the gut are voided as meconium when the anus opens at the completion of larval growth. An original adaptive advantage of this trait may have been physiological conservation of ingested liquids. A secondary but important selective advantage could have been that retaining feces reduced detection by ovipositing orussids, which may use larval feces to locate hosts for oviposition (Powell and Turner 1975).

At the about same time that the closed larval hindgut evolved (branch 3 in figure 1.1), the wasp waist (figure 1.4) also evolved as a major innovation. As with larval leglessness and insectivory, the closed larval hindgut and wasp waist are probably coincident traits rather than a coadapted character complex. The wasp waist is not where superficial inspection suggests it is. The thorax and abdomen are in fact broadly joined as in symphytans, and the constriction is between the first and second segments of the abdomen. Thus the apparent abdomen is not the complete abdomen, so the term "gaster" is commonly used when discussing the third body region of apocritans. ("Metasoma" is the more formal name.) The only problem is, we have no idea what fostered the developmental variability from which selection shaped the morphology, nor are we certain of its adaptive significance.

It has been argued that the adaptive significance of the wasp waist is to facilitate oviposition via a long ovipositor (Vilhelmsen et al. 2001). This seems unlikely. Terminal symphytans (Siricidae, Xiphydriidae, Orussidae) all oviposit into wood, as do basal Apocrita, the latter with the wasp waist but the former without it. The ovipositor of Orussidae is longer than the body and, although sclerotized, it coils within the anterior region of the thorax (!) when in repose

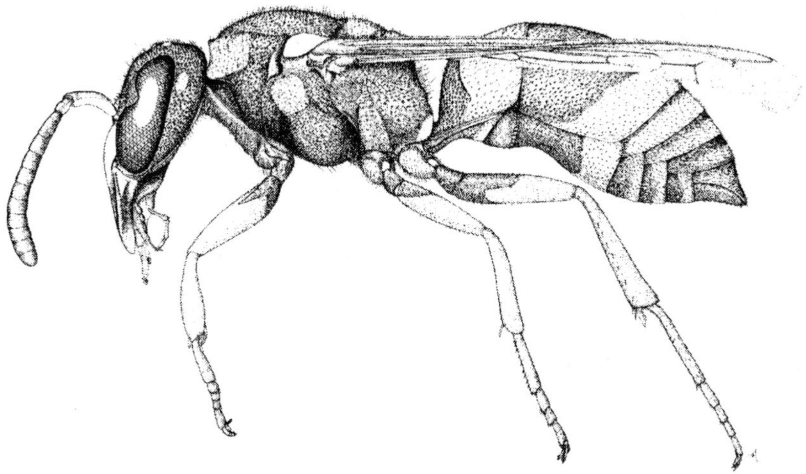

Figure 1.4
A wasp-waisted apocritan, *Ancistrocerus* (Vespidae, Eumeninae). The two wing pairs are on the middle and third segments of the thorax. Posterior to that, the first abdominal segment is broadly fused with the thorax and is an integral part of the middle body region (the mesosoma). The wasp waist is a constriction between the first and second abdominal segments. It is articulated and fully movable, and despite its small diameter, organs of the circulatory, respiratory, nervous, and digestive systems pass through it. From Gauld and Bolton (1988). © Natural History Museum, London. Reprinted with permission of the Natural History Museum, London.

(Vilhelmsen et al. 2001). The wasp waist precludes this anatomical arrangement, and all apocritans have their reproductive system restricted to the gaster. A consequence of this noted by Vilhelmsen et al. (2001, p. 83) is that "basal Apocrita have to struggle with their cumbersome, external ovipositor." Arguments that making the gaster more maneuverable constituted an important preadaptation to the parasitoid lifestyle (Kristensen 1999) might address adaptive refinement of the wasp waist, but this line of reasoning cannot address its origin.

A different possibility for the origin of the wasp waist is that it was selected as a mechanical and/or physiological adaptation reflecting selection for improved flight. Gauld and Bolton (1988, p. 105) note that "the majority of sawflies are rather weak-flying, clumsy insects." Most apocritans, in contrast, are accomplished flyers. Anyone who has tried to out-run aroused social wasps defending their nest can attest that some apocritans fly very well indeed. Flight muscles have very high metabolic demand for both oxygen and fuel (Weis-Fogh 1964a, b; Harrison and Roberts 2000). Insects such as locusts, beetles, and lepidopterans can metabolize fatty acids as an energy source for flight, but a typical apocritan, the honey bee *Apis mellifera*, depends on trehalose (Weis-Fogh 1964b). At the same time, the thorax of hymenopterans is rather rigid, and it changes shape very little during flight (Weis-Fogh 1964a). Although ventilation by thoracic pumping probably plays an important role in the flight of most large insects, ventilation by abdominal pumping is much more important in wasps (Weis-Fogh 1964a). The wasp waist must certainly constrain the ebb and flow of hemolymph during abdominal pumping. Indeed, anyone who has dissected wasps or bees will have noted that some of them—especially older foragers—have very little hemolymph at all. The gut and reproductive organs lie in a gaster with air sacs so large that it seems nearly empty, whereas flight muscles completely fill the thorax. Does the wasp-waist constriction somehow enhance metabolic efficiency of the flight muscles? Miller's (1966) observation that a tethered wasp, *Vespula germanica*, sustained flight wing beat frequency and amplitude for 8 minutes after removal of the gaster lends anecdotal support to this notion.

Diptera is the other insect order characterized by fast and agile flyers. If you see a fast-flying insect or one flying with considerable agility, it is almost certainly a wasp, bee, or fly. Flies are similar to hymenopterans in having a rigid thorax and respiration facilitated by abdominal pumping (Weis-Fogh 1964a), and they also use trehalose as the primary energy supply (Weis-Fogh 1964b). In dipterans, large air sacs that form at adult emergence might serve in part to reduce hemolymph volume with a concomitant increase in concentration of nutrients per unit volume (Miller 1966). Could this be a physiological and functional analogy to the hymenopteran wasp waist? One attraction of the flight efficiency hypothesis is that flight efficiency might be as selectively important, or more so, in males as in females. More scenarios involving diverse agents of selection can be conceived, and at least some of these could be subjected to empirical tests. Neither the flight efficiency hypothesis nor the ovipositional hypothesis, however, gives insight to the multiple constrictions (a "petiolate" gaster) in the abdomens of several parasitoid lineages such as ensign wasps and some fairy flies, all ants, some potter wasps, all hover wasps, and several genera of social Vespidae such as *Mischocyttarus*, *Belonogaster*, *Ropalidia*, and *Agelaia*. Whatever the origin and adaptiveness, however, the wasp waist imposes a specific and severe constraint on all Apocrita.

The cross-sectional anatomy of the wasp waist is shown in figures 1.5–1.7. Through this small, rigid opening must pass the dorsal vessel, ventral nerve cord, two tracheae, and a continuous circulation of hemolymph. Through the midst of all this passes the esophagus, because an adult apocritan's crop, midgut, and hindgut are entirely within the gaster. Although we don't know the origin of the wasp waist, this much is clear: it prevents adult apocritans from ingesting solid food. Adult apocritans, including all social hymenopterans, are constrained to feed only on liquid foods or, at least, on liquid suspensions of particles such as pollen grains that are sufficiently small to pass the hourglass constriction. This constraint, as will be shown, is of profound importance in hymenopteran social evolution. Before reaching that point in the evolutionary scenario, however, additional insights of considerable importance can be gleaned from other traits of Apocrita.

Apocrita is a taxon characterized not only by species richness but also by great diversity in life history, morphology, and behavior (Gauld and Bolton 1988;

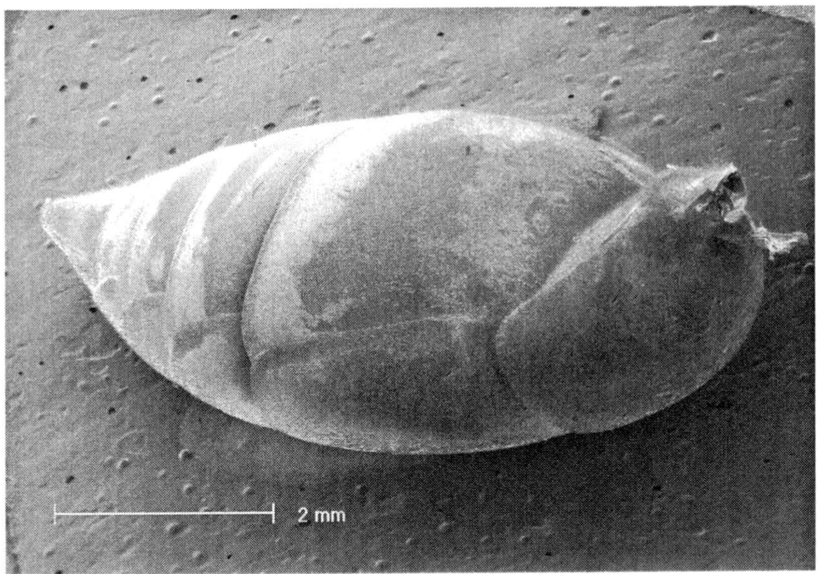

Figure 1.5

Gaster of *Polistes metricus* seen in oblique ventral view. At the upper right, the petiole (wasp waist) has been separated at the articulation with the mesosoma (= the "thorax"). Scanning electron micrograph courtesy of Scott J. Robinson, Beckmann Institute, University of Illinois at Urbana-Champaign.

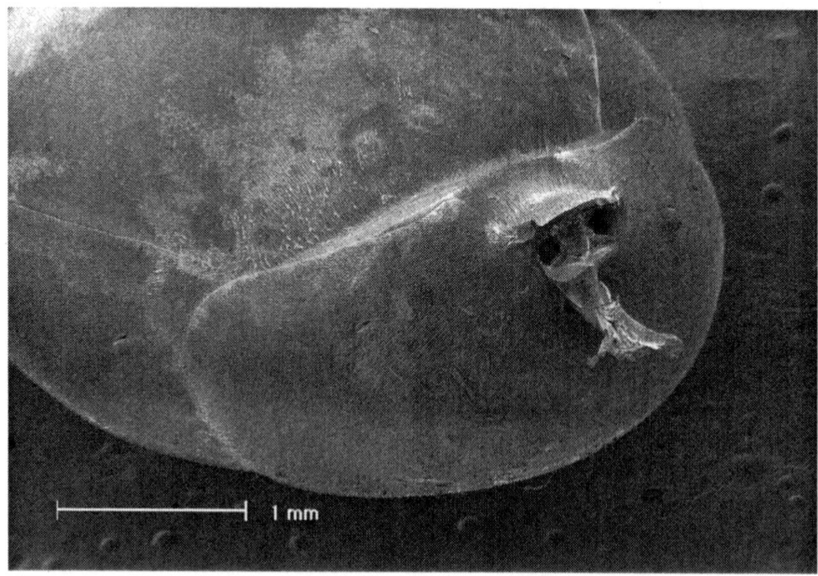

Figure 1.6
Gaster of *Polistes metricus* seen in oblique ventral view. The petiole is a rigid, nondistensible structure through which anything entering the gaster must pass. The two dark openings are tracheae (air trunks); between them lies the esophagus. The tissue extending outward from the exterior of the petiole is the dorsal tendon, one of three tendons that articulate the petiole with the mesosoma. Scanning electron micrograph courtesy of Scott J. Robinson, Beckmann Institute, University of Illinois at Urbana-Champaign.

Hanson and Gauld 1995). Many species are small to very small in size, and loss of morphological characters in concert with size reduction (Dowton and Austin 1994) is among the numerous factors that challenge systematists seeking to establish relationships among subgroups (Gauld 1986). Difficulties notwithstanding, there is general agreement that Aculeata, in which all social hymenopterans are found, is monophyletic (Brothers and Carpenter 1993; Dowton and Austin 1994, 2001; Whitfield 1998). Therefore the exact placement of Aculeata within Apocrita is not critical to unraveling the evolution of sociality, but the approximate placement is informative. Whitfield (1998) found Stephanidae to be sister to all other Apocrita, with Aculeata sister to Ichneumonoidea (figure 1.8, left). More recent analyses (Ronquist et al. 1999; Dowton and Austin 2001) have found Aculeata to be sister to all other Apocrita (figure 1.8, right). Dowton and Austin (2001) point out that, at the very least, the most consistently basal lineages in their diverse analyses were Stephanidae, Ichneumonoidea, and Aculeata. The basal lineages, in turn, of each of these families have larvae that develop by feed-

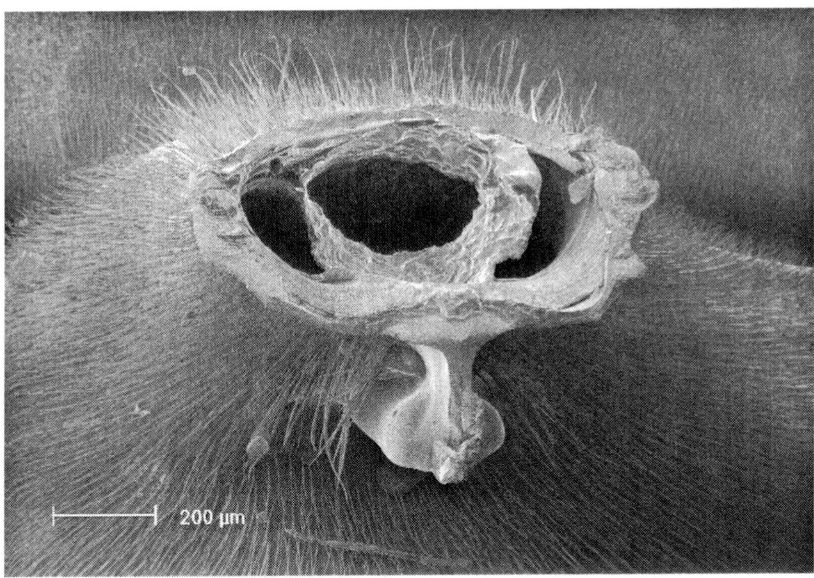

Figure 1.7
Cross-section of the petiole of *Polistes metricus* in inverted view. The esophagus lies between two trachaeae. The dorsal vessel, ventral nerve cord, and a continuous flow of hemolymph also pass through the petiole. The structure on the exterior is a ridge to which the dorsal tendon attaches. Scanning electron micrograph courtesy of Scott J. Robinson, Beckmann Institute, University of Illinois at Urbana-Champaign.

ing on wood-boring beetle larvae (Whitfield 1992a), probably reflecting life history inherited from the common ancestor that Apocrita shares with Orussidae (Dowton and Austin 1994). Many Aculeata, plus all nonaculeate apocritans except gall wasps and fig wasps, retain essential elements of this life history: their larvae develop by feeding on a single host and thereby cause the death of that host. Insects with this type of life history are called parasitoids. Utilization of novel parasitoid hosts opened another door to the adaptive radiation of Apocrita.

Utilization of hosts other than those living in wood fostered the apocritan radiation, but it also introduced a major adaptive challenge due to a particular component of the oviposition procedure. Basal parasitoids sting their host at the time of oviposition, and the sting immobilizes the host larva on which the parasitoid egg is then laid. The host remains alive and fresh rather than decaying, and it does not move. In the confinement of a beetle larva's gallery such immobilization could be adaptive: the parasitoid egg would not be scraped off as the ambulatory beetle larva rubbed against gallery walls. In the case of foliage-feeding hosts, however, immobilization poses the significant problem of exposure. An

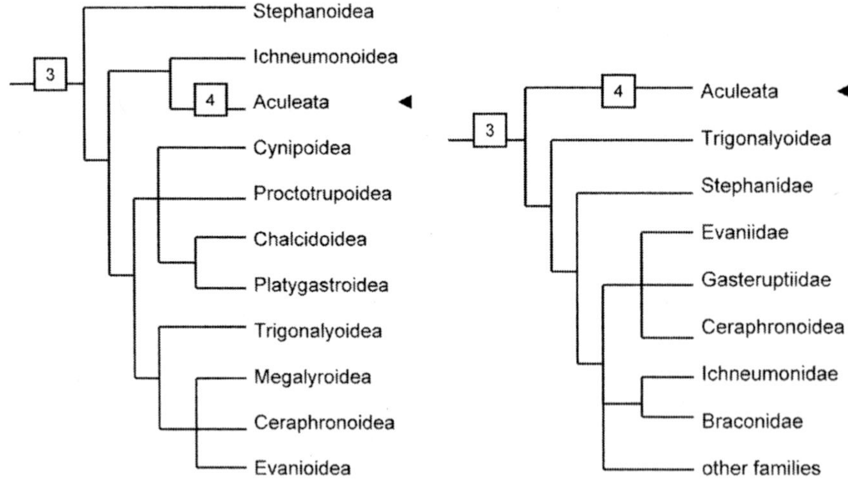

Figure 1.8
Cladograms of Apocrita. Numbered nodes correspond to placement of traits described in the text. The cladogram at left is redrawn from Whitfield (1998), and that on the right is redrawn from Dowton and Austin (2001), although Dowton and Austin "do not consider this a preferred phylogeny." Sociality occurs only in some species of Aculeata, as indicated by the pointer. Traits are: (3) closed larval hindgut, petiolate abdomen; and (4) repositioned oviduct, yolky (lecithal) ova, synovigenesis, adult feeding on host hemolymph, host anesthetization, selection of concealed hosts.

immobilized yet fresh and thus edible host remains in place for days until the parasitoid hatches and additional days as the parasitoid feeds. It is a sitting duck for opportunist feeders (birds and ants are contemporary examples) with obvious consequences for the parasitoid. Although we can't know the specific agents of selection that shaped the course of evolution, we can assess the adaptive suites that they molded. Parasitoids have diverse means for achieving minimum host exposure. The simplest of these means exemplifies what may have been the earliest condition: hosts are as concealed as possible. Thus a host larva under loose bark or in leaf litter might be nearly as secure as one in wood. Hosts more exposed, however, led to selection for new adaptations to reduce exposure. Two suites of traits proved to be particularly successful, as measured by adaptive radiation, and both of these suites of traits evolved multiple times. One of these suites is widespread among nonaculeate parasitoids.

Many nonaculeate parasitoids have a sting that immobilizes the host only briefly or not at all, and the parasitoid egg is generally inserted into the host rather than being deposited on its exterior. The host resumes feeding and growing for at least a while, and the parasitoid typically begins to feed and grow only when the host has secluded itself for pupation during the normal course of its own development. This type of life history is called koinobiosis, and it differs from a life history beginning with host immobilization, called idiobiosis (Haeselbarth 1979; Askew and Shaw 1986). Contrasts of life history features of idiobiont and koinobiont parasitoids have proven to be heuristically valuable (Gauld and Bolton 1988; Hanson and Gauld 1995; Quicke 1997), and the suites of traits have been shown to be generally robust (Mayhew and Blackburn 1999). Numerous traits of parasitoidism are germane to the evolution of sociality, and the idiobiont/koinobiont contrasts of those traits are given in table 1.1. The important point is that idiobiosis appears to characterize basal Aculeata (Whitfield 1992a; Dowton and Austin 1994, 2001; Hanson and Gauld 1995; Ronquist 1999; Belshaw and Quicke 2002). An additional important consideration is that many of the idiobiont traits germane to sociality are traits of adult females.

Adult female idiobionts are characterized by relatively long reproductive lifetimes, relatively low fecundity, and a suite of traits related to these life history characteristics. Females can mate soon after emergence from pupation, storing sperm in a chamber off the oviduct called the spermatheca, but the ova may or may not be ready for oviposition, and they will be few in number. Generally, the wasp-waisted female must feed on flower nectar to gain nourishment to enable full development of the rather large, yolky ("lecithal") ova that she produces. Nectars of all kinds contain amino acids (Baker and Baker 1973) as well as carbohydrates (Maurizio 1975), and variation in nectar nutrient concentrations and compositions suggests selection to address specific nutritional requirements of diverse pollinators (Baker and Baker 1968; Baker and Hurd 1968). Even so, nectar alone may be inadequate to meet the wasp's needs. Perhaps not surprisingly, then, many idiobiont parasitoids feed on hemolymph that oozes from the sting wound that they have just inflicted on

Table 1.1
Life-history correlates of idiobiont and koinobiont parasitoid wasps.[a]

Idiobiont	Koinobiont
Host paralyzed at oviposition	Host continues to develop for a time
Larva feeds as an ectoparasitoid	Larva feeds as an endoparasitoid
Adult female is "synovigenic"	Adult female is "pro-ovigenic"
Oosorption may occur	Oosorption does not occur
Adult females often feed from hosts or prey as well as from nectar sources	Adult females feed primarily, if at all, from nectar sources
Adult lifetimes are relatively long	Adult lifetimes are relatively short
Females often larger than males	Sexes often monomorphic
Wasp may match sex of egg to size of host	No sex–host-size relationship
Generalist parasitoid	Specialist parasitoid

[a]A similar table appears in Quicke (1997). Traits are described in greater detail in Gauld and Bolton (1988), Hanson and Gauld (1995), and Quicke (1997).

the larval host (Jervis and Kidd 1986). Indeed, host feeding of diverse sorts is widespread in parasitoid hymenopterans, including koinobionts (Jervis and Kidd 1986). For idiobionts, this enables synovigenesis (Flanders 1950; Jervis et al. 2001)—the sustained production of ova over the female's lifetime. Because each idiobiont develops on a single host, the second law of thermodynamics dictates that the host must be larger than the wasp it will produce. If the host-searching female (Vinson 1976; van Alphen and Vet 1986) cannot find an appropriate host or if inclement conditions prevail, she may resorb nutrients from mature or maturing ova until she can feed and once again produce mature ova. At the moment the female oviposits, one more life history variable comes into play. Because many basal apocritans are sexually size dimorphic, with females larger than males, the ovipositing female may place unfertilized eggs on hosts that are relatively smaller and fertilized eggs on those that are larger. All these traits characterize basal Aculeata, the lineage in which is found not koinobiosis but, instead, the other major adaptive suite for avoiding host exposure. Aculeata also contains all social hymenopterans.

Before moving on to Aculeata, I must briefly address the assertion that nonaculeate apocritans lack sociality. A key feature of idiobiosis is the oviposition of only a single egg per host, which is called "solitary" in the literature on parasitoid clutch sizes and sex ratios. Some koinobiont parasitoids are "gregarious." Gregarious parasitoids oviposit two or more eggs within a host, and the larvae develop and emerge simultaneously. Phylogenetic analysis reveals that gregariousness has evolved more than 40 times from solitary ancestors (Mayhew 1998). Questions arise as to the fitness consequences of both sex ratio and siblicide, or its absence, when a host contains multiple parasitoid larvae of the same parent(s) (recall that males have only one parent); therefore the origin and maintenance of gregariousness have become foci of attention for evolutionary theorists as well as parasitoid biologists (Waage 1986; Godfray 1987; Rosenheim 1993; Ode and Rosenheim 1998; Mayhew and Van Alphen 1999; Pexton et al. 2003). Yet another layer of complexity is present in some parasitoids that are polyembryonic.

In polyembryony, an egg fissions into many cells, each of which develops into an adult, resulting in a clone of tens to thousands of genetically identical parasitoids within a single host (Strand and Grbic 1997). An exciting discovery was that some polyembryonic species have members of the clone that develop quickly and swim through the host's hemolymph, where they find larvae of other clones or other taxa and kill them. These "defender morphs" do not complete development and so do not reproduce themselves (Cruz 1981). Although heralded at the time of its discovery as sociality in a parasitoid, it is clear that this phenomenon differs in fundamental ways from the sociality of wasps, ants, and bees. Polyembryony is, nonetheless, a deeply interesting phenomenon with many evolutionary complexities and implications (Grbic et al. 1992; Hardy et al. 1993; Craig et al. 1997; Strand and Grbic 1999). Gregariousness and polyembryony, both of which occur in larvae but not in adults, are the closest approaches to sociality among parasitoids.

Aculeata

Aculeata have traditionally been distinguished and unified by a morphological trait—the sting (Latin: *aculeus*; figures 1.9, 1.10). Whereas in all other Apocrita the ovipositor serves in both stinging hosts and laying eggs, in Aculeata it serves only a single function. Ova are passed through an opening that is ventral to the ovipositor, which is retained but is now exclusively a stinger. Branch 4 in figures 1.5 and 1.6 marks the origin of this trait. Most or all idiobiont traits evolved before branch 4, but the idiobiont suite of traits also is placed there because the origin of the various traits cannot be pinpointed and because they may not have evolved as a coadapted character complex (Whitfield 1992b). Thus, placing the suite on branch 4 is a heuristic convenience to make clear that the traits exist at that point and that they typify basal Aculeata.

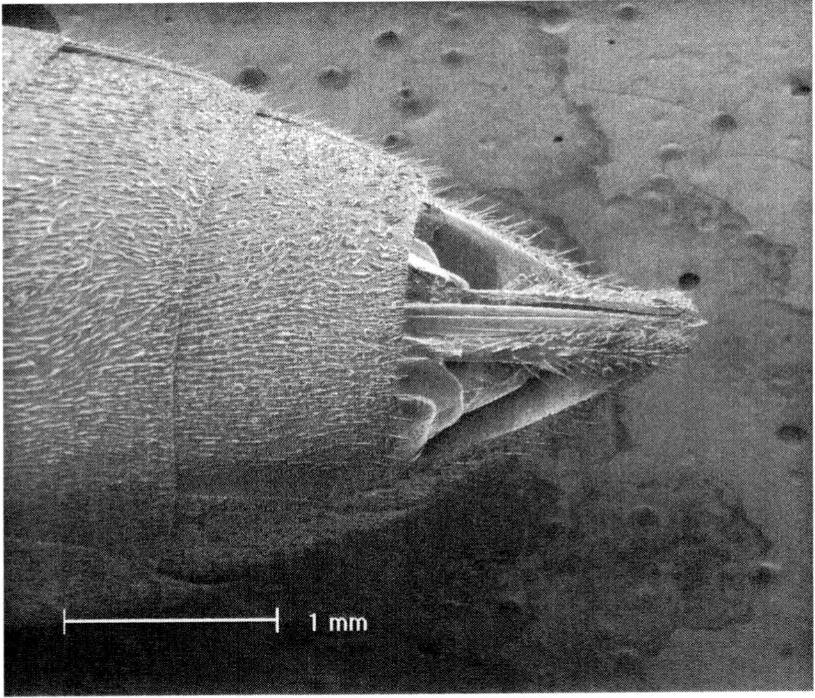

Figure 1.9
The sting of *Polistes metricus* in ventral view. The terminal sternite (ventral sclerite) has been removed to reveal the sting chamber, within which the sting lies in repose. The rigid sting lies between two more flexible parts, the sting sheath lobes. The sting is extended beyond the sting chamber during stinging. Scanning electron micrograph courtesy of Scott J. Robinson, Beckmann Institute, University of Illinois at Urbana-Champaign.

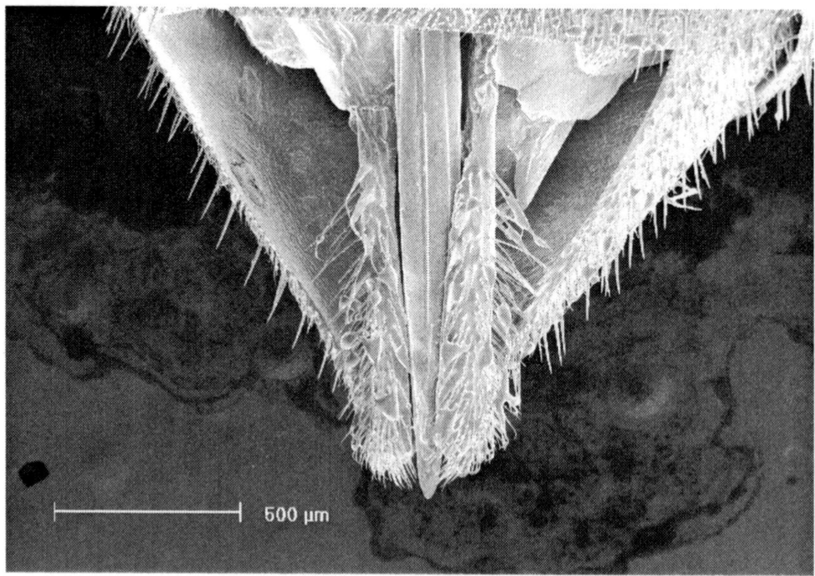

Figure 1.10
The sting of *Polistes metricus* in ventral view. The terminal sternite has been removed to reveal the sting between the two sting sheath lobes. The sting itself consists of three parts, the lancet and two stylets, that overlap tightly to form a tube through which venom flows from the venom gland (not shown) into the victim. Scanning electron micrograph courtesy of Scott J. Robinson, Beckmann Institute, University of Illinois at Urbana-Champaign.

Aculeata radiated into three major subgroups (figure 1.11), which are given superfamily rank (Gauld and Bolton 1988; Brothers 1999). Chrysidoidea contains primarily parasitoid forms plus some, called cleptoparasites, that develop by feeding on provisions placed in a nest cell by a hunting wasp. Apoidea contains parasitoids, hunting wasps, and bees (Apidae). Vespoidea contains parasitoids, hunting wasps, and ants (Formicidae). Basal or relatively basal taxa in all three superfamilies are idiobiont parasitoids of beetle larvae in wood, leaf litter, or soil (Hanson and Gauld 1995; Brothers 1995). The simplest life histories of hunting wasps are essentially the same as those of parasitoids that use unconcealed prey but with the addition of measures to conceal the prey item before ovipositing on it. A spider wasp (Pompilidae) dragging a paralyzed spider across the ground is a dramatic example. Elaboration of prey concealment behaviors led to the second of the two major suites of traits whereby prey concealment fostered adaptive radiation: many aculeates build nests (Gauld and Bolton 1988; Hanson and Gauld 1995; O'Neill 2001). Among spider wasps, for example, although many drag a spider to a naturally occurring concealment site, some drag the spider to a site where they then excavate a simple nest, and some drag the spider to a nest that has been previously prepared (Wasbauer 1995). Construction of nests to which larval provisions are taken is a hallmark of five of the families in figure 1.11. Multiple placements of this trait indicate multiple independent origins of nesting behavior. Most Sphecidae, but not all (Hanson and Gauld 1995), build nests, as do all Crabronidae (which have traditionally been included in Sphecidae *sensu lato*) and all Apidae, which are bees. Nesting characterizes all Formicidae (ants) and Vespidae, but it is absent in the closely related Scoliidae, which are idiobiont parasitoids of beetle

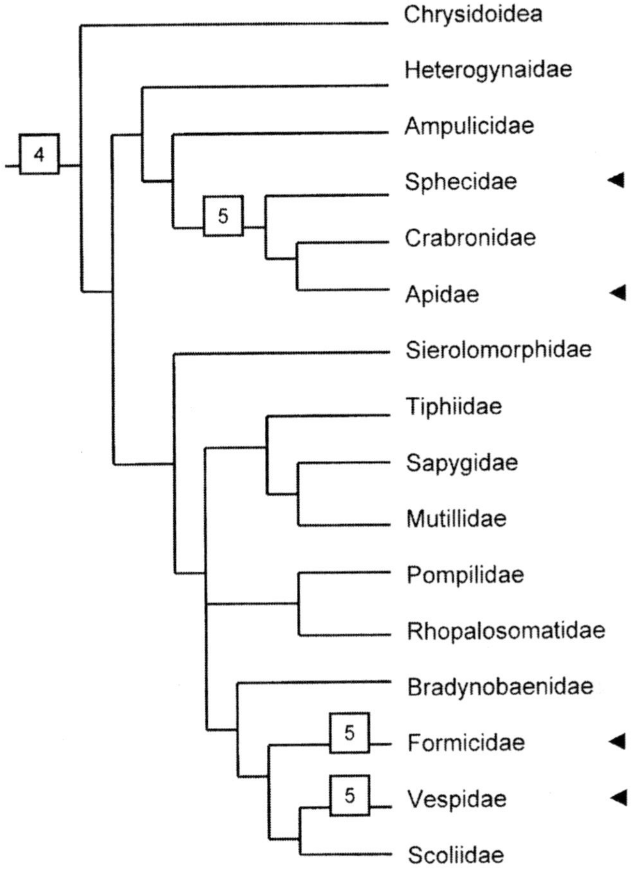

Figure 1.11
Cladogram of Aculeata. Numbered nodes correspond to placement of traits described in the text. Social taxa are found in the four families indicated by pointers. Traits are: (4) repositioned oviduct, yolky (lecithal) ova, synovigenesis, adult feeding on host hemolymph, host anesthetization, selection of concealed hosts; and (5) nesting behavior. Redrawn from Brothers (1999).

larvae in soil. Therefore, the precise number of origins of nesting behavior is still unknown, but it is crystal clear that social forms are found only in families in which nesting behavior is well developed: Sphecidae and Apidae in the Apoidea, and Formicidae and Vespidae in the Vespoidea. Pointers in figure 1.11 mark those families.

Sociality in Sphecidae has been confirmed for only a single species of tiny wasps, *Microstigmus comes* (Matthews 1968; Ross and Matthews 1989a, b), although observations of nest architecture (West-Eberhard 1977) and behavior (Melo and Campos 1993) suggest that similar sociality occurs in other *Microstigmus* species. Sociality may also occur in related genera of small sphecid wasps, *Arpactophilus* (Matthews and Naumann 1988; Matthews 1991) and *Spilomena* (McCorquodale and Naumann 1988). Sociality has evolved several times among Apidae (Michener 1974), which is a clade of nesting aculeates in which pollen, rather than immobilized arthropods, serves as larval food. We call these fuzzy, pollen-gathering aculeates "bees." Most bee species are solitary. Sociality in bees ranges from simple, small societies to the extraordinary complexities of honey bees (Michener 1974). The multiple origins and absence of transitional forms for the highly social taxa make full resolution of social evolution in bees a continuing challenge. A challenge of a different sort attends unraveling the origin of sociality in Formicidae: all ants are social (Wheeler 1910; Hölldobler and Wilson 1990). Basal ant taxa are soil-nesting hunters or scavengers, but absence of presocial forms exhibiting transitional states of behavior and morphology may make it impossible to elucidate the pathway to sociality in ants.

None of these problems attends the Vespidae. This family contains a full spectrum from solitary to advanced forms of sociality, with living taxa exemplifying all levels of social organization. Consequently, Vespidae offers a clear pathway to understanding the origin of sociality in Hymenoptera. The biology of solitary, presocial, and facultatively social vespids is reviewed in the next chapter, and the biology of the social vespids is reviewed in chapter 3.

Questions Arising

What is the respiratory efficiency of symphytans versus apocritans? If the respiratory efficiency hypothesis for adaptiveness of the wasp waist has merit, there should be differences between the two.

Does the wasp waist differ anatomically or functionally between basal and derived apocritans? If precision in oviposition has, in fact, been a selective agent affecting the wasp waist, there should be correlations between variables of anatomy and function.

What are the developmental anatomy and gene regulation of the wasp waist during metamorphosis? Perhaps the most certain route to understanding the wasp waist is to unravel its development.

Are the closed larval hindgut and wasp waist in fact a coadapted character complex? It doesn't seem likely that they are, but they do fall together in character mapping.

Do Vespidae and Formicidae share a nest-building common ancestor? Or, to put it another way, are the phylogenetic relationships among families of Vespoidea fully resolved?

2 Pollen Wasps, Potter Wasps, and Hover Wasps

Hornets, yellowjackets, and paper wasps are among the most familiar insects. There are common names for them everywhere they occur. Everyone has seen them, and most people fear them. They can inflict painful stings and advertise their ability to do so with warning colors of black and yellow. When dining outdoors on a late summer day, the watchwords are *caveat potor*—drinker beware: there could be a wasp in your can of soda or beer. These wasps, all of which are social, are in the family Vespidae. So, too, are the less well-known forms addressed in this chapter, which include pollen wasps, potter wasps, and hover wasps. These are solitary or have a different form of sociality from that of the more familiar species. All Vespidae are currently arrayed in six subfamilies of living forms (Carpenter 1982, 1991) plus one Mesozoic genus accorded subfamily status (Carpenter and Rasnitsyn 1990). Relationships among living subfamilies have been challenged (Schmitz and Moritz 1998) but defended (Carpenter 2003), and the relationships shown in figure 2.1 remain the prevailing hypothesis.

Euparagiinae

Euparagiinae, the living subfamily that is sister to all other living vespid taxa, is represented by a single genus containing nine species that live in arid and semiarid areas of southwestern North America (Bohart 1989). These wasps are rarely encountered, and so they have no common name. Adult female euparagiines visit water sources to drink, and then they regurgitate the water from their crop to soften soil as they excavate nests (Longair 1985), a behavior inferred to characterize basal

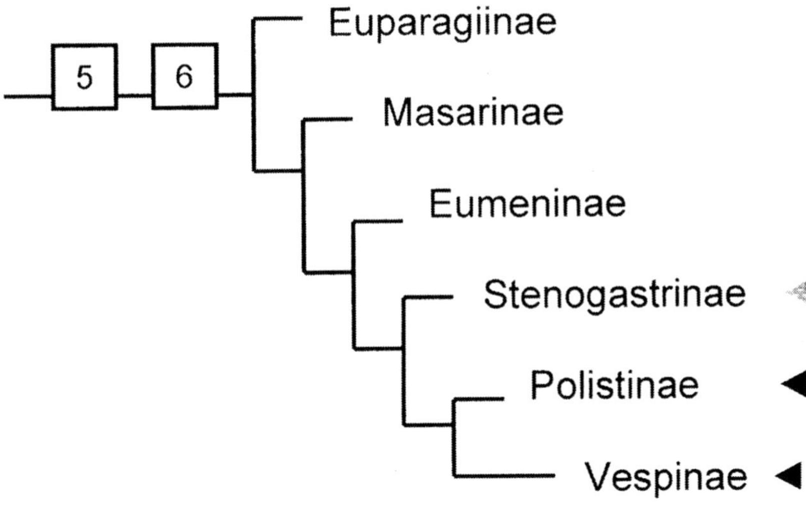

Figure 2.1
Cladogram of subfamilies of Vespidae, from Carpenter (1982). Traits are: (5) nesting behavior; and (6) oviposition preceding provisioning, mass provisioning with multiple weevil larvae via central-pace foraging. Pointers indicate the subfamilies that contain only social forms. The gray pointer for Stenogastrinae denotes facultative sociality.

Vespidae (Gess and Gess 1992; Gess 1996). Two traits associated with nesting behavior that typify Vespidae and are of relevance to the origin of sociality are placed at the base of the clade (node 6) in figure 2.1. First, euparagiines are hunting wasps that seek out beetle (weevil) larvae as prey that they sting, anesthetize, and carry via flight to their nests (Williams 1927). Of particular importance is the fact that they bring up to 30 weevil larvae over multiple foraging trips to provision a single cell (Clement and Grissell 1968; Trostle and Torchio 1986). This means that the link between prey/provision size and wasp size has been broken. Whereas a parasitoid wasp must be reared on a single prey item with greater mass than the wasp that will develop by feeding on it, a hunting wasp can be reared on prey items that are individually smaller yet collectively larger than the wasp that develops. Use of small prey may foster ecological efficiencies in comparison to searching for single, large prey items, and these efficiencies probably enable successful central place foraging, which is a necessary concomitant of nesting. Provisioning in euparagiines is completed and the nest cell sealed before the larva ecloses from its egg (= "mass provisioning").

The second important trait is that euparagiines lay an egg in the empty nest cell before provisioning begins (Moore 1975; Trostle and Torchio 1986). This means that whether an egg is fertilized (female) or not (male) is not determined by the ovipositing female's assessment of the size of a prey item as occurs in idiobiont parasitoids. Selection thus can operate on brood gender sequence in successively provisioned nest cells as an adaptive life history trait. Little more is known about the behavior, life cycle, and ecology of these wasps.

Masarinae

Masarinae represents an extraordinary independent origin of the principal trait that characterizes bees: they provision their larvae not with prey but with pollen. Like many bees, pollen wasps have long "tongues"—mouthparts with which they imbibe nectar from flowers with deep corollas (Gess 1996). In contrast to the majority of bees, which transport pollen externally, masarines take pollen mixed with nectar into their crop and regurgitate it into the nest. F. W. and S. K. Gess have been the most prolific contributors to our knowledge of the nesting behaviors and life history of Masarinae, and almost all of what is known about pollen wasp biology is summarized in a volume by S. K. Gess (1996).

The 300 or so masarine species are found primarily in Earth's five Mediterranean climate zones and nearby semiarid regions. Many excavate nests in nonfriable soil that they soften with water; others construct mud nests above ground, also using crop-transported water. Nest cells have smooth internal walls and are bonded by water, nectar, or in one remarkable genus by silk spun from glands at the mouths of adults (Gess and Gess 1992). Pollen for larval provisioning is gathered from a narrow range of plant species over successive foraging trips and molded into a single mass in a cell. The mass-provisioned nest cell is then sealed before the larva ecloses from its egg. Most species produce one generation of generally 10 or fewer offspring per year, and larvae pass the unfavorable season in diapause as postdefecation final-instar larvae ("prepupae") in their nest cells. Some species have two generations annually, but the relatively short lifetimes of adults and long larval development times for the first generation result in little or no overlap of generations. Although aggregations of nests are common, and reuse of nests is known, nest-sharing by adult females has been ascribed only to a single tropical species (Zucchi et al. 1976), which, together with one other species (Brauns 1910), also exhibits successive placement of pollen loads into open nest cells containing growing larvae ("progressive provisioning"). This, apparently, is the closest approach to sociality among the pollen wasps. The pollen-gathering life history of all masarines makes clear, however, that these cases evolved independently and do not foreshadow sociality in other vespids.

Eumeninae

Eumeninae is the largest currently recognized subfamily of Vespidae. It contains more than 3000 described species, some of which have life history and reproductive biology traits that clearly foreshadow those of social vespids (Evans 1977a). Lamentably, this biologically rich and fascinating subfamily has received only fragmentary and inconclusive phylogenetic analysis (Carpenter and

Cumming 1985; Vernier 1997). Much of the twentieth-century taxonomy of eumenines featured splitting genera and naming new ones (Carpenter and Cumming 1985), leading to a classification of 211 genera (J. M. Carpenter, personal communication) that has been characterized as "chaotic" (Parker 1966) and "irrational" (Menke and Stange 1986). Current researchers (Carpenter and Garcete-Barrett 2002) acknowledge this state of affairs but defer its resolution until a later date. I nonetheless present life history information on the subfamily in a way that suggests sequential evolutionary transitions. Indeed, eminently logical transitional sequences for traits germane to the evolution of sociality are easy to draw from the natural history literature. It must be stressed, however, that such sequences are wholly speculative in the absence of a comprehensive phylogeny of eumenine genera.

Some eumenines nest in soil (Gess and Gess 1976; Iwata 1976). Notable among these, several species of *Odynerus* transport water in their crops that they use to moisten and excavate clay soil (Isely 1913; Bohart et al. 1982). Nest cells, constructed individually, receive an egg (Cowan 1991) before being mass provisioned with multiple weevil larvae (Iwata 1976; Miotk 1979; Bohart et al. 1982). This nesting pattern is identical to that of the Euparagiinae. A few other eumenine genera also prey on beetle larvae, among which the use of larval Chrysomelidae (Iwata 1976) is noteworthy. Chrysomelids are leaf-feeding beetles. Some eumenine nests have been found with chrysomelid larvae plus leaf-feeding larval Lepidoptera (caterpillars) in a single nest cell (Evans 1956; Krombein 1967). The majority of Eumeninae provision their larvae exclusively with caterpillars (Krombein 1967; Iwata 1976; Cowan 1991). Many eumenine species are generalists and use caterpillars from several families of Lepidoptera (Iwata 1976).

Most soil nests (figure 2.2) are excavated by the female that reproduces in them. Another nesting pattern was described by Markin and Gittins (1967), who observed a soil-nesting *Stenodynerus* at the start of its nesting season. New wasps exited their natal cells by means of the previous year's tunnel, which remained largely intact. After several days, a single female remained in possession of the tunnel, to which she added her own new nest cells. Thus, this single female, from among the total brood, nested as a "renter" (Iwata 1976) rather than as an "excavator"—she used a preexisting construction. Other examples of renting soil nests include use of preexisting burrows excavated by tiger beetle larvae (Knisley 1985) and a burrow excavated by a sphecid wasp (Evans 1977a). Behaviors such as these could exemplify the antecedent state of renting nest cavities above ground, a behavior that is common among eumenines. Most renting involves hollow twigs of shrubs and trees, and vacated cells of nests constructed by other species of wasps or bees are also used. Renters also will occupy, with apparent ease, devices such as bamboo tubes or holes drilled into blocks of wood, collectively called nest traps (figure 2.3). Wasps that use nest traps are called trap nesters, and they have been extensively studied (Krombein 1967). Evans (1956) noted nesting plasticity in a species that is primarily an excavator but that Hartman (1905) observed renting crevices in brick walls and fence posts and closing the nests with mud. In a reverse case of nesting plasticity, Cooper (1979)

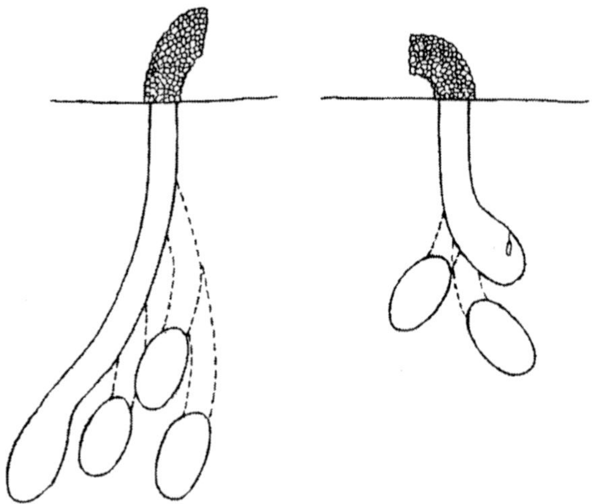

Figure 2.2
Excavated nests of a soil-nesting eumenine, *Rygchium annulatum arvense* (Saussure). Cells are constructed and provisioned individually. The nest at left shows one open cell and three complete cells, with the access to the latter now filled with dirt. The nest at right shows a cell with a newly oviposited egg suspended from the ceiling. Each nest is surmounted by a turret constructed of mud pellets. Figure by H. E. Evans from Proceedings of the Entomological Society of Washington (1956, vol. 58:266).

reported a single individual of a species, previously known only as a renter, that had excavated nests in a piece of Styrofoam inside an above-ground closed box. The excavated nests, with partitions and closures made of mud, were strikingly similar to those of some ground-nesting, excavating eumenines. West-Eberhard (1987a) reported the probable plasticity of a eumenine that both constructs nests and rents existing cavities.

Most excavating and renting Eumeninae use mud for construction. Excavators often add a mud turret above ground at the nest entrance (figure 2.2), and renters commonly use mud to make partitions between cells inside a tubular cavity and to cover the opening of the tube (figure 2.3). Either of these might exemplify the behavioral state antecedent to a nesting behavior as common as, or more common than, renting: many eumenines use mud to construct nest cells above ground. Some nests built by "builders" (Iwata 1976) are tubular, with each tube containing a linear series of cells (Iwata 1938a; Jayakar and Spurway 1966), which makes these nests identical in some ways to nests of tube-nesting renters. Species in several genera build mud nests that are a cluster of elongate cells, and the cluster is sometimes covered by a heavy coat of mud (Roubaud 1916; Claude-Joseph 1930; Evans 1973). *Stenodynerus* and *Ancistrocerus* are genera in which excavator, renter, and builder species all occur (Evans 1956; Evans and Matthews 1974).

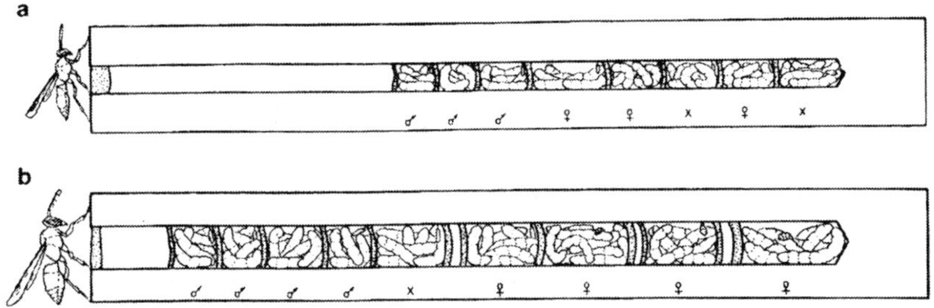

Figure 2.3
Trap nests of two eumenines, *Ancistrocerus adiabatus* (a) and *Euodynerus foraminatus* (b). Both are mass provisioners. Female offspring are in the more interior cells, and males in cells closer to the opening; X indicates an offspring that did not successfully develop. From Figure 1 in D. P. Cowan, 1981, "Parental investment in two solitary wasps *Ancistrocerus adiabatus* and *Euodynerus foraminatus* (Eumenidae: Hymenoptera)," *Behavioral Ecology and Sociobiology* 9:95–102, © Springer-Verlag 1981. With kind permission of Springer Science and Business Media.

Another common form of built nest is an independent cell similar in its interior to the cell of an excavated nest. These single-celled mud nests are often elegant (figure 2.4), and they bestow the name "potter wasps" on their architects. A female potter wasp will build a new one-celled nest for each of her offspring, and the nests of a single female are often found near one another in a row or cluster. In Guyana I once saw 25 independent pots evenly spaced in a row—apparently the work of a single female. Such nests most often remain independent of one another, but in some species the cells are massed together (Iwata 1953). The cells may receive additional exterior mud (Iwata 1953; Jayakar and

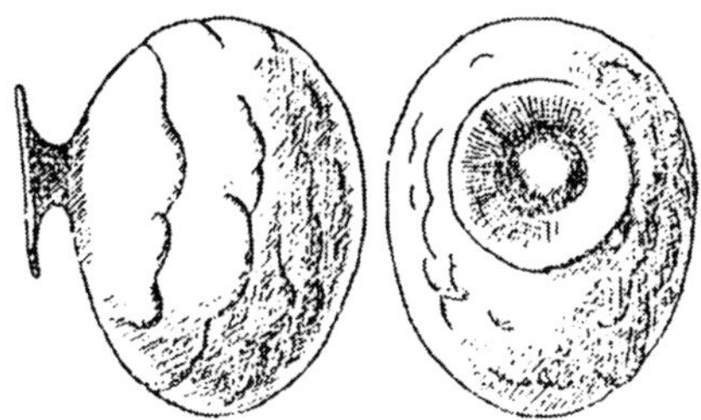

Figure 2.4
A mud pot single-cell nest of a potter wasp, *Eumenes micado*, shown from the side (left) and in frontal view. After the cell is constructed, the wasp inserts her petiolate gaster through the narrow opening and attaches an egg to the roof of the cell, which she then fills with caterpillar provisions before sealing the opening. From Iwata (1953). Reprinted with permission.

Spurway 1965), which in closely adjacent or massed cells can result in an aggregate nest that is more or less amorphous (Hingston 1926; Claude-Joseph 1930). Some tropical potter wasp species that build cluster nests may have more than one adult female at the nest, and more than one nest cell may be open and provisioned simultaneously (Evans 1973; West-Eberhard 1978a, 1987a).

Mud, which may be mixed with saliva (Roubaud 1911) as well as water, is the primary, but not the only, plastic material used by eumenines. Cowan (1991) reviewed the literature on nest materials such as leaf fragments, macerated leaves, and plant gums. Iwata (1939, p. 85) reported a remarkable *Eumenes* that coats its pot nests with wood-pulp paper: "Often I observed females collecting fibre with her mandibles, just as many *Polistes* do, on the weather-beaten surface of a pole."

Diverse parasitoids exploit larval Eumeninae as hosts, including ichneumonids (Hymenoptera, "parasitica"), cuckoo wasps (Hymenoptera, Aculeata), phorid flies, and rhipiphorid beetles. Tubes at the entrances to eumenines' excavated nests, plugs closing the entrance to renting nests, and heavy coats of impenetrable mud or camouflage coatings of plant material on aerial nests may have been selected for their efficacy in defense against such parasitoids. Some behaviors may have similar adaptive significance. Most eumenines mass provision their nest cells: they fill a cell with caterpillars that have been stung and anesthetized and then seal the cell. Some mass-provisioning wasps may remove the caterpillars already placed into a cell and carefully examine each one before stuffing the caterpillars back into the cell (Cowan 1981). A caterpillar is sometimes discarded during such inspections (Rau and Rau 1918), presumably because a parasitoid egg or larva has been detected (Cowan 1981). Such inspections resemble provisioning behaviors of some eumenines in which a female wasp will provision slowly or after an initial delay, thereby placing caterpillars into a cell that contains a wasp larva that has eclosed from its egg and is feeding on the caterpillars already placed there by its mother (Cowan 1991). This behavior is called progressive provisioning. Uneaten caterpillars may occasionally be inspected and, presumably, parasitoids removed (Itino 1986) before provisioning is completed and the cell is sealed.

Cells remain unsealed until the end of larval growth in the few species that practice "fully progressive provisioning" (Cowan 1991), a behavior found only in tropical locales (Krombein 1978). Progressively provisioning wasps mandibulate the caterpillars that they bring to nests (Iwata 1938a), which may kill parasitoid eggs or larvae inside the caterpillars and may also kill the caterpillars (Cowan 1991). *Synagris cornuta* brings a caterpillar to the nest only after mandibulating it into an unrecognizable pulp, which is placed directly on the mouthparts of the larva (Roubaud 1911; figure 2.5). Other *Synagris* species mass provision or progressively provision with intact caterpillars (Roubaud 1908). The progressively provisioning *Antepipona tropicalis* constructs, oviposits into, and begins to provision a second cell while continuing to provision a partially grown larva in a preceding cell (Roubaud 1916).

Mean fecundity for mass provisioning species rarely exceeds 15, and maximum fecundity rarely exceeds 30. A reported maximum fecundity of 60 (Cowan

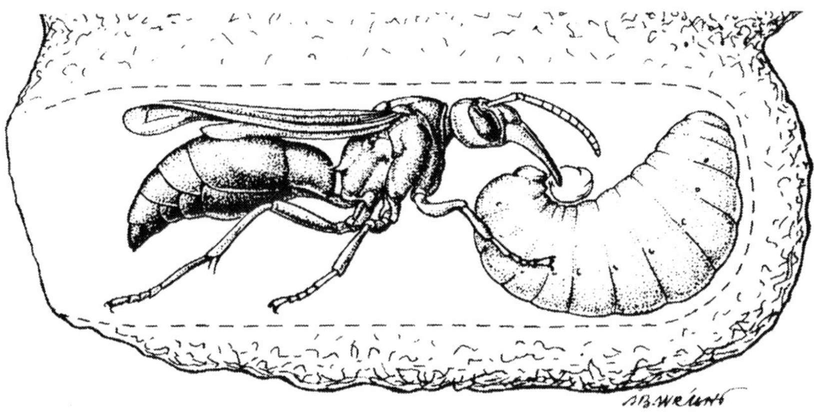

Figure 2.5
Provisioning of a late instar larva of *Synagris cornuta*. The caterpillar provision has been thoroughly mandibulated by the provisioning female, and it is placed just ventral to the mouthparts of the larva and left there for the larva to feed upon. From Cowan (1991).

1991) is exceptional. Progressively provisioning species have lower lifetime fecundities of only four to eight (Cowan 1991), which reflects the time requirement of tending an open nest cell throughout larval development. A female *Synagris cornuta* spends 1 month raising a single offspring (Roubaud 1911). Longer egg maturation time in progressively provisioning species is one component of their lower fecundity (Cowan 1991).

Larvae eclose from their egg covering (the chorion) by chewing through one end. Some crawl free immediately (figure 2.6), but others remain attached to the egg filament through the first instar and then fall free sometime after the first molt (figure 2.7). As larvae then feed and grow, they pass through five instars

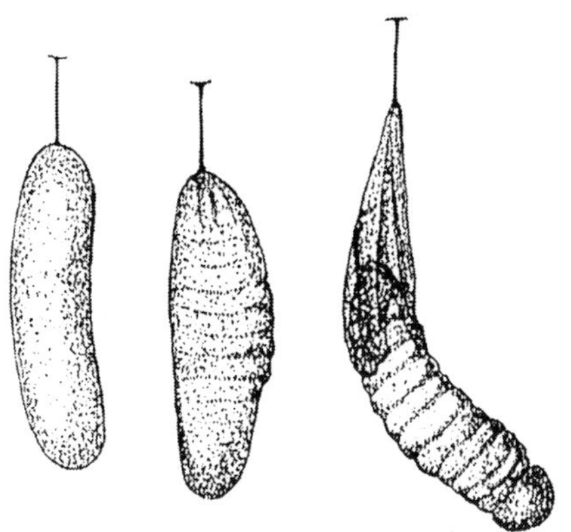

Figure 2.6
Eclosion of a new larva of *Ancistrocerus* sp. from the chorion that enclosed its egg. The egg had been suspended from the top of the cell within a nest trap. After chewing through and then wriggling from the chorion, the first-instar larva fell free and began to feed on caterpillars that had been mass provisioned in the cell. From Taylor (1922).

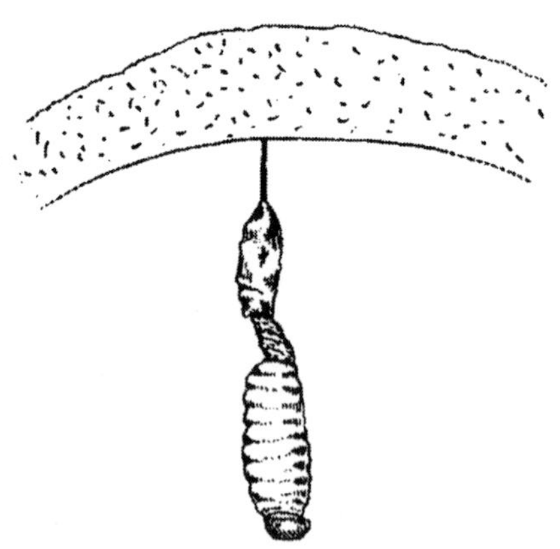

Figure 2.7
A second-instar larva of *Eumenes fulvipennis*. The larva is attached to the shed exoskeleton (exuvium) of the first-instar larva, which is attached to the empty chorion of the egg, which is suspended by a filament from the roof of the nest cell. Thus suspended, the larva feeds on stored provisions (caterpillars) until it falls free (at the next molt?). From Williams (1919).

(Cooper 1966). Increases in ambient temperature can cause accelerated larval development (Jayakar and Spurway 1968). Pupae of potter wasps have a distinctive, folded posture that is coincident with an elongate, petiolate gaster in adults (figure 2.8).

Eumenines at upper latitudes, high altitudes, or seasonally arid locales may have only one generation per year, and wasps in these environments pass the

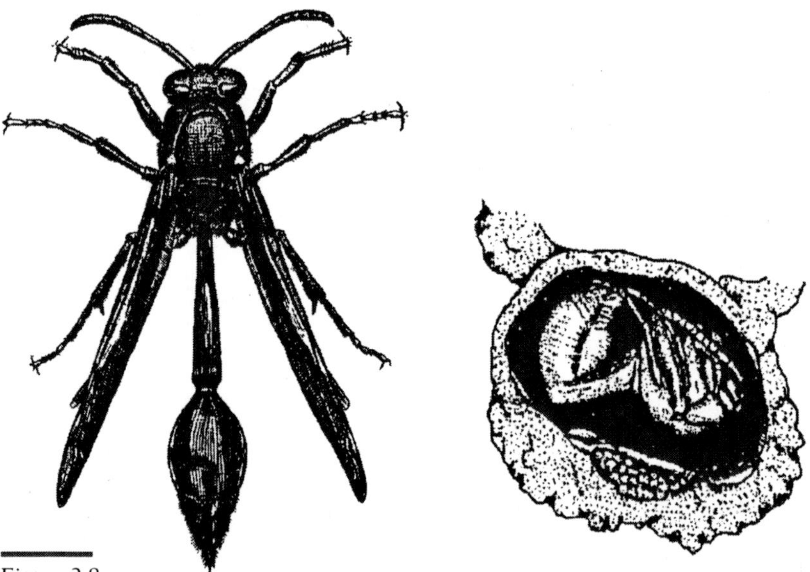

Figure 2.8
Adult female *Eumenes curvata* and a cross-section of a nest cell showing a pupa within. In addition to the usual abdominal constriction of all Aculeata, there is a jointed constriction between abdominal segments 2 and 3 (= gastral segments 1 and 2) at which the abdominal flexion occurs. From Williams (1919).

unfavorable season within their natal nest cells as last instar larvae in prepupal diapause. Diapause is a developmental arrest that is genetically programmed, rather than being a response to local environmental cues (Flannagan et al. 1998; Denlinger 2002). Many eumenine species in seasonal locales have two generations per year: the first generation has continuous development, and the second enters prepupal diapause (Iwata 1938a; Medler and Fye 1956; Medler 1964; Itino 1986). Some tropical species produce offspring throughout the year without diapause (Taffe 1978).

Individuals of the same sex vary in size as a function of the amount of provisions they received as a larva (Iwata 1953), but experimental overfeeding showed that large larvae eventually stop feeding even when provided additional food (Cooper 1957). A large female of *Monobia quadridens* differed from a smaller female in the presence of storage proteins residual in the fat body after metamorphosis (Hunt et al. 2003). Larger adult females tend to have higher fecundity than smaller conspecific females (Cowan 1981).

For eumenines in trap nests or those that build tubular multicell nests, the innermost cells are almost always longer, receive more provisions, and produce female offspring, whereas cells nearer the opening are shorter, less abundantly provisioned, and produce male offspring (Taylor 1922; Iwata 1938a; Rau 1945a; figure 2.3). Because oviposition precedes provisioning and because females are generally larger than males, an adaptive link must exist between fertilization of the egg (yielding a female) and subsequent provisioning cues (Medler 1964; Jayasingh 1980). A consequence of the gender sequence in a trap nest is that males, which are closer to the opening, emerge before females. The gender sequence for excavating and building eumenines has not been well studied, but some builders in tropical locales oviposit unfertilized eggs and produce males from their first nest cells (Jayakar 1963; Brooke 1981). Emergence of males before females therefore seems to be a general phenomenon in eumenines. One consequence is that newly emerged males can remain at their natal nest and mate with nestmate females as soon as they emerge from pupation (Taylor 1922; Jayakar and Spurway 1966; Cowan 1979, 1986, 1991). Males may also patrol for females at nesting areas, at water and/or mud sources, and at flowers (Rau 1935; Cowan and Waldbauer 1984; Cowan 1991). Trap nests have been found that contain only male brood, only female brood, and brood of both sexes. There is no evidence that a female eumenine might go unmated, and mating probably occurs soon after female emergence or at least before the onset of nesting.

Flower visitation is commonly seen in eumenines, which seem to be generalists rather than specialists in their nectar choices (Markin and Gittins 1967; Heithaus 1979; Haeseler 1980). On rare occasions females have been observed to feed on the hemolymph of caterpillars that they then discard rather than use as provision for their larvae (Iwata 1938b; Rau 1945b). Although females sting the caterpillars that they take as larval provisions (Steiner 1983; Veenendaal and Piek 1988), eumenines rarely sting in defense of their nests, although West-Eberhard (1987a) has been stung at the nest by two species. Eumenine stings are painful to humans who mishandle captured wasps and, therefore, presumably also to vertebrate predators such as birds, which may play a role in the seemingly aposematic coloration of many eumenines (Waldbauer and Cowan 1985). Foraging eumenines fall prey to diverse invertebrate predators, including robber flies, praying mantises, and spiders (Isely 1913; Iwata 1953).

Stenogastrinae

Hover wasps are the will-o'-the-wisps of the Vespidae. These slender, delicate, and secretive wasps, the Stenogastrinae, are characteristic of rainforest habitats from south India across the Indochina peninsula and Indonesia to the Philippines and New Guinea. The smallest are around 1 centimeter or slightly less in length, and the largest are only about two centimeters (Ohgushi et al. 1983; Turillazzi 1991). More than 50 named species (Carpenter and Kojima 1996; Turillazzi 1999) occur in 7 genera (Carpenter and Starr 2000; figure 2.9). Numerous aspects of stenogastrine life history are unlike those of any other Vespidae or, for that matter, any other Hymenoptera.

Outstanding among the unique traits of Stenogastrinae are behaviors of oviposition. Adult female hymenopterans have a ducted gland that, in Aculeata, opens near the base of the sting into the chamber through which eggs pass during oviposition (Downing 1991). Called Dufour's gland, it serves diverse roles in ants and bees, but its role in wasps is uncertain except in Stenogastrinae (Downing 1991). The Dufour's gland of hover wasps produces a gelatinous, milky secretion (Jacobson 1935) that has come to be called "abdominal substance" (Turillazzi 1991). Hansell (1982), Turillazzi (1985a, b, c, 1991), and Sledge et al. (2000) have described the role of the abdominal substance in oviposition and larval provisioning. As the wasp begins oviposition, she flexes her petiolate gaster and expels abdominal substance, which she grasps with her mouthparts and forelegs. After working the material into a wad, she flexes her gaster forward a second time and expels an egg onto the wad of abdominal substance that she holds in her mandibles and forelegs (see photographs in Turillazzi 1985c, 1991). Holding the wad and attached egg in her mandibles, she places the sausage-

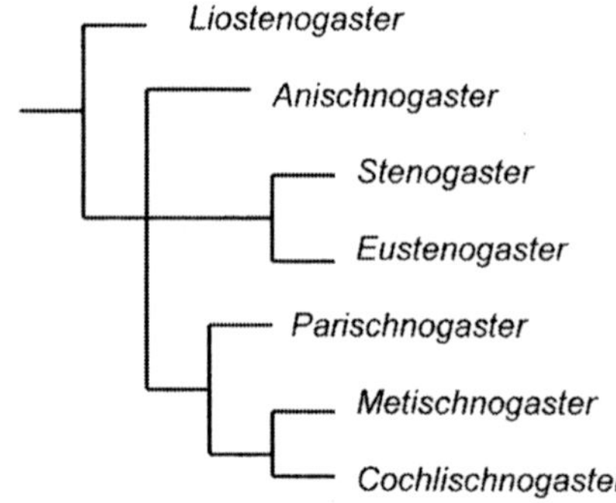

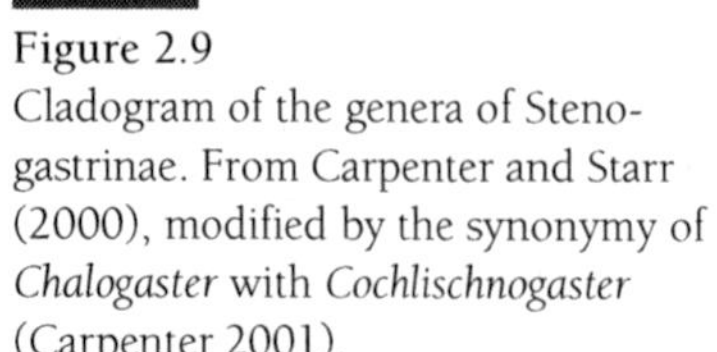

Figure 2.9
Cladogram of the genera of Stenogastrinae. From Carpenter and Starr (2000), modified by the synonymy of *Chalogaster* with *Cochlischnogaster* (Carpenter 2001).

shaped egg into a nest cell, where adhesive on the egg causes its convex surface to adhere to the cell apex. The wad of abdominal substance is left adhering to the concave surface of the egg. Later, more abdominal substance is added to the wad (figure 2.10). After about 1 week of embryonic development (Turillazzi 1991), the new larva exits the egg chorion and worms its way onto the wad of abdominal substance until it is coiled around it (figure 2.11). A provisioning female places regurgitated liquid onto the egg before it hatches, and she continues to place both liquid and, after the larva is 1 day old, solid food onto the wad of abdominal substance (Turillazzi 1985a). There is no direct (i.e., mouth-to-mouth) provisioning (Turillazzi 1985a). The larva consumes the provisions and small amounts of abdominal substance (S. Turillazzi, personal communication), although the abdominal substance is non-nutritive (Sledge et al. 2000).

This sequence is extraordinary and unique. Placement of eggs into the nest cell using the mandibles but without using abdominal substance has been observed in *Anischnogaster* (Turillazzi and Hansell 1991) and may also occur in *Stenogaster* (Spradbery 1975). Turillazzi (1989) has suggested that the energetic cost of producing abdominal substance could be a constraint on stenogastrines' ability to rear numerous larvae simultaneously.

Other distinctive features of Stenogastrinae are the nature and, especially, the diversity of their nests (Pagden 1958; Iwata 1967; Ohgushi et al. 1983, 1986; Wenzel 1991). All nests are built above ground, where they are placed on leaves, small branches, plant rootlets exposed along stream banks or road cuts, beneath boulders, or on the roofs and walls of caves or structures such as buildings and

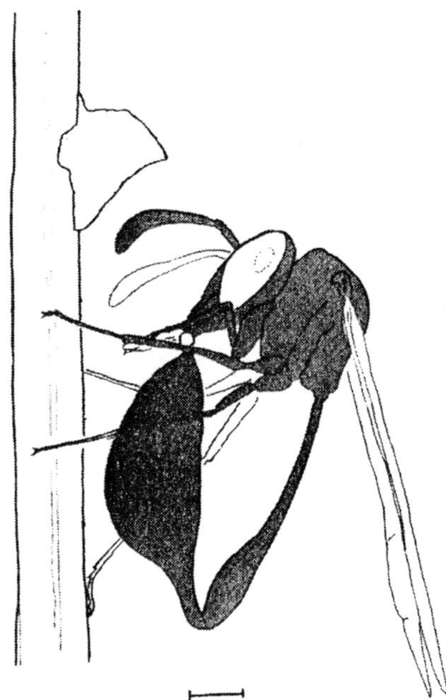

Figure 2.10
Secretion and collection of abdominal substance by *Parischnogaster mellyi* prior to placing it into the nest cell, which is just above the wasp's antennae. From Hansell (1982).

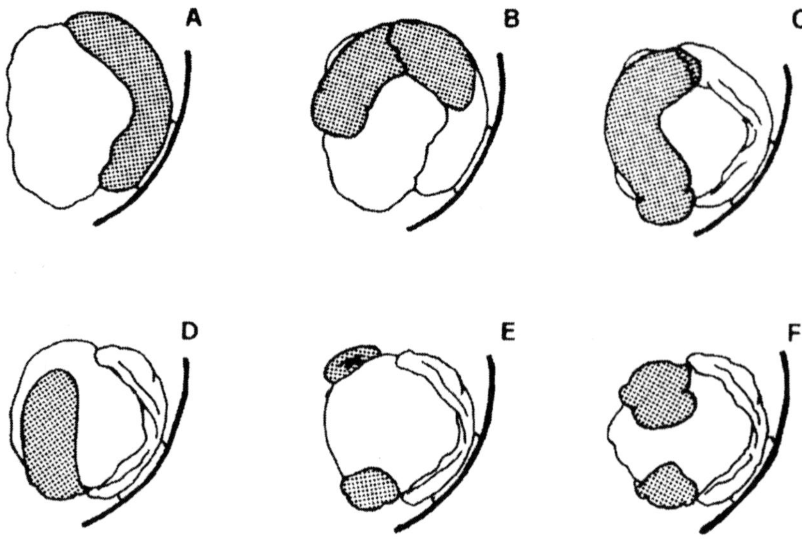

Figure 2.11
Eclosion of a new larva of *Parischnogaster mellyi* from its chorion. The larva (shaded) is shown in A within the chorion, which is attached by adhesive to the cell wall and with a wad of abdominal substance on the concave surface. The larva then chews through one end of the chorion (B) and crawls onto and over the wad of abdominal substance (B–E) until it has assumed a position wrapped around the wad (F). The chorion remains after the larva has crawled free from it. From Turillazzi (1985b). Reprinted with permission.

bridges. Architectural diversity impressed pioneering investigators (Pagden 1958), and discovery of new species has expanded the catalog of known nest types (Turillazzi 1999). No other group of wasps shows as much variety. Bringing conceptual order to this diversity is a nut that has not yet been cracked, for although the genera of Stenogastrinae have been subjected to cladistic analysis (figure 2.9), the species have not. Therefore, the same caveat applied to Eumeninae must be applied to Stenogastrinae: in the absence of cladograms, evolutionary scenarios for changes in life history variables for species within a genus are speculative. Indeed, some of the scenarios and contrasts presented below are based on observations of species in different genera. However, the natural history data show that whereas nests are diverse, essential features of life history and development are similar across species and genera.

Nest architecture in *Liostenogaster*, the sister genus to other stenogastrine genera, is nearly as diverse as that of the subfamily as a whole (Turillazzi 1991, 1999). A *Liostenogaster* nest made of mud and consisting of parallel, contiguous, cylindrical cells (Williams 1919) could exemplify the basic form for the subfamily (Turillazzi 1991) and is similar to nests of some Eumeninae. Mud nests of more complex architecture are constructed by some *Liostenogaster* species as well as by species in other stenogastrine genera. A *Parischnogaster* species uses mixtures of mud and plant material (Coster-Longman and Turillazzi 1995). Some *Liostenogaster* and species in other genera build exclusively with plant material, which can be "fine crumbs and small flakes" (Hansell 1981) and can be bonded together with oral secretion (Kudô et al. 1996). One nest made of "brittle carton" was green and resembled a leaf (Ohgushi et al. 1986). Turillazzi and Pardi (1982) observed a female *Parischnogaster nigricans* gathering nest material by scraping wood fibers in the manner of paper wasps. Hansell (1981, 1985, 1987a) has argued that heavy mud and paper of poor quality have constrained the evolution of large colony size in Stenogastrinae.

Structural diversity among nests is extreme. In *Liostenogaster* alone nests may consist of rows of cells, two tiers of cells, multiple rows of tiered cells, or the extension of a row or tier of cells into a semicircle with the openings of the cells facing inward. Complete circles of cells are known, looking rather like a donut with the cell openings facing the hole. Cells may be attached to the substrate at their closed ends rather than along one side, and such cells can be in a row or cluster. Most clusters have the closed ends of all cells against the substrate, but some may have only the initial cell attached, with additional cells being longitudinally offset such that there is a gap between their closed end and the substrate (Turillazzi 1999). Clusters of cells in full contact with the substrate may be surrounded by a perimeter wall that extends beyond the cell openings. The perimeter wall may be a full enclosure with the cells opening into the resulting chamber, which has an opening for entry and exit (Turillazzi and Carfi 1996). In *Liostenogaster* and at least three other genera, nests pendant on rootlets or branches also have deposits of abdominal substance between the nest and the attachment point of the supporting plant structure (Keegans et al. 1993). These deposits of abdominal substance have been interpreted as chemical and physical barriers against foraging ants that might prey upon the brood (Pagden 1958; Turillazzi and Pardi 1981; Turillazzi 1985b, 1994).

Most nests of *Liostenogaster* are small, with the largest consisting of about 100 cells (Turillazzi 1989). Two *Liostenogaster* species and one of *Parischnogaster* may form aggregations of hundreds of nests (Coster-Longman et al. 2002). Species in other genera have small nests with diverse architectures that include scattered cells; irregular clusters of cells; compact clusters hanging from a rootlet or small branch; pendant compact clusters enclosed within elegant walls; separate cells arrayed serially, hanging from a slender branch or rootlet; and a single cell hanging from the tip of a rootlet with successive cells hanging individually on the cell built before. One species lines its nest cells with fungal hyphae (Spradbery 1975). When summed across the subfamily (Wenzel 1991; figure 2.12), the nesting diversity of Stenogastrinae is truly astonishing.

Liostenogaster flavolineata has a complex life history that can follow a diverse array of paths (Samuel 1987). Essential elements and similar diversity of life history may typify most hover wasp species (Yoshikawa et al. 1969; Yamane et al. 1983; Turillazzi 1985d, 1989, 1991). Solitary females initiate most new nests (Turillazzi 1991). Other females may take up residence with the foundress at a

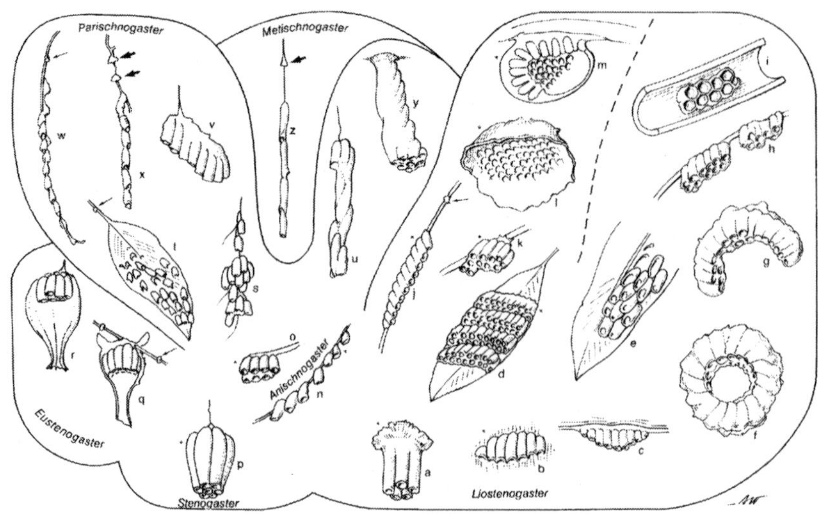

Figure 2.12
Some nest diversity in Stenogastrinae. Asterisks indicate nests made of mud, and arrows indicate ant guards of abdominal substance. For species identities see Turillazzi (1991). From Turillazzi (1991).

recently initiated nest. In *Parischnogaster alternata*, nests occasionally may be founded by small groups of females, such as after destruction of an active mature nest (Turillazzi 1985d) or when nests are closely aggregated (Turillazzi 1991).

Larvae of different ages, each in a separate nest cell, are progressively provisioned simultaneously. The first few offspring are always females. Once female offspring begin flying, they also begin to forage and to participate in provisioning larvae in their natal nest. In time, female offspring mate and pursue any of several nesting activities. They can found a new nest alone or with another female, occupy an existing but unoccupied nest, or displace a single female from her nest and usurp it. They can join a nest that has one or more females already present, although this occurs rarely (J. Field, personal communication). The most common route to nesting is to become an egg layer on the natal nest by succeeding to the uppermost position in a dominance queue.

Usurper females will provision and rear older larvae in a nest they have usurped (Hansell 1987b), but nests with only young larvae are generally not usurped, even if no adult is present (Field et al. 1998). Nest cells may be reused, so nests can remain active beyond the lifetime of particular individuals (Turillazzi 1991). Although *Anischnogaster* species have only one or two females per nest, they exhibit many of the reproductive behaviors described above (Hansell 1987b; Turillazzi and Hansell 1991).

Kojima (1990) asserts that stenogastrine larvae pass through five instars, but larger data sets reveal only four instars (Turillazzi 1985a, 1990; Hansell

1986). Stenogastrine larvae spin incomplete cocoons (S. Turillazzi, personal communication) that do not cover the cell opening. Instead, adults seal nest cell openings completely, or in some species only partially, at the end of larval growth (see Carpenter 1988). After the enclosed larva opens its anus and voids its meconium, an adult will open the cell, remove and discard the meconium, and then reseal the cell (Turillazzi and Pardi 1982). Larvae enter metamorphosis without prepupal diapause. The naked pupa has a characteristic folded posture (Williams 1919; Iwata 1967; Spradbery 1975; figure 2.13), with the fold occurring at the joint between abdominal segments 2 and 3 (gastral segments 1 and 2). Full development from egg to adult can take from about 40 to more than 100 days (Hansell 1981; Turillazzi 1985a, 1991). Newly emerged females spend less time on the natal nest as they age, with foraging commencing as early as 4 days of age (Field et al. 1999).

Small, young larvae of Stenogastrinae are provisioned with liquid that presumably is regurgitated from the provisioner's crop (Turillazzi 1985a). Larger, older larvae are provisioned with solid food that has been minced by the provisioner's mandibles (Turillazzi 1985a). The mincing is less thorough than that of paper wasps (J. Field, personal communication). Liquid and solid provisions are both placed on the wad of abdominal substance, where the larva then

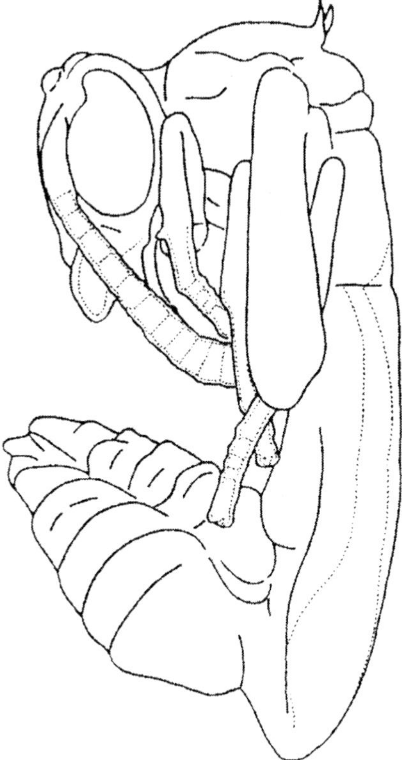

Figure 2.13
Pupa of *Parischnogaster mellyi* showing the folded posture characteristic of all stenogastrines. Dotted lines show outlines of the adult within the pupal cuticle. From Iwata (1967). Reprinted with permission.

feeds on them over the course of a day or more (Spradbery 1975; Turillazzi 1989). Older larvae lie in a closed coil at the base of the nest cell, and they open the coil to allow provisions to be placed in the center (Spradbery 1975; Turillazzi 1985a). This mode of provisioning (figure 2.14) resembles that of the eumenine *Synagris cornuta* (figure 2.5).

Although solid provisions are generally unrecognizable, various stenogastrines use their delicate, hovering flight, which begat their common name (Carpenter 1988), to forage at spider webs, plucking out ensnared insects (Williams 1919; Pagden 1962; Turillazzi 1983a). Iwata (1967) found a whole small caterpillar, which is unlikely to have come from a spider web, and fragments of spiders and adult insects among provisions. Predation of termite sexuals (Samuel 1987) and midges (small flies) in flight has been observed (S. Turillazzi, personal communication). Provision may be moved from one nest cell to another within a nest or taken by adults for their own nourishment (Turillazzi 1985a).

Adult females can take provisions for their own nourishment from returning foragers (Turillazzi 1987) as well as from nest cells. They mandibulate these provisions for extended periods, presumably extracting liquid, and then discard

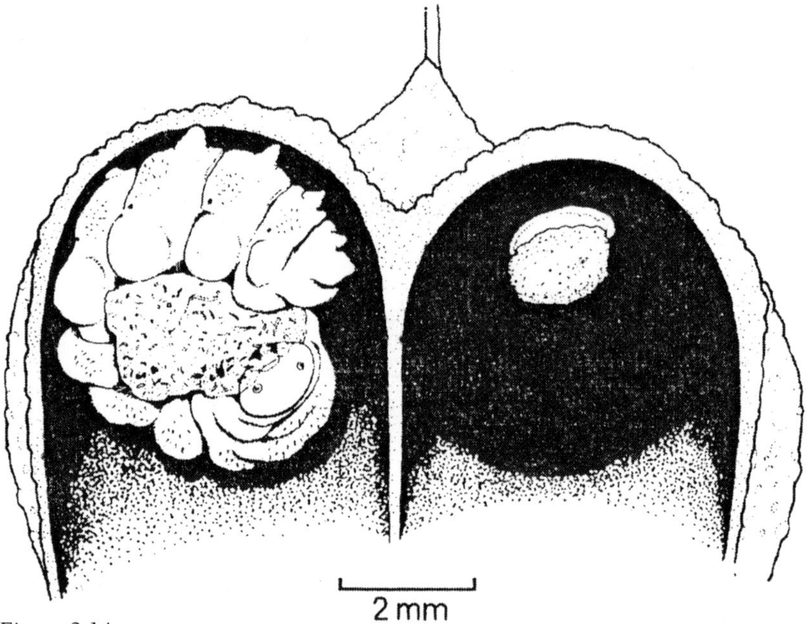

Figure 2.14
Two nest cells of *Stenogaster concinna*. At left, a large, late instar larva feeds on a provision mass that has been placed by an adult female within the sphincter formed by the coiled larva. The cell at right contains an egg adhering to the cell and a wad of abdominal substance adhering to the egg. The nest is suspended from a slender rootlet. From J. P. Spradbery, 1975, "The biology of *Stenogaster concinna*," *Australian Journal of Zoology* 14:309–318, Blackwell Publishing Ltd. Reprinted with permission.

the solids rather than providing them to a larva (Turillazzi 1985a). Adult females may expel a drop of "salivary fluid" (Turillazzi and Pardi 1982), which I think is more likely to be liquid regurgitated from the crop than secreted from a gland, in response to aggressive solicitation by one of their nestmates (Turillazzi 1987). Females may lick a clear liquid from the ventral abdominal surface of their larger larvae (Turillazzi 1991). Turillazzi (1987) describes adults that solicit liquid expressed by larvae. Larvae of *Stenogaster* (Spradbery 1975) and of *Parischnogaster* and *Liostenogaster* (S. Turillazzi, personal communication) have enlarged salivary glands, which are the probable source of liquid that the adults solicit and drink. The nutritional value of the liquid has not been documented.

Egg eating (Hansell 1982; Turillazzi and Pardi 1982) occurs frequently (J. Field, personal communication). The egg layer may eat her own eggs immediately after oviposition (Turillazzi and Pardi 1982), and an invading female may eat eggs in a nest that she has just taken over (Turillazzi 1985d). Eggs may also be preyed upon by the same or another species of stenogastrine (Turillazzi 1985a; Turillazzi et al. 1997).

When male offspring appear on a nest, they take provisions from foragers (Turillazzi 1987) or from nest cells. They feed on these provisions by keeping them in the mouth for extended periods, presumably extracting liquids, and then dropping the remaining solids rather than provisioning a larva (Turillazzi 1985a). Males also will feed on eggs taken from nest cells (Hansell et al. 1982; Turillazzi and Francescato 1989). Both J. Field (personal communication) and S. Turillazzi (personal communication) have seen stenogastrines visit flowers, but flowers are scarce in rainforests, and there are no published observations of flower visitation as a common behavior.

On average there are two to four females per nest, with a maximum in the low teens (Turillazzi 1989; Field et al. 1998). Females on a nest may form a dominance hierarchy (Yoshikawa et al. 1969; Turillazzi and Pardi 1982; Turillazzi 1986), or at least a principal egg layer will emerge that spends more time at the nest and less time away from it than her nestmates (Yoshikawa et al. 1969; Turillazzi and Hansell 1991). Mature colonies may have multiple females capable of laying eggs, although some may resorb their eggs rather than lay them (Turillazzi 1985d). A single female lays most of the eggs (Sumner et al. 2002). Offspring females remain for a time at the nest in a pattern of serial occupancy in which older offspring eventually leave the nest or succeed to the top of the dominance queue and become the principal reproductive (Hansell et al. 1982), while new offspring continue to emerge (Turillazzi 1985d, 1989). All females have the capacity to become egg layers (Field and Foster 1999). Young females' ovaries reach reproductive capacity in about 2 months (Turillazzi 1985d, 1989). Mating can occur at any age (J. Field, personal communication), and mated females have more developed ovaries than unmated females of the same age (Turillazzi 1985d). Females may leave their natal nest without immediately founding or joining another nest, thus forming a pool of potential founders or joiners (Turillazzi 1985d). The assertion of Samuel and Hansell (cited in Turillazzi 1991) that some *Liostenogaster flavolineata* females never mate and remain permanently at their natal nest as "true workers" is now interpreted as the death of individuals before they succeed to the reproductive position at the top of the dominance queue (Field et al. 1999; Shreeves and Field 2002).

Mating has been seen rarely and never during nest observations, so it presumably takes place only away from nests. Males may aggregate on roots or leaves (Williams 1919; Spradbery 1975; Turillazzi 1983b) or on nests (Turillazzi and Pardi 1982). Females may subsequently initiate new nests at male aggregation sites (Turillazzi and Francescato 1989). Males leave clusters at specific times of day to fly in open areas (Turillazzi 1983b), where they perform "patrolling" flights (Pagden 1962; Turillazzi 1983b; Beani and Turillazzi 1999; Beani et al. 2002). Patrolling may involve pheromone emission (Turillazzi and Francescato 1990), so patrolling aggregations could be leks to which females come for mating (Pagden 1962; Turillazzi 1983b). Males are often present on nests (Pagden 1958), but they do not participate in nest construction or larval provisioning. Because nesting cycles in local populations are asynchronous (overlapping) and aseasonal, males probably are continuously present in a population. The disruptive effect of numerous males on a nest intercepting incoming provisions and eating eggs can curtail nest growth and possibly even terminate reproduction on that nest (Turillazzi 1985d).

Some parasitoids use hover wasp larvae as hosts. Hornets in the genus *Vespa*, especially *V. tropica*, and various ants are important predators of hover wasp brood in nest cells. C. K. Starr (personal communication) found evidence of predation by a bat on a nest of a cave-dwelling *Parischnogaster*. Hover wasps will fall or fly from the nest if disturbed by humans, and they rarely sting (Turillazzi and Pardi 1982), although stings are painful to investigators who mishandle captured wasps (J. Field, personal communication).

Questions Arising

Are the character sequences given in this chapter for Eumeninae concordant with a comprehensive phylogeny of the genera? Traits in living forms that can be placed into a logical series, from soil nesting with mass provisioning to aerial nesting with progressive provisioning, were easily drawn from the literature, but this does not make the sequence valid; it is only conjecture. Eumeninae desperately needs phylogenetic analysis.

If single eumenine genera exhibit multiple nest architectures or modes of larval provisioning, can these characters be phylogenetically informative for the subfamily? Difficulty in assessing characters such as these, which could show high levels of homoplasy, touches on my concern about characters used in existing phylogenetic analyses for all Vespidae. New data, drawn from well-conserved genes and analyzed independently of existing data, would provide independent tests of the current phylogenies and generate phylogenies for taxa such as genera of Eumeninae and species of Stenogastrinae for which phylogenies do not yet exist.

How does a female eumenine match provision mass to egg gender in a nest cell that she mass provisions after ovipositing? The ability of an idiobiont parasitoid to assess the size of its host seems plausible simply by means of visual cues. What cues, however, enable a solitary, mass provisioning vespid to provision a greater mass for females than males? Are they the same cues that affect the decision to oviposit a female versus male egg in the first place?

Does the egg chorion in Stenogastrinae differ from that of other Vespidae? Illustrations in the literature show stenogastrine larvae crawling out of the chorion, whereas in *Polistes* (chapter 3) the chorion seemingly disappears.

Is flower nectar (or honeydew) a significant source of nourishment for adult stenogastrines? *Polistes* conspicuously visit flowers, but rainforest stenogastrines apparently rarely do so. Does the absence of flowers as a nourishment source play a role in stenogastrines remaining at their natal nest to take nourishment from larvae or other adults?

What nourishment, if any, do larval stenogastrines derive from the liquid that adult females regurgitate onto the egg and, after hatching, onto the wad of abdominal substance? What nourishment, if any, do adult stenogastrines derive from the liquid that they lick from the ventral surface of larvae? Nutrient analyses of these liquids remain to be done.

3 Paper Wasps and Vespines

A wasp alights on the unpainted fence in my backyard. I stop what I am doing and step nearer. She touches her mouthparts to the wood, and a dark spot appears where she expels a drop of water from her crop. I lean close. She extends her head and then rotates it rearward, biting with her mandibles on the backstroke. I can hear squeaks as she scrapes fibers off the weathered board. She repeats this a few times until she has collected a small, dark bolus, then she flies to her nest beneath my back steps. She selects a cell, places her mouthparts to its rim, and mandibulates the fibers into a thin sheet of paper that extends the cell a millimeter longer.

Papermaking by wasps was first accurately described in the eighteenth century by René-Antoine Ferchault de Réaumur (for a beautiful article on wasp paper, see Hansell 1989). The wasp that I was watching was *Polistes*—a paper wasp. Réaumur knew of the paper nests of *Polistes*, but his description concerned the sheets of paper that envelop the nests of European yellowjackets (Réaumur 1742a). Paper wasps and yellowjackets represent two subfamilies of Vespidae, the Polistinae and Vespinae, for which making nests of paper is a hallmark. Collectively they all could be called paper wasps, but that name usually is reserved for polistines only. The vespines have other common names: yellowjackets, hornets, and night wasps. Wasps of these two subfamilies share more than making paper. Most of the fundamental aspects of their life cycles and individual development are the same, and all of them are social.

Independent-Founding Polistinae

Among polistines, a distinction is usefully drawn between two modes of colony founding (Jeanne 1980), because many life history features correspond to those founding modes. In "independent founding," nests are founded by one or several inseminated females. In seasonal localities there also is a break in nesting activity that roughly coincides with generations. That is, mature colonies produce female offspring that spend some time apart from any nest before founding or cofounding a new nest. In "swarm founding," nests are founded by relatively large numbers of uninseminated females together with a smaller number of inseminated females. Usually there is little or no time lapse between leaving the old nest and initiating the new, and swarm founding occurs only in tropical and subtropical locales. Nests of swarm founders may not coincide with wasp generations, and in many cases nests clearly are occupied by serial and overlapping generations of occupants. Independent founding characterizes the genera *Polistes*, *Mischocyttarus*, *Belonogaster*, *Parapolybia*, and some but not all species of *Ropalidia* (figure 3.1). *Polistes*, the sister genus to other polistines, is the only cosmopolitan genus of social wasps, and its 200 or so species (Carpenter 1996a) are found in tropical and temperate zones both continentally and on many islands. *Polistes* has been bountiful in yielding its secrets, and it is the focus of the following overview.

Polistês, from the Greek, means city founder.[1] *Polistes* wasps build cities. Their paper nests are always aerial and may be sited on plant stems and twigs, on sheltered faces of rocks and boulders, or on human artifacts such as buildings and abandoned automobiles. Each new nest begins with construction of a stalk, also called a pedicel or petiole, from which the cluster of usually downward-facing nest cells then is suspended (figure 3.2). The construction material is wood fibers mixed with saliva. The stalk is rather rigid and black due to secretions from oral glands applied by "licking" (Downing 1991). The paper cells are somewhat pliable and grey. As soon as the first nest cells are mere cuplike receptacles, an egg is placed in each. Successively initiated nest cells receive eggs in turn, so that several eggs of different ages are present before the first-laid egg ecloses as a larva. As larvae are fed and grow, their nest cells receive additional paper at the rim and are lengthened. Cell extension stops when cells are large enough to accommodate full-grown larvae, but new cell initiation continues at the margin of the nest. Cell diameter is mediated by the constructing wasp's anatomy and behavior (figure 3.3; Downing and Jeanne 1990).

1. Latreille (1802) did not give the etymology of his new genus name. Hamilton (1996) incorrectly said that *polistes* is the plural of the word for a Greek city-state, which is *poleis*. Instead, *polistês* means founder of a city (Liddell et al. 1978), a meaning and spelling that can also be had by making a noun of the verb *polizô*, to found a city, as used by Herodotus (C. Rapp, personal communication).

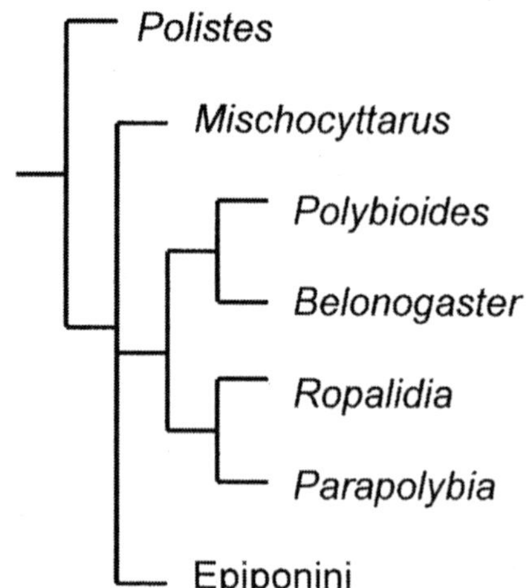

Figure 3.1
Consensus tree of Polistinae. Independent founding characterizes *Polistes, Mischocyttarus, Belonogaster, Parapolybia*, and some but not all *Ropalidia*. Swarm founding occurs in *Polybioides*, some *Ropalidia*, and all 20 genera of the tribe Epiponini. Redrawn from Carpenter (1991).

Figure 3.2
A single foundress *Polistes annularis* on her nest. The wasp is malaxating a caterpillar to provision larvae in central cells of the nest. Eggs can be seen in marginal nest cells. No larvae have yet reached maturity and spun cocoons. The nest cells will be lengthened by the foundress as the larvae grow. From H. E. Evans and M. J. West Eberhard, *The Wasps*, University of Michigan Press, © University of Michigan 1970. Reprinted with permission of the University of Michigan Press.

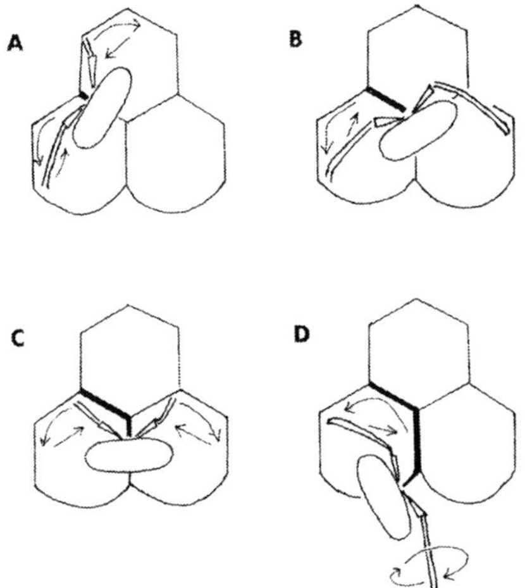

Figure 3.3
Schematics in sequence, A through D, showing the positions of a wasp's head and antennae relative to three nest cells during the application of new paper (thick lines) to the wall that divides the lower left cell from its two neighbors. In the A to B step, the wasp's right antenna moved from the upper cell to the lower right cell. Continuous antennal contact with neighboring cells mediates both cell size and shape. When an antenna encounters no neighboring cell, as in D, the construction at the nest margin is rounded rather than hexagonal. From West Eberhard (1969). Reprinted with permission of the University of Michigan Museum of Zoology.

Nests may be symmetrical with the pedicel in the center, asymmetrical with the pedicel at one margin, or elongate with only two rows of parallel cells with the pedicel at the top (figure 3.4). The variation is modest and can be modeled by a construction algorithm in which change of a single parameter can result in all known forms (Karsai and Penzes 1998). Multiple wasps perform independent acts of construction behavior, and enlargement of the nest as a unified whole is mediated by clues drawn from the existing construction ("stigmergy") (Karsai 1999).

As the nest increases in size and weight (due largely to the brood), the wasps apply oral secretions to the pedicel to increase its diameter and thereby strengthen it. Oral secretions are also added to the upper nest surface (Réaumur 1742b). The amount of secretions added correlates with exposure to rain (Kudô et al. 1998). Wasps rub the nest pedicel with the underside of their gaster, from which cuticular glands add a substance that has been shown to repel ants. This was first described in *Mischocyttarus* (Jeanne 1970) but is also found in *Polistes* (Turillazzi and Ugolini 1979) and the other independent-founding genera (Downing 1991). Wasps also rub the face of the nest comb with the underside of the gaster (West Eberhard 1969), and the secretion applied, which differs from

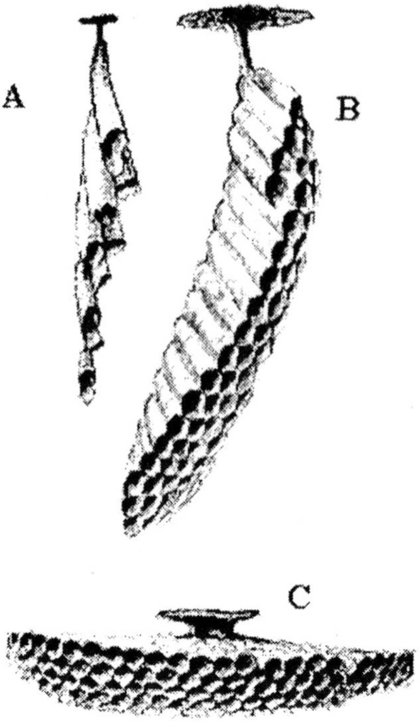

Figure 3.4
Nest forms in *Polistes*. (A) *P. goeldii*—eccentric pedicel and two rows of cells. (B) *P. infuscatus*—eccentric pedicel and oval, inclined comb. (C) *P. flavus*—central pedicel and round, horizontal comb. From Karsai (1999), therein taken from Evans and West Eberhard (1970). Reprinted with permission of M. J. West-Eberhard.

the ant-repelling secretion, may act as a pheromone that mediates behaviors among nestmates (Downing 1991; Van Hooser et al. 2002).

Nests of many species can reach sizes of several hundred cells, and many cells may serve two or three offspring in sequence. A colony at its peak on such a nest can have more than 100 adults and hundreds of immatures present simultaneously. Some colonies may consist of more than one nest comb in a cluster (Reed et al. 1988). The largest *Polistes* nests to come to my attention are one of *P. olivaceous* that was 32 centimeters in diameter and contained 1456 cells (Alam 1958) and nests of *P. annularis* with 1575 cells (Pierce 1909) and one that had at least 1500 cells that produced a pupa (Wenzel 1989). At the peak of the colony population, these nests would have been home to hundreds of adult wasps and more than 1000 immatures. Gobbi (unpublished; cited in Gobbi et al. 1993) reports giant nests of *P. simillimus* in Brazil that had 1000 to 4000 cells. Cities indeed!

Eggs are laid directly from the oviduct into nest cells. Cement at one end attaches the egg into the rounded closed end or side wall of the cell. After embryogenesis, the larva must chew an opening in the egg chorion at eclosion, and the chorion must slip off the larva to the posterior attachment. However, this is an inconspicuous event (Rau 1930a), as is also the case in the honey bee *Apis mellifera* (Collins 2004), and the tiny new larva can be distinguished from the egg, with difficulty, primarily by the annulations that mark its segments. A viscous secretion (Buysson 1903) and/or the incompletely shed egg chorion (Edwards 1980) attach the larva's posterior end to the nest cell, continuing the attachment that held the egg in place. Larval attachment continues through three (Ishay 1975a; Edwards 1980) or four instars (Matsuura and Yamane 1990) or into the early fifth instar in the case of small individuals (Hunt, personal observation of laboratory-reared *Polistes metricus*). At each molt the previous instar's cuticle splits open and slips over the new cuticle to the posterior attachment (Hunt, personal observation of *P. metricus*). Incomplete shedding of the cuticle is said to continue the attachment to the nest cell (Edwards 1980; Matsuura and Yamane 1990). Third- and fourth-instar larvae develop a segmental series of rounded protuberances ("dorsal welts and lateral bosses": Wheeler and Wheeler 1979; "transverse dorsal ridges and pleural lobes": Chao and Hermann 1983). By means of these protuberances, large fourth-instar larvae can hold themselves in the downward-facing cell via friction against the cell walls, and the posterior attachment can be ended. Fifth-instar larvae generally have a girth that fills the cell diameter.

First-instar larvae must be fed exclusively on liquids, which could be nectar or prey hemolymph. In some cases, and perhaps typically, newly eclosed larvae are provisioned with recently laid eggs (Mead et al. 1994). Liquid provisions, whatever their nature, are regurgitated from the crop of an attending female and fed directly to the larva via mouth-to-mouth contact. Only enough food is provided for a single feeding, and provisioning is repeated several times daily. Late instar larvae are provisioned with solid food consisting primarily of caterpillars (Rabb 1960; Hunt, personal observations) that have been thoroughly "malaxated" (from the French *malaxer*, to knead) by the provisioning adult. Undigested solids accumulate in the midgut, where they are contained within a membrane called the peritrophic sac. These wastes are often pink and can be seen through the body wall of second- to fourth-instar larvae (Hunt, personal observation).

When larval growth is completed, the larva secretes silk from its labial glands, which have ducts at the larva's mouth, and spins a cocoon that first covers the open end of the cell and then is spun in the cell interior (Chao and Hermann 1983). Adult wasps play no role in cell closure. Once cocooned, the larva's anus opens, and the meconium is voided into the closed end of the cell, where it dries and remains. The resulting less corpulent larva, now called a prepupa, becomes turgid and ivory colored, whereas the predefecation larva had been plump, soft, and milky white. The turgidity and yellowing reflect not only the passing of the meconium but also the synthesis of storage proteins (Berlese 1900; Roubaud 1911; Hunt et al. 2003) that serve as the source of amino acids for tissue building during metamorphosis. Metamorphosis then ensues, with the pupa molting from the prepupal cuticle and the adult molting from the pupal cuticle before the new adult bites a perimeter incision around the silk cocoon cap and emerges onto the face of the nest (Chao and Hermann 1983). There is no prepupal diapause.

The first offspring are always females. In time, males (from unfertilized eggs) will be among the brood, although female brood production generally continues.

Developmental variability is a result of environmental factors, including temperature and the quantity or quality of provisions (West Eberhard 1969). Interindividual size variation in female *Polistes* follows a general pattern of initial smaller offspring, then larger offspring (West Eberhard 1969) up to a species-typical maximum (Karsai and Hunt 2002), followed by somewhat smaller brood again (Haggard and Gamboa 1980) or, at least, brood with increased size variance (Hunt, personal observation of *P. metricus*) near the end of the nesting period.

Polistes olivaceous in India makes two sizes of nest cells: shorter cells of smaller diameter in the nest center and taller cells of larger diameter at the periphery (Alam 1958). Nonreproductive females are produced from the smaller cells, and males and reproductive females are produced from the larger cells (Alam 1958; Kundu 1967). Continuous allometric variation from small to large females has been proposed for *P. olivaceous* in Micronesia, but this proposal was based on a small sample size that contains only a single individual intermediate between more numerous smaller and larger individuals (Miyano 1994). Dimorphism in the species seems possible.

Individual variation among male *Polistes* has been little studied, but males have greater size variation than do contemporaneously reared females (Eickwort 1969a; Haggard and Gamboa 1980). Size declines from initial large males to later emerging smaller ones (Miyano 1983). Experimentally supplementing provisions led to larger males (Seal and Hunt 2004).

Although mating has been observed on nests (Kundu 1967; Hook 1982; O'Donnell 1994), most mating probably occurs at aggregation sites away from nests (West Eberhard 1969; Kasuya 1981a). "Early males" may be sterile diploids (chapter 9) incapable of reproduction (Tsuchida et al. 2002, 2004; Liebert et al. 2004), although some early males are fertile (Strassmann and Meyer 1983). In seasonal locales, all colonies in a population are roughly synchronous in their time of founding, growth, and production of reproductives, and most mating takes place in aggregations at the end of a nesting season (Reed and Landolt 1991). Most wasps in these aggregations are inseminated females. The foundresses/queens and uninseminated alloparental females of that nesting cycle have died by then. The inseminated females and occasionally some males become quiescent in a sheltered location (Rau 1930b), to which they may migrate some distance (Beall 1942; Hunt et al. 1999; González et al. 2005). In this manner they pass the inclement season—either the temperate winter or the tropical dry season. At the onset of the ensuing favorable season the inseminated females, no longer accompanied by males (cf. Brimley 1908; Hermann et al. 1974), emerge from quiescence, and the nesting cycle begins anew. In tropical locales without strong seasonality, species lack quiescence and have year-round nesting activity that can be asynchronous among colonies (Young 1986), although some seasonal patterning may be discerned (García A. 1974).

Females newly emerged from quiescence feed on flower nectar as they initiate nesting. Nectar ingested in excess of immediate needs may be regurgitated as small drops of honey on the interior of nest cell walls (Rau 1928; Hunt et al.

1998; Prezoto and Gobbi 2003). Nectar and honey may be provisioned to larvae throughout the nesting season (Grinfel'd 1977). Once larvae are present, females also begin to take caterpillars as larval provisions. Without first stinging the prey item, the female captures a small caterpillar with her mandibles, or she bites and severs a fragment from a larger caterpillar. In either case, malaxation begins at the site of capture, and the provision arrives at the nest as an unrecognizable mandibulated pulp. Malaxation continues at the nest, and liquids are drawn from the morsel into the provisioner's crop (Hunt 1984). The female then uses the remaining morsel to feed larvae, allowing each of several larvae to bite off a fragment until the morsel is gone. The provisioning female then grooms herself before regurgitating the caterpillar hemolymph in her crop to smaller larvae. The end of provisioning is marked by a second bout of grooming. Each time a female provisions larvae in this manner she retains at least some of the caterpillar hemolymph for her own nourishment (Hunt 1984; Greenstone and Hunt 1993).

When more than one female is present on a nest, one may take part or all of a provision load from another female and then engage in the larval provisioning behaviors just described. Interactions among adult females often lead to one wasp regurgitating crop liquid that then is imbibed by another. First described in yellowjackets by Réaumur (1742a), this behavior can be called interadult trophallaxis (Hunt 1982). At times other than when provisioning, females touch their antennae and mouthparts to the mouthparts of a late instar larva and drink a nutritious saliva (Hunt et al. 1982) that is secreted by the labial glands (Maschwitz 1966) onto the larva's mouthparts in response to being touched. This larva–adult trophallaxis (Hunt 1982) is a significant source of adult nourishment (Roubaud 1916).

At the termination of nesting, recently produced females feed at flowers before entering quiescence. When males appear on a nest, those males nourish themselves by intercepting provision loads, by soliciting regurgitation from female nestmates, and by drinking larval saliva. Males also visit flowers at the end of a nesting cycle. Females may emerge from quiescence on warm winter days and seek nourishment such as stores of honey remaining in the natal nest (Strassmann 1979). Although a tropical *Polistes* apparently does not feed when occasionally active during quiescence that it passes in cool cloud forests (Hunt et al. 1999), this same species and other *Polistes* species visit flowers when active in nearby seasonal lowlands (Heithaus 1979).

When inseminated females emerge from quiescence, they may return to their natal nest, if it remains from the previous year, before initiating a new nest nearby (Rau 1930a). Old nests are occasionally reused (Starr 1976, 1978; Queller and Strassmann 1988). New nests may be founded by a single female ("haplometrosis") that may continue nesting alone, or the foundress may be joined by one or more females. Intraspecific variation in foundress number may reflect site history (Klahn 1979) or site characteristics (Cervo and Turillazzi 1985). A single foundress may be displaced by a female that usurps the nest and brood (Klahn 1988; Makino and Sayama 1991). Usurping wasps may have lost their own initial nest, left a foundress association, or delayed any nesting activity until nests of other foundresses contain large larvae (Nonacs and Reeve 1995; Starks 1998). Three *Polistes* species are social parasites (inquilines) whose exclusive mode of reproduction is to displace a foundress of a different species from her nest and take over the brood and sometimes even the initial offspring (Cervo and Dani 1996). Usurping and inquilinous wasps eat eggs and small larvae in the nests they take over, whereas larger larvae are reared to adulthood and behave alloparentally to the unrelated larvae of the invader.

Small groups of inseminated females may found nests together ("pleometrosis"). The largest number of cofoundresses I have seen on a preemergence nest (before any offspring mature) was 17, not counting the ones that got away, on a nest of *Polistes carolina*. Strassmann (1989a) recorded as many as 28 cofoundresses in *P. annularis*. Pleometrosis is more common in the tropics than in temperate zones (Richards and Richards 1951), but it may be less common in Old World tropics than New World tropics (Suzuki and Ramesh 1992). Cofounding or, more likely, joining by two species on a single nest occurs occasionally (Snelling 1952; Hunt and Gamboa 1978; O'Donnell and Jeanne 1991).

When more than one female is present on a preemergence nest, they may form a dominance hierarchy (Pardi 1942, 1946, 1948; West Eberhard 1969). The dominant female spends more time at the nest, less time foraging, intercepts more provision loads, and receives more droplets of regurgitated liquid than her nestmates. She also initiates more new nest cells and lays more of the eggs. Egg laying by subordinate wasps (Pardi 1942; Gervet 1964) generally ends by the time offspring emerge (West Eberhard 1969) and is accompanied by ovarian regression in subordinates (Pardi and Cavalcanti 1951). In some cases, however, more than one female will continue to lay eggs simultaneously on postemergence colonies (Hoshikawa 1979; Kasuya 1981b; Reed et al. 1988; figure 8 of Hunt et al. 2003).

The females that emerge early in the brood sequence are small and have low nutrient reserves. These females forage for wood pulp, larval provisions, nectar, and water. At the nest they engage in construction, larval provisioning, nest cooling during high temperatures (Rau 1931), and defense against predators and parasitoids. The brood in the nest is not their own, so they are performing alloparental behaviors. Most offspring that undertake alloparental behavior under these circumstances will never mate, never produce their own offspring, and will die before the end of the nesting season. The presence of such females has been considered the distinguishing characteristic of sociality in paper wasps (Hunt 1991).

Although most remain unmated and perform alloparental care at the natal nest, female offspring of the early brood may follow various life paths. If the queen dies after emergence of the first offspring, one of the offspring can become a replacement queen (Strassmann et al. 2004). In some *Polistes* species in which early males (Reeve 1991) are among the brood, it is not uncommon for the queen to be replaced by an inseminated offspring (Strassmann and Meyer 1983; Suzuki 1985, 1997, 1998). Some offspring establish "satellite nests" near their natal nest, where they undertake reproduction (Strassmann 1981; Page et al. 1989), although these mid-season nests rarely become large or produce offspring before the end of the favorable season (Rau 1941; Hunt, personal observation of *P. metricus* and *P. fuscatus*). Other females that depart the nest mid-season may enter early quiescence (Reeve et al. 1998; Hunt and Dove 2002). Females emerging later in the nesting season, together with male offspring, may remain at the nest, where they intercept provisions and feed on larval saliva. Some males that intercept provisions will, after prolonged malaxation, give the remaining solids to a larva (Kojima 1993). This has been called brood care (Cameron 1986), but such care is incidental to males' self-nourishment (Jeanne 1972; Hunt and Noonan 1979; Makino 1983, 1993). When many provision loads are intercepted, larval growth slows. As foragers die and are not replaced, fewer provision loads are brought to the nest. Nonforaging females and males may cannibalize remaining brood. The empty nest is abandoned.

Paper wasps take provisions with their mandibles, not with their stings. Stinging is used only in defense of nests against vertebrate predators, and the venom has components that can cause considerable pain (Starr 1985). Nonetheless, nests are taken (for the larvae in them) by birds, mice, and omnivores such as raccoons (Hunt and Dove 2002). Ants may strip nests of brood despite the ant-repellent nest pedicel (Young 1979). In the Old World, paper wasp brood is preyed upon by hornets of the genus *Vespa*. Diverse parasitoids use larval paper wasps as hosts (Nelson 1968), and adult paper wasps may fall prey to invertebrate predators such as mantises and spiders.

Independent-founding paper wasps in genera other than *Polistes* have individual development, colony cycle, and life history that generally differ from those of *Polistes* only in small details (Gadagkar 1991b). A few differences merit special attention. Larvae of the New World genus *Mischocyttarus* all have a unique and distinctive lobe on the ventral surface of the first abdominal segment (figure 3.5). A fifth-instar larva can flex and withdraw its head so that its mouthparts are concealed, and doing so increases hydrostatic pressure of the hemolymph and causes the abdominal lobe to become engorged, turgid, and erect beyond the head (figure 3.6). This dramatic behavior occurs when the larva has been solicited for saliva but does not secrete it. Larvae in preemergence nests apparently always secrete saliva in response to solicitation and do not perform the behavior, but larvae in nests with alloparental adult offspring may perform the behavior and not secrete saliva (Hunt 1988).

Belonogaster females are nearly dimorphic, with small females emerging earlier and large females later (Pardi and Marino Piccioli 1981; Keeping 2002). *Belonogaster* nest cells are conical rather than cylindrical (Keeping 1991), and nest architecture that changes as the nest is expanded (Marino Piccioli and Pardi 1978) may play a role in female dimorphism. In *Ropalidia galimatia*, an independent founder, the nest comb consists initially of small cells from which alloparental females emerge, but then larger cells are built from which larger, morphologically distinct females emerge that are potential foundresses of the next nesting cycle (Wenzel 1992). This is the clearest case known to date of

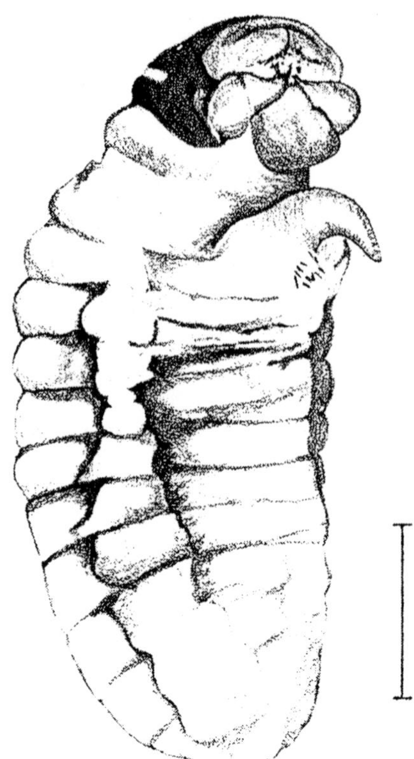

Figure 3.5
A preserved fifth-instar larva of *Mischocyttarus immarginatus* in oblique ventral view. Osmotic dehydration in preservative has caused the larva to contract to less than life size (scale bar = 2 mm), and the resultant hydrostatic pressure of the larva's hemolymph has caused the ventral abdominal lobe to become engorged and turgid. Originally published as Figure 1 in J. H. Hunt, 1988, "Lobe erection behavior and its possible social role in larvae of *Mischocyttarus* paper wasps," *Journal of Insect Behavior* 1:379–386. © Plenum Publishing Corporation. With kind permission of Springer Science and Business Media.

morphological castes ("worker" and "queen" in the usual terminology) in independent-founding paper wasps.

Like *Polistes*, some tropical *Mischocyttarus* and independent-founding *Ropalidia* species found nests seasonally, but other species found nests throughout the year (Gadagkar 1991b). Even for the year-round founders, however, founding frequency may show an annual pattern (O'Donnell and Joyce 2001). Most of the asynchronous (year-round founding) tropical paper wasps studied to date have determinate colony growth (Jeanne 1991), in which colonies, regardless of when they are founded, have similar trajectories and decline at about the same age (Jeanne 1972). *Ropalidia marginata*, found in tropical India, is the only species of independent-founding polistines known to date in which nesting cycles are indeterminate as well as asynchronous (Gadagkar 1991b). The first emerged offspring in *R. marginata* are all females, but not all of them necessarily become alloparental. Long-lived colonies of *R. marginata* may have multiple egg layers one after another in a pattern called serial polygyny (Gadagkar 2001). Males may always be present in populations of tropical asynchronous paper wasps (Gadagkar 1991b; O'Donnell and Joyce 2001), and females may be uninseminated when they join foundress groups but subsequently become inseminated (West Eberhard 1969).

Figure 3.6
Larvae of *Mischocyttarus immarginatus* in a mature colony. A domed pupal cocoon covers a pupa; two cells contain an egg. Four large larvae (one fourth-instar at upper right and three fifth-instar; note the head capsule size) can be seen. Three of these are in normal repose, with the head and mouthparts near the cell opening. The fifth-instar larva in the center has just been gently probed on its mouthparts with a pipette, whereupon it withdrew its head, causing its abdominal lobe to stand erect as the most anterior part of the larva. The larva is now part way back to repose. Larvae that perform this behavior do not give saliva in response to solicitation.

In an aggregation of 240 nests of *Ropalidia formosa* beneath a large boulder in Madagascar, nests were small, the average number of adult wasps per nest was scarcely more than one, and males emerged as early as the second or third offspring (Wenzel 1987). In a single colony of *Ropalidia rufoplagiata* in India, 33 of the 46 adult females on the nest oviposited, and more than half of a sample of these were inseminated. Data suggest a possible temporal pattern in which younger females forage and older females are nonforaging egg layers (Sinha et al. 1993). These are the only two cases reported to date in which an independent-founding polistine may not have permanently nonreproductive alloparental offspring.

The Old World genera *Belonogaster*, *Ropalidia*, and *Parapolybia*, and *Polybioides* all extract the larval meconium from nest cells. Adults do this by chewing a hole from the dorsal face of the nest comb into the apex of a nest cell and withdrawing the meconium through the hole (Marino Piccioli 1968; Kojima 1983; Turillazzi and Francescato 1994).

Swarm-Founding Polistinae

Genus *Polybioides*, the tribe Epiponini, and some species of *Ropalidia* (figure 3.1) represent three independent origins of a complex suite of behaviors. The key trait is that new nests are founded by groups of uninseminated workers accompanied by a smaller number of inseminated, egg-laying queens that collectively make up a swarm. Swarm-founding wasps (Jeanne 1991) occur only in the tropics and subtropics—*Ropalidia* and *Polybioides* in the Old World, and Epiponini in the New. Epiponini is a clade of 20 genera (figure 3.7), which constitute the dominant hunting wasps of the New World tropics (Jeanne 1991).

Apoica, the sister genus to other Epiponini, is the only swarm-founding polistine that builds a nest consisting of a single comb that is exposed rather than being enclosed by an envelope or concealed among leaves or within a cavity. Whereas nests of independent-founding polistines are made of wood fibers and suspended from the substrate by a pedicel, *Apoica* nests are made of plant hairs and are broadly attached to the substrate.

Nests of other swarm-founders have diverse architectures (Wenzel 1991) that usually, but not always (Hunt and Carpenter 2004), are uniform at the species level. Some are concealed among leaves (attached to plants) or within cavities such as hollow trees. Most are enclosed by envelopes. Construction materials include wood fibers, plant hairs, and, in some neotropical *Polybia*, mud. Construction material is mixed with saliva. Although some epiponine genera appear to use small amounts of saliva, others use substantial amounts,

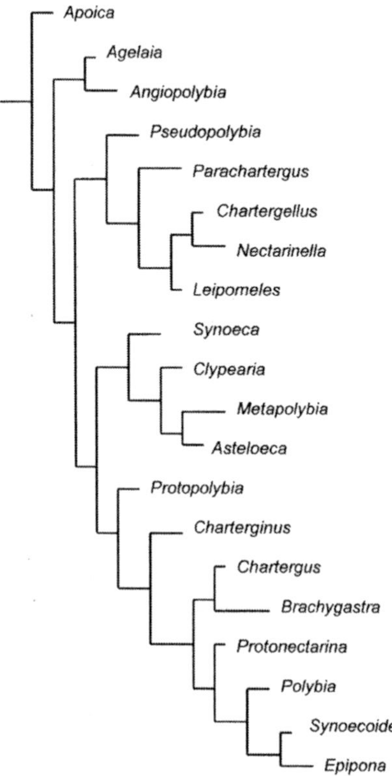

Figure 3.7
Cladogram of genera in Epiponini. From Wenzel and Carpenter (1994), modified by the synonymy of *Occipitalia* with *Clypearia* (Carpenter et al. 1996) and *Marimbonda* with *Leipomeles* (Carpenter 2004).

and a few shelter their nests with transparent sheets of pure saliva (Wenzel 1991). Most nests have more than one comb. Some nests contain thousands of nest cells, each of which may be used to rear several successive offspring. Mature colony sizes range from a few tens of wasps in some epiponines, to 60,000 in *Ropalidia montana* (Jeanne and Hunt 1992), to several million (!) in *Agelaia vicina* in Brazil (Zucchi et al. 1995; F. B. Noll, personal communication). The city has become a megalopolis.

Swarm movement from an old nest to a new site is mediated at least in part by pheromones. Glands on the underside of the gaster (Smith et al. 2002) are rubbed on landmarks such as leaves, rocks, and fence posts on which swarm members alight momentarily during their swarming flight (Naumann 1975a; Jeanne 1981). *Apoica* lacks the rubbing behavior found in other swarm founders and instead coordinates swarm assembly by exposing glands on the underside of the elevated gaster (Howard et al. 2002)—a behavior first observed and described by R. L. Jeanne (in Hunt et al. 1995).

Swarms can be of two types. Reproductive swarms comprise some females of a mature colony that depart the still-active colony, whereas absconding swarms comprise all members of a colony. Reproductive swarming therefore represents the establishment of a new colony, whereas absconding reflects continuation of the existing colony. Absconding is an adaptive response to nest destruction, and several taxa also abscond in the course of seasonal nest relocation (Naumann 1975b; Jeanne 1991; Hunt et al. 1995, 2001b). The few swarms found containing males could be absconding swarms, although Bouwma et al. (2000) found that males remain behind even in absconding swarms.

Foragers collect water (for construction and cooling), nest construction material, nectar, and larval provisions (Jeanne 1986). Nectar collection has not been studied extensively. Storage of excess nectar as honey droplets, commonly seen in temperate *Polistes*, is rare in swarm founders, except in some species of *Polybia* and *Brachygastra* in subtropical or seasonally arid locations, which may store substantial quantities of honey in nests during the cool, dry season when many nest cells contain no brood (Hunt et al. 1998). *Polybia* and *Agelaia* take diverse insects, including caterpillars (Gobbi and Machado 1985; Machado et al. 1987). *Apoica* is apparently also a generalist forager (Hunt et al. 1995). A few species forage at carrion (Dejean et al. 1994; O'Donnell 1995a), but none recruit nestmate foragers to rich, ephemeral food sources (Jeanne et al. 1995b). Nutrient dynamics in *Polybia occidentalis* (Hunt et al. 1987) are probably typical of all swarm founders. Adults at the nest intercept incoming loads of prey and nectar. Larval provisioning has not been documented but almost certainly is identical to that of independent founders. Adults regurgitate crop contents to one another. Larvae produce saliva that adults drink. Larvae are sometimes cannibalized.

Oviposition, egg attachment, and larval development are identical to those of independent-founding polistines. Large numbers of females are produced before any males are produced. Males are apparently few or at least are infrequently encountered in swarm-founding wasps (Carpenter and Mateus 2004; J. M. Carpenter, personal communication), with occasional exceptions (Hunt et al. 2001b). Dimorphism of workers and queens characterizes swarm-founding *Ropalidia* (Sô. Yamane et al. 1983; Fukuda et al. 2003), *Polybioides* (Turillazzi et al. 1994), and the relatively basal epiponine genera *Apoica* (Jeanne et al. 1995a) and *Agelaia* (Jeanne and Fagen 1974). In *Apoica* the morphological differences are in proportions only, but other epiponines can have dimorphic females in which workers and queens have different body proportions and also are different sizes (e.g., Hunt et al. 1996). Morphological differences between queens and workers may diverge as colonies age (Noll and Zucchi 2000). Some epiponines have little or no difference in size or shape between "workers" and "queens," and so these terms often apply to behavioral and reproductive classes rather than morphological distinctions. Noll et al. (2004) argue that female monomorphism is the ancestral state in Epiponini.

Queen production may be seasonal, and it precedes reproductive swarming, but details of queen differentiation are unknown. Queen number is high in new nests founded by reproductive swarms, and it declines over time. New queens may be reared when the number of queens is reduced—a phenomenon called "cyclical oligogyny" (Queller et al. 1993). In young colonies of epiponines that have no morphological differences among females, queen numbers may diminish to one. When the sole queen dies, a replacement queen may emerge from among the adult females present on the nest (West-Eberhard 1978b). Uninseminated male-producing eggs are laid by inseminated queens (Henshaw et al. 2000), but factors regulating male production are not known. Mating may take place at sites where males patrol away from nests (Jeanne 1991).

Nests may remain active for more than a year, and some are very long lasting (Richards 1978). Perennial colonies can have seasonal patterns of brood rearing (Kojima 1996) and reproductive swarming (Jeanne 1991).

Apoica is unique among polistines in being nocturnal. The entire colony of adults spends the day sitting in a radial array on the face of the nest, and they depart at dusk for a night of foraging, building, and brood care (Hunt et al. 1995).

There are few records of vertebrate predation on nests of swarm-founding wasps. Some swarm founders have stings that can cause excruciating pain (Hunt, personal observation), but some smaller epiponine species with cryptic nests do not sting even when a nest is under attack (Strassmann et al. 1990). Venom spraying by such wasps (Jeanne and Keeping 1995) may be effective against arthropod predators. In the Old World tropics, swarm-founding *Ropalidia* can be victims of depredation by foraging hornets (*Vespa*) that take larvae from nests over periods of days or weeks. In the New World tropics, army ants of the genus *Eciton* can quickly strip nests of all brood (O'Donnell and Jeanne 1990). Although the envelopes that surround nests of most swarm founders may have evolved as adaptations to cold or rain (see Wenzel 1991), one current function of envelopes is to reduce exposure of the brood to parasitoids (London and Jeanne 1998).

Vespinae

Vespinae is a monophyletic group of four genera (figure 3.8). *Vespula* and *Dolichovespula*, collectively called yellowjackets, are broadly distributed in subarctic and temperate zones of the Northern Hemisphere, extending no farther into the tropics than high elevations at the temperate/tropical boundary (Hunt et al. 2001a). *Vespula germanica* is invasive in temperate zones of the Southern Hemisphere. The wasp in your summertime can of soda or beer is a yellowjacket. Hornets, which constitute the genus *Vespa*, occur in temperate Palearctic and tropical Indo-Pacific regions. *Vespa crabro* is invasive in North America (Hunt 2000). The three species of *Provespa* occur only in the tropical Orient. Pagden (1958) calls these "night wasps," a common name evocative of their nocturnality. Excellent short reviews of vespine biology are those of Matsuura (1991) and Greene (1991).

Vespine nests share basic features with those of independent-founding polistines. A stalk or pedicel is attached to a substrate, and from it are suspended cells made of paper. There are, however, several elaborations beyond that basic similarity (Wenzel 1991). Nests are enclosed by an envelope of multiple overlapping sheets of paper, and multiple nest combs will be built, parallel to and suspended from one another (figure 3.9). Nests of many vespines are aerial on branches, but some hornets and yellowjackets nest in voids such as hollow trees or underground. Subterranean nests are begun in an existing small hollow that then is expanded as the colony increases and the nest is enlarged.

Yellowjacket and hornet nests are founded by a single inseminated female, which can become the sole colony queen. However, foundresses may be usurped, and usurpation in some yellowjackets may be more common than nest initiation (Greene 1991). Some yellowjackets nest only as social parasites, which means that they usurp nests of another species and then produce their own offspring. A few yellowjackets are inquilines that usurp the nest and brood of another species but have no worker brood of their own (Greene 1991). Night wasp nests are founded by a single queen accompanied by a swarm of several tens of workers. This constitutes an independent evolution of swarm founding, and it differs from polistine swarm founding by the presence of only a single queen.

Hornets and yellowjackets have small-colony and large-colony species (Weyrauch 1935). Mature colony sizes may be no larger than a few tens of work-

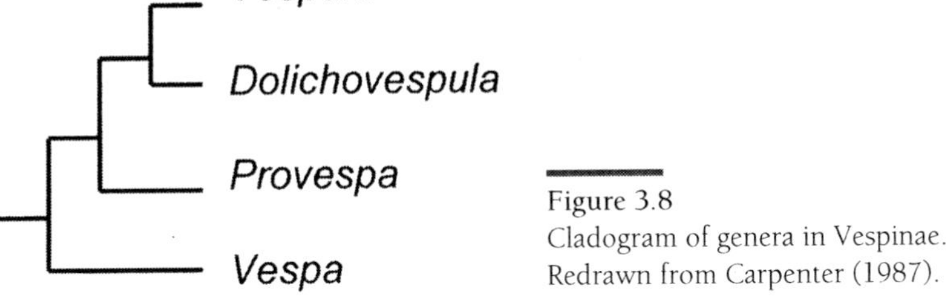

Figure 3.8
Cladogram of genera in Vespinae. Redrawn from Carpenter (1987).

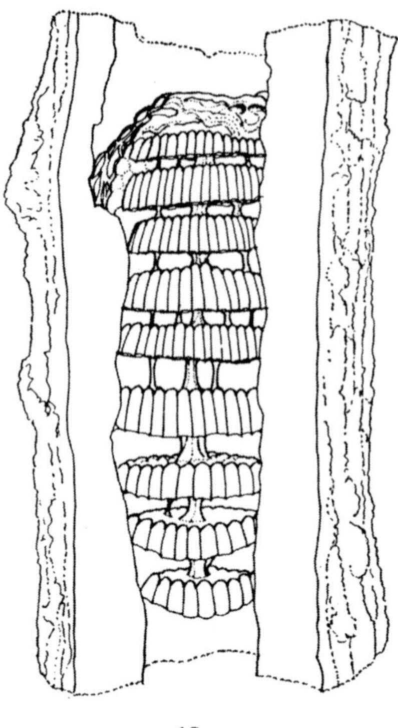

Figure 3.9
View of a nest of *Vespa crabro* in a hollow tree showing the arrangement of combs in Vespinae. From Spradbery (1973). © J. Philip Spradbery. Reprinted with permission of the author.

ers in some subarctic yellowjackets, and maximum sizes in temperate small-colony yellowjackets are generally in the hundreds. Large-colony hornets can have several thousand adults, and large-colony yellowjackets can attain populations of tens of thousands. Some large-colony yellowjackets have populations of hundreds of thousands in subtropical habitats where nests remain active through winter and therefore last longer than one year (Pickett et al. 2001). These perennial nests come to have multiple queens as females that mate and return to their natal nest are integrated as egg layers (Greene 1991). Some tropical hornets have multiple queens that arise when several inseminated females join a young nest and its foundress shortly before emergence of the first offspring (Matsuura 1991; Kojima et al. 2001).

Oviposition and larval provisioning in vespines are identical to those of polistines. Most vespines are generalist predators of insects for larval provisions, and the large-colony yellowjackets are also scavengers. *Vespa tropica* is a specialist predator on the brood of stenogastrine and polistine social wasps (Matsuura 1991). *Vespa orientalis* in the Middle East and *V. mandarinia* in the Orient can be major predators of honey bees. *V. mandarinia* is unique among vespines in sometimes attacking en masse the nests of other social hymenopterans (Matsuura 1984, 1991). Adult vespines visit flowers for nectar. Large-colony hornet species also collect plant sap, often by chewing on or stripping bark from plants.

Adult large-colony hornets and yellowjackets also feed on ripe fruit. Adults regurgitate crop contents to other adults. Larval provisions are malaxated as in polistines. Larvae produce saliva that adults drink (Maschwitz 1966).

Larval development in vespines proceeds as in polistines, but with an additional complexity. Vespine nest cells vary in size, and the sizes of adults vary correspondingly. The initial nest cells of yellowjackets are small, and worker females are reared in them. In lower, later-built nest combs, larger cells are constructed in which reproductives are reared (see chapter 5). Worker cells of small-colony species may show graded variation, and size difference between worker cells and reproductive cells is not great. Interindividual sizes are correspondingly graded among workers and are modest between working and nonworking females (figure 3.10). Large-colony hornets and yellowjackets have two distinct size classes of female offspring (figure 3.10), workers and "gynes" (prefoundresses in the season of their development, preceding the season in which they might become foundresses and queens). Gynes/queens of *Vespa mandarinia* are the largest social wasps. Stocky and 4.5 centimeters or more in length, they are the size of your thumb. It is no mystery where some Japanese insect sci-fi films draw their inspiration.

The males and gynes of temperate species mate in the fall, and inseminated gynes overwinter alone in sheltered locations such as the leaf debris of forest soils or in soft, rotting logs. Tropical *Vespa* have asynchronous, determinate nesting cycles, probably without gyne quiescence (Matsuura 1991). *Provespa* is

Figure 3.10
Gynes (left) and workers (right) of a small-colony yellowjacket, *Dolichovespula maculata* (top), and a large-colony yellowjacket, *Vespula squamosa* (bottom), illustrating the range of caste dimorphism in yellowjackets.

asynchronous and aseasonal and has determinate nesting cycles, an unusual trait for a swarm-founding social wasp (Matsuura 1991).

Numbers of gynes and males produced by a colony range from a few tens in the subarctic yellowjackets, to hundreds in hornets and small-colony yellowjackets, to thousands in large-colony yellowjackets, to many thousands in yellowjacket colonies that last longer than one season (Greene 1991). In the single-season colonies, although worker production continues until the end of the colony cycle, the production of males and gynes generally marks the end of the cycle. After a *Provespa* colony begins to produce males and gynes, it continues to do so over the latter half of the colony cycle, and new queens and swarms leave the source colony intermittently until that colony declines (Matsuura 1991).

Parasitoids of Vespinae are less diverse than those of *Polistes*, and no parasitoids are known that severely restrict colony productivity. The stinging ability of vespines is legendary; therefore most predation on nests by vertebrates is on foundress nests (Greene 1991) or on colonies in decline toward the end of the season.

Questions Arising

What role does nectar play in larval provisioning? Few holometabolous insect larvae feed on nectar, but nectar apparently is fed to *Polistes* larvae, especially by foundresses at the start of nesting, but perhaps also throughout the colony cycle.

Do vespid wasps produce invertase and glucose oxidase? These are the hallmarks of converting nectar to true honey.

Is honey from nest stores fed to larvae? As with nectar, the nutritional roles of honey at individual and colony levels are unknown.

What is the relative nutritional value of wasp honey versus larval saliva? It seems possible that honey may play a nutritional role for adult females principally during times when there are few or no larvae producing trophallactic saliva.

Do independent foundresses take caterpillars for self-nourishment before larvae are present? Or, more generally, can foundresses produce eggs using only food reserves and nectar, or must they feed on protein-rich sources?

Why are the larval peritrophic sac contents sometimes pink?

Does *Polistes olivaceous* in fact have size-dimorphic females? The intriguing reports on this are sketchy and need to be confirmed with quantitative data.

Are independent-founding tropical paper wasps that found nests year-round truly aseasonal? Although tropical locales may have little or no temperature seasonality, most or all tropical locales have wet/dry seasonality. Do the paper wasps there show patterns of nest founding or male production, for example, that correspond to wet/dry seasonality? Individual longevities (of males, for example) in relatively benign environments might obscure seasonal patterns of production.

How did swarm founding evolve? It has evolved independently at least four times in wasps (and twice in bees). Can either a comparative approach or a case study be framed in terms of experimental tests of possible mechanistic components of the behavior?

What ecological factors affect male and queen production by swarm founders? For example, would experimental supplementation or diminishment of nourishment have an effect?

Is there a basis for nocturnality occurring only in swarm-founding wasps?

What regulates the nest cell sizes of vespines, and what precipitates the change from small cells to large? *Polistes* uses its antennae to gauge cell diameter. A foundress/queen vespine must use different behaviors when initiating smaller worker cells versus larger gyne cells. Are the differences quantitative or qualitative? Do vespine workers initiate cells that are larger than themselves from which gynes are reared, or do the workers only follow, via stigmergy, a geometry that has been established by the queen?

The Historical Scenario of Social Evolution

4

Darwin labored over how best to illustrate the historical course of evolution. Although a branching coral seemed to him a better metaphor (Bredekamp 2002), a more familiar organism—a tree—is commonly invoked when describing the single figure that appeared in Darwin's *On the Origin of Species*. Stephen Jay Gould sought to substitute a bush for the tree, highlighting ramification rather than a central main course (a tree's trunk) as the dominant pattern of evolution. When trying to understand the evolution of wasp sociality, it is tempting to think in terms of a main course from solitary to social. However, it must be stressed that the pathway was far from straight, even though such an impression might easily be drawn from a diagram (including the one in this chapter) focused on the evolution of a specific trait or behavior. Branching points along the path were numerous, and if branches are weighted by the number of living tips they support, the branches leading to sociality were not always larger. What is important for the historical scenario that follows is that the branch points have living descendants on both sides of the divide. Many branches diverged only slightly from the stem that preceded them, whereas others diverged more. By using phylogenetics to hypothesize the pattern of coalescence from the tips to the root and by looking at the branches that diverged only slightly but that have living descendents, one can build a realistic scenario for the history of sociality in vespid wasps. It is important to keep in mind that the less diverged branches exemplify character states shared with a common ancestor of the (eventually) social forms, but the living taxa on those less diverged branches are not ancestral to the social forms.

Phylogenetic hypotheses have been built by others, making it possible to begin the scenario at a branch well down in the interior of the bush and work toward the tips that are social Vespidae. Vespid sociality is at least 63 million years old (Wenzel 1990), so the main elements of the scenario unfolded when dinosaurs roamed the earth.

Overview

A central element of the scenario to be drawn from the first three chapters of this book is that vespid sociality evolved in a context of nesting and parental care. Reflecting her idiobiont ancestry, a nesting adult female requires continuous access to proteinaceous nourishment, with which she sustains the production of large ova. Within the nesting milieu, means of such nourishment are diverse, and some of them have consequences for the larvae being provisioned. Details differ among the social vespids. Specific traits and the roles that they play in this scenario are encapsulated in tables 4.1 and 4.2 and placed on branches of a phylogenetic hypothesis in figure 4.1. Several aspects of the evolutionary scenario merit discussion. The first of these, discussed below, is particularly important.

Diphyletic Sociality in Vespidae

When I began to write this book I had no doubt that sociality in Vespidae was monophyletic. By the time I reached the end of chapter 3, I had no doubt that it is diphyletic. The extraordinary amount of new natural history knowledge produced in the past 20 years, especially on Stenogastrinae, led me to shape two different views of sociality in Vespidae, one for Stenogastrinae and one for Polistinae+Vespinae. These two views, described in the following sections, are what caused me to question the hypothesis of monophyletic sociality in Vespidae. The question began to loom large in my mind, and so I delved into the data and arguments on which the hypothesis of a single origin of vespid sociality is based. I concluded that the data that support the hypothesis of monophyletic sociality are unconvincing.

The cladogram in figure 4.1 presents the current phylogenetic hypothesis. I think the hypothesis is probably wrong. Others before me have argued that sociality in Stenogastrinae is separate from that of Polistinae, and they based their arguments on the same reasons that initiated my thinking—that there are numerous and often dramatic trait differences between stenogastrines and other vespids, as exemplified by ovipositioning, provisioning, and nest architecture (Richards 1971; Spradbery 1975; Vecht 1977; Pardi and Turillazzi 1982; Turillazzi 1989). Carpenter (1988) refuted this line of reasoning by pointing out that traits unique to a taxon ("autapomorphies"), regardless of their number or distinctness, are uninformative as to shared ancestry between that group and any other. Carpenter is correct on this point of phylogenetic interpretation. However, I would point out that there are numerous and distinctive

Table 4.1
Traits germane to the evolution of sociality in Vespidae and the role played in vespid sociality by each.[a]

Node	Trait	Role in vespid sociality
1	Biting mandibles	Prey/provision transport and malaxation; nest construction
	Wings joined by hamuli	Heavy wing loading for prey/provision transport
	Sclerotized ovipositor	Stinging in defense of colony
	Prominent antennae	Detection of pheromones; nest and nestmate recognition
	Haplodiploidy	Enables protogyny
2	Larval carnivory	High-protein diet
	Larval leglessness	Restriction to nonlocomotory feeding mode
3	Meconium voided at pupation	Larvae do not soil nest
	Petiolate abdomen	Adults restricted to a liquid diet
4	Long reproductive lifetime	Node 4 traits, collectively, are the idiobiont suite of traits; feeding on host hemolymph provides nutritional support for the other traits
	Synovigenesis	
	Lecithal ova	
	Relatively low fecundity	
	Host hemolymph feeding	
5	Nest excavation in compact soil	Brood protection; necessitates central place foraging
6	Mass provisioning with multiple weevil larvae	Eliminates need for single large prey item; enables ecological efficiency
	Oviposition precedes provisioning	Breaks link between prey size and larval sex
A, B	No traits at these nodes are antecedent to sociality	
C	Nests constructed above ground	These traits found among Eumeninae are seemingly antecedent to traits of social forms.
	Plant material incorporated into construction	
	Caterpillars as primary prey	
	Progressive provisioning	
7	Traits at this node are the putative synapomorphies of Carpenter (1991) as listed and discussed in the text.	Behavioral or life history traits at this node are found in multiple social hymenopterans and include the traits used to define "eusocial."
8	Traits of table 4.2, left-hand column	These are the life-history autapomorphies of Stenogastrinae
9	Traits of Table 4.2, right-hand column	These are the life-history autapomorphies of Polistinae+Vespinae

[a]Numbers refer to branches and nodes on the phylogenetic hypothesis in figure 4.1 and correspond to numbered branches on phylogenetic diagrams in previous chapters. A and B mark nodes in figure 4.1 for which no life-history traits antecedent to sociality are apparent. C is not a node but is placed outside Eumeninae in figure 4.1 to indicate the origination of several traits of relevance to sociality within the subfamily.

Table 4.2
Distinctive features of Stenogastrinae and of Polistinae+Vespinae.[a]

Stenogastrinae	Polistinae+Vespinae
Nests constructed of mud or of paper made from plant chips or short fibers	Nests constructed of long fiber wood pulp paper
No nests have a pedicel	Nests suspended from a pedicel
Great architectural diversity of nests	Only minor architectural diversity of nests among independent-founding Polistinae
Egg placed into cell via mandibles and adheres to cell via convex surface	Egg oviposited directly into cell and attached via one end
Larva crawls free of chorion at eclosion	Larva remains attached to nest cell through three or more instars
Abdominal substance	No abdominal substance
Provisions placed where larva can feed over time	Provisions fed directly to larva, mouth to mouth
Adults drink liquid from ventral surface of larva	Adults drink saliva from larva, mouth to mouth
Larvae not provisioned with nectar; adults do not store honey	Larvae provisioned with nectar; adults store honey
Four larval instars	Five larval instars
Larva spins incomplete cocoon and adult seals nest cell at pupation	Larva spins complete silk cocoon and seals own nest cell
Adult opens nest cell and removes meconium	Meconium remains in nest cell
Pupa lies flexed at second constriction of the abdominal petiole	Pupa elongate
Wings not folded lengthwise	Wings folded lengthwise
Delicate, hovering flight	Strong flight
Generalist predator of diverse arthropods as larval provisions	Specialist predator of caterpillars (*Polistes*)

[a]Some features are autapomorphies of the group in which they occur, and others are shared with some Eumeninae. Some features are characteristic of basal members of the taxon but not of all members of the taxon. None of the features, however, are shared between the two taxa.

autapomorphies of Polistinae+Vespinae as well as of Stenogastrinae. Polistinae+Vespinae are the only Vespidae that construct a pedicle from which the nest is suspended, build nests of long-fiber wood pulp paper, have larvae that remain attached to the nest cell wall through three or more instars, provision larvae via mouth-to-mouth contact, feed their larvae with liquid as well as solid provisions, have larvae that close their own nest cells completely with silk cocoons, and in which adults play no role in nest cell closure. The many autapomorphies of Stenogastrinae and those of Polistinae+Vespinae as well as other life history differences between them (table 4.2) include major aspects of morphology, development, and life cycle. The number and importance of the differences so

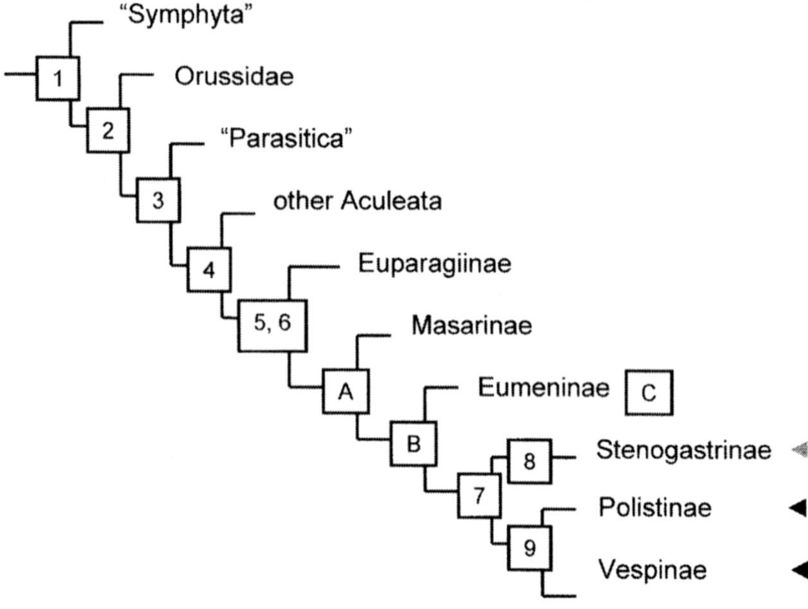

Figure 4.1
A condensed phylogeny of Hymenoptera combined with the cladogram of Vespidae, both as presented previously in this book. Numbers and letters correspond to previous figures and to table 4.1 and illustrate placement of key traits of vespid sociality. Pointers indicate social subfamilies; the gray pointer signifies facultative eusociality. The sequential placement of traits germane to sociality on successive nodes fails at the node subtending Masarinae and again at the node subtending Eumeninae and the social subfamilies. In the latter case (node B) at least, this almost certainly is because Eumeninae is paraphyletic with regard to the social subfamilies. Many traits that seemingly presage sociality are found within the Eumeninae, and so these are placed apart in box C.

greatly exceed the number and importance of the synapomorphies (table 4.1, node 7) that reexamination of the argument for recent common ancestry seems called for.

Carpenter (1982, 1988) based the common ancestry of Stenogastrinae and Polistinae+Vespinae on three shared derived traits ("synapomorphies"): forewing marginal cell pointed onto costa, larval labrum not narrowed where it joins the clypeus and narrower than maximum width of the clypeus, and the behavior of simultaneous progressive provisioning. Carpenter called monophyly of the three subfamilies "problematic" and noted that "all the synapomorphies show homoplasy" (Carpenter 1982, p. 28), which is to say these traits are shared with other taxa on the basis of convergence from unrelated ancestry. Carpenter (1991) identified additional behavioral characters as synapomorphies of Stenogastrinae and Polistinae+Vespinae: simultaneous progressive provisioning with a masticated paste from a broad range of adult and immature arthropod prey, progressive provisioning probably only after egg hatch, care extended until adult eclosion, construction of complete nests that hang free from the substrate, nest construction probably primitively (ancestrally) of plant material, cell reuse, nest sharing among adults, adult–adult trophallaxis, cooperative brood care, and temporary reproductive division of labor. One of the original morphological traits, the larval labium (a mouthpart), is unsupported by the data of Kojima (1998). No nests of Stenogastrinae have pedicels, and so they do not "hang free from the substrate" in the same way as those of Polistinae and Vespinae. To propose that the ancestral construction material for stenogastrine nests is probably plant material is speculation. I disagree with Carpenter's interpretation of most of the behavioral traits, and some of the behavioral traits proposed to be synapomorphies are, in fact, the very social behaviors used to define "eusocial": overlap of generations ("care extended until adult eclosion"), cooperative brood care, and reproductive division of labor (which I would argue is not temporary in Polistinae and Vespinae). To use these traits as evidence of common ancestry for taxa categorized as "eusocial" constitutes a fallacy of affirming the consequent. Thus, I think the best that can be said for the hypothesis of common ancestry of Stenogastrinae and Polistinae+Vespinae is that its support is weak in terms of both morphological and behavioral evidence.

The same can be said for molecular evidence. Both 16S mitochondrial ribosomal DNA and 28S nuclear ribosomal DNA have been sequenced from species of Eumeninae, Stenogastrinae, Polistinae, Vespinae, and honey bees. Phylogenetic analysis of these data led Schmitz and Moritz (1998) to conclude that Eumeninae is more closely related to Polistinae+Vespinae than any of these are to Stenogastrinae; thus sociality must have evolved twice independently in Vespidae. Carpenter's (2003) reanalysis of the Schmitz and Moritz data supports the hypothesis he first advanced in 1982. Both the Schmitz and Moritz analysis and Carpenter's reanalysis, however, are fraught with problems, and nothing can be gained by arguing over which one is right. The data in contention are too meager to withstand repeated reanalysis, and I suspect that neither has got it right. There is a clear and compelling need for new molecular data, from different genes and from many more taxa, and for these new data to be analyzed independently of the existing data sets.

Thus, I reject the hypothesis of a monophyletic taxon (Stenogastrinae + (Polistinae+Vespinae)) not merely because of the number and seeming significance of the differences between stenogastrines and polistines+vespines but also, and primarily, because the data that support the monophyletic hypothesis are few and weak and because those few, weak data are badly argued. I believe, therefore, that it is not merely plausible but responsible to advocate two propositions. First, the monophyletic hypothesis should be rigorously tested by new and independent data. Second, until this is done, researchers should thoughtfully consider hypotheses and research that are not based on the monophyletic hypothesis.

Sociality in Stenogastrinae

Stenogastrinae are characteristic of tropical rainforests. Many traits of their sociality can be interpreted as adaptations to life in warm, wet, dark, and aseasonal forest interiors, and they most likely evolved there. In these forests stenogastrines rear larvae on a diet of diverse arthropods. Minced prey is placed where larvae can feed on it. Adults probably feed on hemolymph from prey items, although this has not been documented. Adults also probably drink saliva that larvae secrete, although the behavior of doing so differs from that in Polistinae+Vespinae. Adults rarely visit flowers for nectar, so adult nourishment outside the context of brood rearing may never be abundant. When females emerge from pupation, they probably have low nutrient reserves, and, like all Vespidae, they have undeveloped ovaries. They can gain nourishment via handling larval provisions, by taking provisions from larvae or from storage on wads of abdominal substance in nest cells, by feeding on larval saliva, and by imbibing regurgitated crop contents of other adults. Thus, females gain the nourishment needed to develop their ovaries and produce eggs primarily in the milieu of colonies in which larvae are being progressively provisioned. Access to nourishment is greater when larvae are being cooperatively reared.

Males are always present in the asynchronously nesting populations in the aseasonal rainforests, and most or all females become inseminated. Inseminated females may pursue any of several pathways of direct reproduction, including spending time in a dominance queue on the natal nest before succeeding to the dominant rank, by independent nest founding, by nest joining, or by nest usurpation. Colonies never become large. Permanently sterile workers are nonexistent. Females vary widely in reproductive success (see chapter 7), although variance among nests may no more than moderate (J. Field, personal communication).

Sociality in Polistinae (+ Vespinae)

Polistine sociality most likely evolved in a seasonal environment. I first developed this perspective based colony cycle and demography data that can be logically interpreted as adaptations to an annual seasonal cycle (Hunt 1991). My entry point to this perspective was the role played by males in polistine wasp societies. Male *Polistes* do not survive temperate winters. Because inseminated females generally do not work at their natal nest (e.g., Yanega 1988, 1989), I thought that an annual cycle in which no males are present to inseminate first-brood females would be an important component in inducing those females to work at their natal nest (Hunt 1991). However, it has been shown that insemination status per se does not affect ability or inclination to work in *Polistes* (Downing 2004). Therefore, the significance of the absence of males among the early brood is not that they would inseminate their sisters, but that males do not work (notwithstanding Hunt and Noonan 1979; Cameron 1985, 1986; Hölldobler and Wilson 1990, p. 185; O'Donnell 1995b; see also Makino 1993). Males among the early brood of a colony would consume resources that could instead produce a worker, and early males would thereby reduce potential colony growth. Production of females as exclusive first brood (protogyny) yields only offspring that can, and perhaps will, perform alloparental care. The consequence of a major investment in alloparental care by the early brood is the production of a large brood of reproductives (gynes and males) toward the end of the colony cycle. The resulting demographic similarities between the life cycle of *Polistes* and those of some annual plants (Hunt 1991) are so striking that interpreting the *Polistes* life cycle as reflecting origin in a seasonal locale seems inescapable. A tropical locale (West Eberhard 1969; Carpenter 1996b) with wet/dry seasonality would suffice. To envision the origin of polistine sociality in the framework of a seasonal environment (chapter 8) is key to understanding polistine sociality.

Another key to understanding polistine sociality (indeed, insect sociality in general: Hunt and Nalepa 1994; Hunt 1994) is to know the role and significance of unequal distribution of nourishment among nestmates, both as immatures and as adults. A major source of nourishment for adult female polistines is prey hemolymph imbibed during malaxation of larval provision items. Drinking larval saliva is another source of proteinaceous nourishment for adults. Variable provisioning of larvae and drinking larval saliva by the foundress(es) caused first offspring to be smaller or, at least, poorly endowed with internal nutrient stores as adults. These first offspring all were females, and these poorly endowed females performed parental care behaviors at their natal nest. The alloparental care that these first female offspring provided enabled that colony to produce a large number of offspring, including males and well-endowed females later in the nesting cycle/season. These well-endowed females, following insemination and passage of the inclement season, then synchronously initiated the next generation. Selection favored traits of individuals and properties of colonies that led to maintenance of this pattern. Usurping, joining, and cofounding all reflect selection for behaviors that evolved from independent founding as the initial condition (chapter 7). Different behavioral roles among adult females in foundress groups emerge in the context of dominance interactions and are usually reflected in differential egg laying. Variance among females in reproductive success is extremely high (chapter 7).

A Further Argument for Diphyletic Vespid Sociality

If the nature of sociality in Stenogastrinae and Polistinae is as different as it seems, are there additional arguments in support of the proposition that these subfamilies evolved sociality independently? I believe that one such argument is the implausibility of a key component of the current phylogenetic hypothesis. Carpenter (1991, p. 9) asserts that origin of Stenogastrinae from among Eumeninae, as proposed by van der Vecht (1977), is logically impossible because subfamily Eumeninae is united as a monophyletic group by four autapomorphies not found in other Vespidae: "parategula, hindcoxal carina, bifid claws, and bisinuate larval labrum." The proposed monophyly of Eumeninae raises a major difficulty. If Eumeninae is monophyletic and separated from other vespid subfamilies, then examination of Eumeninae for traits antecedent to sociality in other subfamilies is a misguided exercise. Instead, only traits in the common ancestor of Eumeninae and the social subfamilies are germane. However, diversity among eumenines ranges from taxa with life histories identical to the Euparagiinae to taxa that have behaviors and/or life-history traits seemingly antecedent to sociality. Accordingly, numerous investigators have turned to eumenines over the past century for insight into the origins of sociality. If, despite the richness of seemingly informative life histories, Eumeninae is truly monophyletic, then any eumenine traits that seem to presage sociality are nothing more than convergences with traits of the stenogastrine and/or polistine+vespid social forms.

According to the current cladogram, the hypothetical social ancestor of (Stenogastrinae + (Polistinae + Vespidae)) (node 7 in figure 4.1) evolved its sociality from the hypothetical ancestor that it has in common with Eumeninae (node B in figure 4.1). If this is true, then we can never know the nature of the hypothetical social ancestor at node 7 in the figure (was it more like Stenogastrinae or Polistinae?), nor the sequence of changes it went through en route from solitary to social. This is because there are no living descendants of forms intermediate between the hypothetical ancestor at node B, which can plausibly be argued to have been solitary, and the hypothetical ancestor at node 7, which can strongly be argued (based on the current phylogeny) to have been social. That there are abundant forms that show life history traits intermediate between solitary and social is immaterial, according to the current phylogeny, because all the intermediate forms are in the monophyletic Eumeninae. Thus the biology of Eumeninae, encompassing the many striking and informative presocial behaviors described in chapter 2, can have no bearing whatsoever on our understanding of the evolution of the living social forms. I think this is implausible, and data on the systematics of Eumeninae are insufficient to be the basis of a firm conclusion.

A consensus opinion on the number of origins of sociality in Vespidae will require more data. Sampling of eumenines must be comprehensive in future analyses. Sequence data, especially of well-conserved nuclear genes, seem the most promising source of informative new data, but more precise definition and coding of behavioral traits might also bring clarity to the question. The quality of data compilation and of assessing results has lagged behind the technical aspects of data analysis in metazoan phylogenetics (Jenner 2004), and the same might be said of the phylogenetics of Vespidae. The behavioral, developmental, and ecological details are compelling: sociality evolved twice in Vespidae. Stenogastrinae evolved in a tropical rainforest, whereas Polistinae evolved in a seasonal locale. The living nonsocial sister taxa of each of the two social lineages are currently classified as Eumeninae.

Beyond the Sociality Threshold

There are two forms of sociality in Vespidae that arise from social antecedents: swarm founding and vespine sociality. Swarm-founding polistines represent three internal clades of Polistinae—Epiponini, *Polybioides*, and part of *Ropalidia*—and the derivation of swarm founding from independent-founding ancestors is uncontested. The natural history basis for the origin of swarm founding has been contemplated (West-Eberhard 1982) and sought in the field (Hunt et al. 1995; Smith et al. 2002), but it remains an open problem.

From a cladistic perspective, Vespinae and Polistinae share an ancestor that would logically be inferred to have been social. Several traits shared by the two subfamilies provide strong support for the hypothesis of shared ancestry (table 4.2). From a natural history perspective, however, it seems plausible that vespines evolved from a lineage of independent-founding social wasps already well along a pathway exemplified by the living independent-founding polistines that have dimorphic female offspring, which is not a basal trait of Polistinae. Evidence that the social ancestor of Vespinae may even have been nonpolistine, however, is that vespine larvae have mandible morphology and ingestion behaviors that more closely resemble those of Eumeninae than of *Polistes* (Reid 1942; Yamane 1976; Matsuura and Yamane 1990). H. E. Evans (1958, p. 456) believed it likely that Vespinae "probably evolved from some early vespid independently from the other subfamilies." Thoughts along these lines have been suppressed by the six-subfamily phylogenetic hypothesis of Carpenter (1982, 1991, 2003), but I think it would be worthwhile to retain some of the spirit of Evans' conjecture. I say this in part because developmental evidence presented in chapter 5 suggests that the origin of castes in Vespinae is likely to more interesting than the current phylogenetic hypothesis indicates. Nonetheless, because the focus of this book is on the evolution of sociality, I do not pursue social evolution beyond the origin of sociality in independent-founding polistines, although great interest attends the more complex forms. In the remainder of this book, however, vespines and swarm-founding polistines are sometimes brought forward as examples or to provide details on topics such as individual development and demography.

PART II

Dynamics

The wasp in a soda can has an entirely new twist in Japan. There, marathon runners and mountain climbers can select as their preferred libation a product named VAAM. The marketing strategy is ingenious. Recall that queens of the giant hornet, Vespa mandarinia, *are the size of your thumb, and such a queen, once her first daughters begin alloparental care, can produce thousands of eggs over the course of several months without ever again leaving the nest. The nourishment that sustains her endurance and performance is the saliva of her larvae. The nourishment that sustained my endurance and performance as I hiked to the top of Mt. Fuji was VAAM—Vespa Amino Acid Mix. The drink includes a cocktail of amino acids in the proportions found in hornet larval saliva! To be honest, I also took Gatorade and several Snickers bars up the mountain, and I had a Coca-Cola at the top. Somehow, though, the Japanese beverage was fitting to the occasion.*

That queens of large-colony social wasps can lay thousands of eggs reflects their physiological capacity in a social context, and it has the demographic consequence that successful queens leave a great many reproductive offspring. The historical route to vespid sociality was marked by changes in individual traits, colony-level phenomena, and the demographic consequences of interactions between the two. Therefore, these are the three foci of investigation that can best reveal the dynamic maintenance of vespid sociality. I explore these in the next three chapters, and then I draw a synthesis of sociality in the present tense. Individual variation constitutes the basic level and so will be the starting point.

Individuals

5

During the Watergate political scandal, reporter Bob Woodward was advised by Deep Throat that mysteries could be resolved if he would "follow the money." Money, of course, is the mother's milk of politics. In life, protein—a key ingredient of mother's milk—is the currency of growth and reproduction. Therefore I have found it a useful analogy that many, perhaps most, of the mysteries of hymenopteran sociality might be resolved if investigators would "follow the protein." The first place I investigated was adult wasps' midgut, the organ of digestion and absorption (figure 5.1).

Midgut Proteases

If an adult social wasp with idiobiont heritage is going to sustain production of large eggs throughout a colony cycle, she must have protein nourishment. Because the wasp is constrained by the apocritan thread waist, it can use protein only in liquid forms. In progressively provisioning wasps that malaxate caterpillars, a key proteinaceous liquid is the hemolymph of those caterpillars. It is imbibed into the crop before being regurgitated to feed larvae, but some is always retained (Hunt 1984) and passes into the midgut. Insect hemolymph contains free amino acids (Wyatt 1961) as well as whole proteins. A question that attends feeding on hemolymph, therefore, is whether the malaxating female can digest the whole protein that she has ingested. Ikan et al. (1968) asserted that adult females of *Vespa orientalis*, a hornet, did not contain digestive proteases and therefore were incapable of digesting protein. In collaboration with Donald Grogan and Barry Kayes, I found that proteolytic enzymes are present in midguts

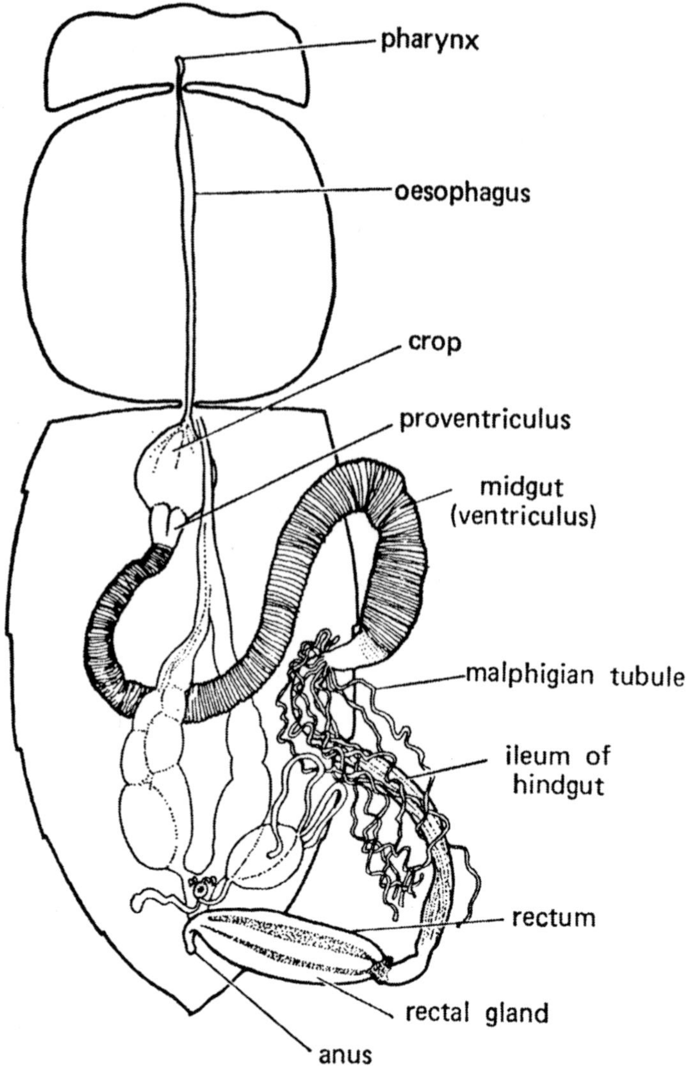

Figure 5.1
Anatomy of the digestive tract of the yellowjacket *Vespula germanica* showing the narrowly constricted thread waist through which ingested food must pass. All of the digestive system except the esophagus and salivary glands is in the gaster. The crop is a distensible storage organ from which liquids may be regurgitated and from which the proventriculus, a muscular valve, regulates flow into the midgut, which is the organ of digestion and nutrient absorption. Malphigian tubules are organs of nitrogenous waste excretion. From Spradbery (1973). © J. Philip Spradbery. Reprinted with permission of the author.

of adult *Vespula* (Grogan and Hunt 1977), *Polistes* (Kayes 1978), the swarm-founding polistine *Polybia* (Hunt et al. 1987), and the hornet *Vespa crabro* (Grogan and Hunt 1986). I was frankly disappointed that the Ikan et al. hypothesis was incorrect. Obligate dependence of adults on liquid foods containing only free amino acids rather than whole proteins would make a wonderful story. The story line took a plot twist, however, because of findings from work that Grogan and I pursued for a time on pollen and bees.

Pollen is the primary proteinaceous food of bees. Bees gather pollen as larval provisions by means of branched setae (hairlike structures) that cover the cuticles of all but cleptoparasitic forms. Adult bees drink nectar to nourish themselves. Nectars of all kinds contain free amino acids (Baker and Baker 1973), and nectar may also contain pollen grains. If so, that nectar will be enriched nutritionally. The enrichment occurs via a surprising means: pollen in aqueous solution will digest itself. The hard shell of pollen grains contains an enzyme with the same cleavage specificity as the mammalian digestive enzyme chymotrypsin, and some pollens also have enzymatic activities of trypsin and/or carboxypeptidases A and B (Grogan and Hunt 1979). The amounts of chymotrypsin and trypsin in midguts of honey bees are within ranges that could be eluted from the quantity of pollen that worker honey bees are known to ingest (Grogan and Hunt 1979). Enzyme activity correlations with both season (Grogan and Hunt 1984) and worker age (Grogan and Hunt 1980) suggest the influence of ingested pollen. It seems possible that worker honey bees could gain all of their midgut proteases via pollen ingestion.

This perspective on honey bee midgut proteases took on added significance in light of the findings of Martin (1987) on the midgut proteases of fungus gardening ants, including the famous leaf cutter ants of the New World tropics. These ants cut and collect fragments of leaves, but rather than eat the leaves directly (except for liquid released by the cutting: Littledyke and Cherrett 1976) they cultivate fungus on them and then feed the fungus to their larvae (Belt 1874; Weber 1966; Wilson 1971; Hölldobler and Wilson 1990). Martin found that the proteases that worker leaf cutter ants defecate on fresh leaves are fungal enzymes rather than insect enzymes. That is, the fecal proteases of adult leaf cutter ants have been ingested. Has natural selection shaped the fungal enzymes for the roles they may play as they pass through the ant midgut? Similarly, might pollen protease activities have been naturally selected, at least in part, for a role in facilitating pollinator nourishment? These notions are plausible. Ikan et al. (1968) may have been incorrect in asserting that proteases are absent in adult oriental hornets, but the idea that adult wasps might obtain midgut proteases from external sources rather than by, or at least in addition to, endogenous production began to gain traction in my mind.

I had known for some time that adult wasps ingest at least some proteases, because proteolytic activity had been documented in wasp larval saliva (Maschwitz 1965; Ishay and Ikan 1968a; Kayes 1978), which the adults certainly eat. Might pollen be a second source of ingested proteases for wasps? Masarinae collect pollen internally in their crop, and such a feeding habit did not arise *ex nihilo*. There had to be an antecedent, simpler behavior. I looked to see if pollen ingestion might occur passively in wasps as they drink floral nectar. Several students and I sought pollen in the midguts of museum specimens of various wasps and bees, and, lo and behold, counts of pollen grains from several species of Vespidae, both solitary and social, approach those obtained from bees (Hunt et al. 1991). It also occurred to me that as wasps malaxate living caterpillars they would obtain proteases that are in the caterpillars' midguts, but I have not investigated this possibility.

Kayes (1978) had found proteolytic activity in the midgut lumen (the hole in the donut, if one takes a transverse slice of midgut) of *Polistes* wasps newly emerged from pupation. It was a simple test, however, and I wanted to confirm the result. I therefore designed an experiment in which I isolated *Polistes metricus* females immediately upon their emergence from pupation in the lab. Some were given a small cube of unflavored gelatin with a bit of table sugar in it (wasps wouldn't take gelatin plain or with beef extract). I let the wasps malaxate the gelatin as they wished, and then I removed and froze the wasps' midguts at 1, 2, 3, and 4 hours after they had started malaxation. Those midguts were subsequently analyzed for protease activity by Gary Felton, then at the University of Arkansas. The data (table 5.1), published for the first time here with Dr. Felton's kind permission, show a striking pattern. Newly emerged wasps and wasps given gelatin and sampled up to 3 hours after the start of malaxation had amounts of chymotrypsin and trypsin in the same range as workers and queens that were present on nests at the time of collection. The 4-hour wasps had significantly higher activity of both enzymes. Curiously, all three wasps in the 4-hour group malaxated the gelatin for at least 3 hours, which is an aberrant behavior. Some wasps of the other time periods had dropped the gelatin before I collected their midguts, and a few consumed it. Nonetheless, the data show two things: that adult wasps isolated at emergence have midgut protease at the same levels as adults that have access to nourishment, and a change in protease levels is inducible. Both lines of evidence argue that the protease is endogenous. I confess, however, a lingering suspicion and curiosity about the matter. For example, although Lepidoptera that feed on proteinaceous foods may have midgut proteases (Terra et al. 1990; Chapman 1998), lack of endogenous proteases may be the rule in butterflies and moths that feed on nectar (Wigglesworth 1972; Chapman 1998). Thus, I see no a priori reason that endogenous proteases must be present in adult wasps. Ingested proteases could play a far more significant role in hymenopteran sociality than now is recognized or even envisioned.

Free Amino Acids

Protein digestion is not an issue where free amino acids are concerned. Amino acids, the building blocks of proteins, occur naturally in the liquid foods that adult wasps eat: floral nectars (Baker and Baker 1968, 1973; Baker and Hurd 1968), insect honeydew (Way 1963), and insect hemolymph (Wyatt 1961). The honey stored

Table 5.1
Activities of chymotrypsin and trypsin from midguts of *Polistes metricus*.[a]

Category (sample size)	Chymotrypsin (mmol/min/mg protein)	Trypsin (mmol/min/mg protein)
Newly emerged (8)	3948±752	7.7±1.7
1-Hour (3)	6556±468	6.0±0.5
2-Hour (3)	3064±674	8.4±2.4
3-Hour (3)	3845±352	3.4±0.2
4-Hour (3)	15550±1328	25.7±1.3
"Workers" (5)	3490±718	11.0±1.5
"Queens" (4)	2894±722	8.7±4.8

[a]Data are means ± standard error; sample sizes are in parentheses. "Workers" and "queens" were taken from field-collected nests on July 5, 1996, and separated by the absence/presence of developed ova. Other wasps emerged in captivity between July 7 and 16 and were held in isolation with only water and from zero to three feedings of dilute sucrose solution until July 16, when gelatin was given to wasps of the time-treatment groups. ANOVA: chymotrypsin, $F = 11.74$, $p < .001$; trypsin, $F = 6.66$, $p < .001$.

by paper wasps has nectar and/or honeydew as its origin, and so it too contains free amino acids (Hunt et al. 1998). While all of these free amino acid sources may be important, the most important is larval saliva. Free amino acids were first shown to be present in larval saliva of *Vespula* (Maschwitz 1965) and *Vespa* (Maschwitz 1966; Ishay and Ikan 1968a). Maschwitz (1966) showed that the saliva comes from the labial gland. Proportional abundance of individual amino acids in the larvae of *Vespa*, *Vespula*, and *Polistes* were quantified by Hunt et al. (1982), and additional analyses for *Vespa* were given by Abe et al. (1991). The quantitative analyses revealed several striking patterns. The saliva of social vespid larvae contains most or all of the nutritive amino acids, whereas most flower nectars contain less than the full complement. The quantity of amino acids is up to 50 times greater than that of nectar from flowers typically pollinated by wasps. The percentage of proline among the other amino acids is unusually high—up to 50%. The high percentage of proline can be an entry point to the mysteries of larval saliva.

Proline is technically an imino acid (containing an NH group) rather than an amino acid (containing an NH_2 group). It acts a structural disruptor for α helixes and as a turning point in β sheets. That is, proline puts kinks into proteins. Thus proline plays an important role in protein structure, but most proteins require it in only modest proportions. Table 5.2 gives the amino acid proportions for three average proteins: the King-Jukes average of 53 mammalian peptidases (King and Jukes 1969), the average of the Swiss-Prot Database (http://us.expasy.org/tools/pscale/A.A.Swiss-Prot.html), and the D-50 average of 50 representative *Drosophila* proteins (Hunt et al. 2003). The latter was compiled by my colleague Diana Wheeler and her student Dan Hahn. Proline constitutes about 5% of all 3 of these average proteins, which is less than the 21% average for larval saliva of 18 species of Polistinae and Vespinae (table 5.2).

Therefore it is reasonable to infer that the full complement of proline in wasp larval saliva is not used for protein synthesis by adult wasps. Instead, it is probably used as an energy source.

Bursell (1963), studying flies, was the first to report that proline in insect hemolymph diminishes during flight. The same was reported for an aculeate, the honey bee, along with the finding that amounts of other amino acids do not change during flight (Micheu et al. 2000). Proline is converted to glutamate, and this conversion occurs in mitochondria of flight muscle cells during flight (Sacktor and Childress 1967). Sugars would seem a better energy source than proline, and larval

Table 5.2
Amino acids, as molar percentages, of three average proteins, larval saliva of vespid wasps, and diverse arthropod silks.[a]

	Proteins			Saliva	Silks		*Polistes annularis*	
Amino acid	K-J	S-Prot	D-50	(18 salivas)	24 I, B	24 Hosts	Saliva	Silk
Alanine	7.4	7.7	7.2	5.4	21.0	28.1	1.4	33.6
Arginine	4.2	5.2	5.6	3.7	3.4	2.3	8.3	8.1
Asparagine	10.3	9.6	10.7	2.7	10.9	7.1	4.2	6.0
Cysteine	3.3	1.6	1.6	2.6	—	—	6.3	—
Glutamic acid	9.5	1.5	10.9	8.6	6.6	5.2	14.5	6.5
Glycine	7.4	6.9	6.0	9.8	18.3	25.7	5.5	7.5
Histidine	2.9	2.3	2.6	1.3	1.1	0.4	0.6	0.1
Isoleucine	3.8	5.9	5.0	3.7	0.8	1.5	2.2	1.6
Leucine	7.6	9.6	9.1	3.7	1.1	2.2	3.0	4.5
Lysine	7.2	6.0	5.6	5.2	1.2	1.2	7.8	3.1
Methionine	1.8	2.4	2.2	2.2	—	—	—	0.3
Phenylalanine	4.0	4.1	3.4	3.6	0.2	0.7	7.3	0.3
Proline	5.0	4.9	5.5	21.3	1.5	3.2	16.7	0.7
Serine	8.1	7.0	8.2	4.6	30.0	13.4	2.4	19.6
Threonine	6.2	5.6	5.7	5.8	1.1	2.6	3.4	3.1
Tryptophan	1.3	1.2	1.0	1.7	—	—	1.4	—
Tyrosine	3.3	3.2	3.2	5.6	1.2	2.9	8.9	0.3
Valine	6.8	6.7	6.5	4.9	1.2	2.6	2.6	4.6

[a] K-J = the King-Jukes average of 53 mammalian peptidases (King and Jukes 1969); S-Prot = the average of the Swiss-Prot Database (http://us.expasy.org/tools/pscale/A.A.Swiss-Prot.html); D-50 = 50 representative *Drosophila* proteins (Hunt et al. 2003). For silks, 24 I, B = 24 Ichneumonidae and Braconidae from Quicke and Shaw (2004); 24 Hosts = diverse hosts of the 24 parasitoids in Quicke and Shaw (2004), including Lepidoptera (20), Hymenoptera (3), and Aranae (1). *Polistes annularis* saliva from Hunt (1982) and silk from Espelie and Himmelsbach (1990). The 18 samples of saliva of social Vespidae are from Hunt (1982): *Polistes annularis, P. carolina, P. exclamans, P. fuscatus, Vespula maculifrons, Dolichovespula maculata*, and *Vespa crabro*; from Hunt et al. (1987): *Polistes canadensis, P. instabilis, Mischocyttarus immarginatus, Brachygastra mellifica, Parachartergus fraternus, Polybia occidentalis*, and *P. diguetana*; and from Abe et al. (1991): *Vespa mandarinia, V. crabro, V. tropica, V. analis*, and *V. xanthoptera*. Values have been rounded to the nearest 0.1 from the original sources. Non-nutritive amino acids reported in the original sources have been omitted.

saliva does contain sugars such as glucose, fructose, sucrose, maltose, trehalose, and melezitose (Maschwitz 1965; Ikan and Ishay 1966; Ishay and Ikan 1968a). At least some of these are synthesized via a metabolic pathway called gluconeogenesis (Krebs 1964), whereby protein-fed larvae convert protein to sugars (Ishay and Ikan 1968b). There are more sugars in saliva than amino acids, although *Polistes* larval saliva is not sweet to my taste, as are most nectars. (Maschwitz [1965] reported a sweet taste and fruity aroma for the saliva of some vespines.) Thus, there is less proline than sugars. Even so, evidence suggests that an energetic role for proline has been strongly selected in larvae of polistine and vespine wasps.

Larval saliva is produced by the labial gland. So is the silk that polistine and vespine larvae use to spin cocoons and close their larval cells (Ochiai 1960). Considerable interest attends silk, both for its traditional role as fabric and for nontraditional roles such as bulletproof body armor. These human and insect uses of silk depend on the tensile strength of its long fibers, which are made of protein. The role played by proline in forming kinks in proteins would seem misplaced where long fibers are desirable, and, indeed, silks of all kinds are low in proline (Lucas et al. 1957; S. Hunt 1970; Prashad et al. 1972; Lai-Fook and Wiley 1976; Pant and Unni 1978; Varman 1978; Lombardi and Kaplan 1990; table 5.2). Wasp silk is no exception. Table 5.2 lists amino acid proportions for the saliva and silk (Espelie and Himmelsbach 1990) of *Polistes annularis*. Although the saliva and silk analyses were conducted separately and although the two analyses have been performed on only a single vespid species, the conclusion is almost certainly generally applicable. Proline concentration is high in larval saliva, and it is low in silk produced by the same gland. The conversion from producing a secretion high in proline to one low in proline takes place in a few hours preceding pupation, during which time the larva produces neither saliva nor silk. Such a dramatic physiological change must have a strongly selected basis.

The discovery that wasp larvae behave differently as they give saliva, or not, to soliciting adults (or me) (Hunt 1988) was my breakthrough in beginning to understand saliva. The lobe erection behavior of *Mischocyttarus* described in chapter 3 of this book makes sense in the context of the idiobiont heritage of aculeates. When nourishment for adult females is in short supply, idiobionts can resorb ova in their oviducts and perhaps reproduce again another day when nourishment conditions are better. Alternatively, a female may seek nourishment from sources other than those typically used. For polistine and vespine wasps whose larvae live in open nest cells, an alternative source is obvious. Maschwitz (1965) referred to larvae as a food reserve within the nest. He was thinking about the saliva that the adults drink. There is a source more basic than that. When the going gets tough, the tough eat the larvae (Crespi 1992a). By doing so, they sustain themselves to reproduce another day. From the larva's perspective, this situation has no clear fitness payoff.[1] Selection thus has shaped an adaptive larval

1. One could argue that larvae that make the "ultimate sacrifice" whereby their tissue sustains reproductive relatives, as is often the case, have positive inclusive fitness. I'll reserve inclusive fitness for part III of this book.

behavior: when confronted with a hungry adult, they appease it. Strong individual selection would attend unsuccessful appeasement, and such selection has led to high levels of both sugars and amino acids, notably proline, in the larval saliva.

Selection would most strongly favor appeasement behavior during times of low nourishment, which characterize the early phase of the *Polistes* nesting cycle (West Eberhard 1969). Appeasement during times of low nourishment would also play the role of a food reserve as proposed by Maschwitz (1965), which also could have selected for the nutritional value of the saliva. Later in the nesting cycle, larvae that are tended by alloparental offspring of the colony are better nourished, and it may not be as risky to withhold saliva, or they may simply be solicited less often or less aggressively. The resulting pattern is that less well-nourished larvae early in the colony cycle may pass more saliva to adults, and better-nourished larvae later in the colony cycle may retain more of their saliva. This could accentuate interindividual developmental differences arising from provisioning differences alone.

There is a possible loose end to the saliva story. Brian and Brian (1952) said, incorrectly, that the saliva is not attractive to adult wasps. They proposed, therefore, that the saliva is largely an excretory product to rid the larva of excess water. Because they were wrong on the first point, I always thought that they also were wrong on the second, but there is a chance that in one sense they may be correct. I know of no vespid wasps other than Polistinae and Vespinae that provision their larvae with nectar. This provisioning behavior could well have been selected to play a role in producing the copious quantities of saliva that larvae pass to adults. However, excepting a possible, partial role in water management for larvae provisioned with liquids, the saliva cannot reasonably be considered waste (Maschwitz 1965). When I said to my colleague Irene Baker that the saliva had been proposed to be waste, her astonished response was, "But, oh, what a waste!"

Fat Body and Storage Protein

In insects, the principal organ of intermediary metabolism is the fat body (Chapman 1998). More a tissue than a discrete organ, fat body may be distributed throughout the internal body cavity, but it is most abundant in the abdomen. It is a site of protein synthesis, and it serves as a storage tissue for proteins, fats, and carbohydrates. In *Polistes*, quantity and color of fat body in the gaster have been used as caste indicators. Scant, yellow fat body characterizes workers, and abundant, white fat body characterizes gynes (prequeens) (Eickwort 1969b; Strassmann et al. 1984a; cf. Strambi et al. 1982). The yellow color is probably due to fat. The white color may be due, at least in part, to presence of protein (Berlese 1900; Pardi 1939) and not merely an absence of fat, as fat globules are sometimes scattered through white fat body (Eickwort 1969b). The quantity and quality of fat body reflect nutritional conditions, including nutrition during larval development.

Anthony Rossi and I investigated the role played by honey storage in preemergence colonies of *Polistes metricus* (Rossi and Hunt 1988). In springtime *P. metricus* foundresses will freely initiate single-foundress nests in bottomless wooden boxes placed on poles in oldfields. Rossi used such nests and placed small droplets of honey into nest cells to imitate and augment the honey storage that occurs naturally in *P. metricus* nests in the spring (Hunt et al. 1998). He then collected the first-emerged offspring from treatment and control colonies and tested several variables. In the usual pattern, the first-emerged offspring were smaller than the foundresses. However, the fat content of the gaster (and therefore the fat body) was significantly different among the three groups. Offspring of unsupplemented colonies had low amounts of fat, foundresses had intermediate amounts of fat, and offspring of supplemented colonies had the highest fat levels (Rossi and Hunt 1988).

In a laboratory study, István Karsai and I tested effects of different levels of caterpillar provisioning in *P. metricus* (Karsai and Hunt 2002). Once again, first-emerged offspring were smaller than foundresses. Size of later-emerged offspring varied according to treatment. Colonies receiving few caterpillars produced smaller offspring than those of colonies given caterpillars ad libitum. Another set of colonies received ad libitum caterpillars plus daily hand feeding to satiation of fourth- and fifth-instar larvae. These hand-fed larvae were no larger than the ad libitum group, but they had significantly heavier gasters. This means that they had more fat body, which we interpreted as more fat. Knowledge gained subsequent to that experiment now suggests an additional interpretation.

The fat body of insects stores protein at different times during the life cycle, and this protein can serve diverse functions. The most widespread of these functions is a developmental role in insects with complete metamorphosis, the holometabolous insects. Holometabolous insects grow through a series of larval instars, pass through pupation and metamorphosis, and emerge as adults. Most or all holometabolous insect larvae, at the end of feeding and growth, synthesize in their fat body a suite of large proteins called storage proteins. These proteins then serve as the source of amino acids for the great amount of tissue building and reorganization that occurs during metamorphosis. By the end of metamorphosis, most or all storage proteins generally have been depleted. Storage proteins do, however, play occasional roles in adult insects, and in adult vespid wasps one role may be directly related to sociality.

Colleagues Norm Buck, Diana Wheeler, and I (Hunt et al. 2003) looked for storage proteins in several immatures and two adults of the solitary, trap-nesting vespid *Monobia quadridens*. The immatures conform to the general developmental pattern, but two adult females show a striking difference. Both emerged in the lab on the same day in July. One was smaller than the other, reflecting its having been provisioned with a lesser quantity of caterpillars. The smaller wasp also emerged with little or no storage protein residual from metamorphosis, as is the general pattern for holometabolous insects. The larger wasp, however, emerged with storage protein in its fat body that was residual from metamorphosis. If not sacrificed for science, these two wasps would have entered into reproductive activities before dying in the fall. In expression of their idiobiont heritage, the wasps would have had to develop eggs in their oviducts, and the smaller wasp would have been able to do this only by feeding. The larger wasp, with its reserve of storage protein, certainly would feed as well, but it clearly would have had a head start over the smaller wasp. Although there are no data that can be brought to bear, it seems reasonable to assume that the larger wasp with storage protein residual from metamorphosis would be likely, all else being equal, to have higher fitness.

A similar phenomenon occurs in *Polistes*. The typical metamorphic role for storage proteins can result in depletion during metamorphosis with no residual protein, but it may also result in incomplete depletion with residual storage protein in the adult wasp. We found no storage protein in any adult female that had emerged in the early part of the nesting cycle, but storage protein was present in some (but not all) females present on nests later in the nesting cycle (Hunt et al. 2003). The storage protein in the adults was not confirmed as residual from metamorphosis, although carryover from metamorphosis is likely. As with *Monobia*, the reproductive fates of wasps with and without the storage protein probably differ, and in *Polistes* the difference is tightly linked to sociality. The wasps with no storage protein emerged early in the colony cycle, and those with storage protein emerged later. The former may become alloparental caregivers; the latter become gynes.

Storage protein seems implicated in caste differentiation in *Polistes*, but such a role is apparently not universal in social vespids. No adult yellowjackets, *Vespula maculifrons*, had storage protein (Hunt et al. 2003). Gynes of *Vespula* are larger and strikingly different in appearance from workers, and the gynes' gasters that we assayed were replete with white fat body. However, there was no storage protein in their fat bodies. The presence of storage protein in gynes of *Polistes* but its absence from gynes of *Vespula* constitutes a blip on the sonar screen alerting us that something is lurking out there. This observation contributes significantly to my speculation at the end of chapter 4 that the origin of Vespinae may be much more intriguing than currently is envisioned.

Polyphenism

Polistes species in the north temperate zone apparently all show a similar pattern. Early emerging females are smaller than those that emerge later, except that size may decrease at the end of the brood period (Yamane 1969; Haggard and Gamboa 1980; Turillazzi 1980; Miyano 1983; figure 5.2). Workers and gynes, usually identified behaviorally (or sometimes arbitrarily divided by the emergence of the first male), always overlap in size. Tropical species have been less well examined, and although most may be similar to temperate species, evidence cited in chapter 3 suggests that *Polistes olivaceous* in India could be size dimorphic.

The basis of the potential size dimorphism of *P. olivaceous* is the occurrence of two sizes of cells in a single nest. The unequivocal dimorphism of *Ropalidia*

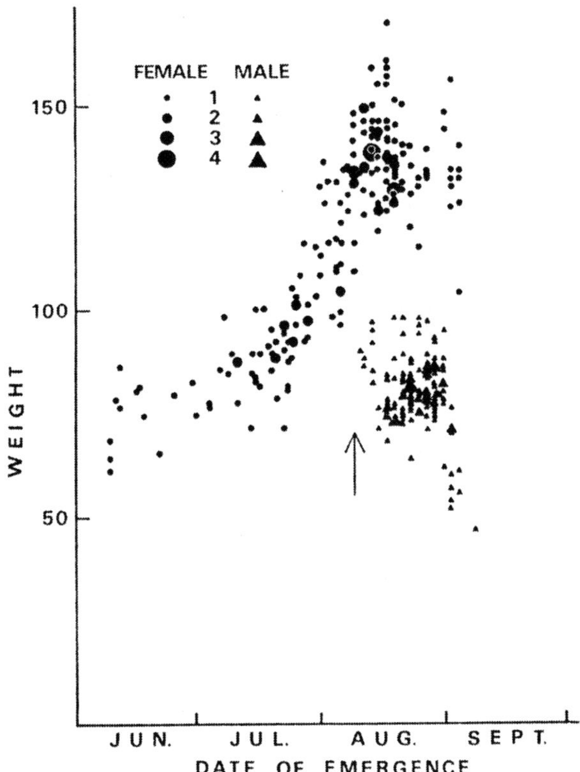

Figure 5.2
Wet weight in milligrams of emerging offspring on a single nest of *Polistes chinensis antennalis* over the course of a season. For females, fresh weights are smallest for the first offspring, increase through the mid-season, and drop slightly at colony dissolution. For males, fresh weight declines from first to last emergence. Different sizes of spots and triangles indicate the number of wasps (1–4) of that weight that emerged on a single day. The arrow indicates the emergence of the first female that did not engage in alloparental care. Originally published as Figure 3 in S. Miyano, 1983, "Number of offspring and seasonal changes of their body weight in a paperwasp, *Polistes chinensis antennalis* Pérez (Hymenoptera: Vespidae), with reference to male production by workers," *Researches on Population Ecology* 25:198–209. With kind permission of Springer Science and Business Media.

galimatia is based on two sizes of nest cells (Wenzel 1992). The basis for dimorphism in *Belonogaster* (figure 5.3) may be similar. The nest of *Belonogaster* is unusual among independent-founding polistines in that the initial cells are not reused but instead the nest walls are torn down, and the paper is remade into walls of new cells. Cell addition is at only one margin of the nest, and the continuing tearing down of old cells and adding new ones results in a J-shaped form

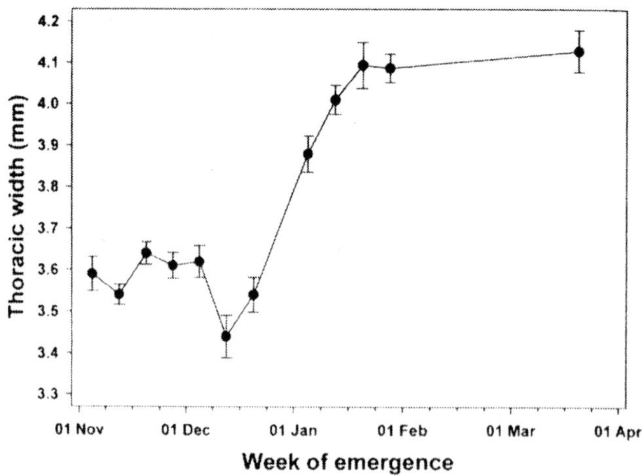

Figure 5.3
Size, measured as thoracic width, of female offspring of *Belonogaster petiolata* that emerged throughout a season in South Africa. Early females are smaller than later females, and the size difference is nearly dimorphic. Reprinted from *Journal of Insect Physiology*, 48, M. G. Keeping, "Reproductive and worker castes in the primitively eusocial wasp *Belonogaster petiolata* (DeGeer) (Hymenoptera: Vespidae): evidence for pre-imaginal differentiation," Pages No. 867–879, © 2002, with permission from Elsevier.

with downward-facing cells in a fanlike array (figure 5.4). The back surface of the nest is cupped. The fanlike array and cupping are due to the shape of the later-constructed cells, which are conical rather than cylindrical. The later, more conical cells, on the left in figure 5.4, are larger than the initial cells and are the ones from which larger wasps emerge.

An intriguing observation on *Polistes fuscatus* and *P. erythrocephalus* is that "the older the cell the larger the diameter of its opening. As a result, new (peripheral) cells are smaller than older (more central) cells on the same nest" (West Eberhard 1969, p. 88). Nest cells in *Polistes* are reused, and they are lengthened by paper addition and/or by rings of cocoon silk added during cocoon spinning by previous occupants (Chao and Hermann 1983; for an extreme case of this, see Yamane and Okazawa 1977). Thus, older cells would become larger and, therefore, would be the ones from which larger offspring emerge. Nests of *Polistes annularis* that enlarge in a cuplike manner (Wenzel 1989) have cells that are conical rather than cylindrical (M. J. West-Eberhard, personal communication). *P. snelleni* also builds cuplike nests (Yamane 1969), and cell volume coincides with caste of the occupant. Males and first-brood workers emerge from smaller cells, whereas late-brood workers and gynes emerge from larger cells (Inagawa et al. 2001; figure 5.5).

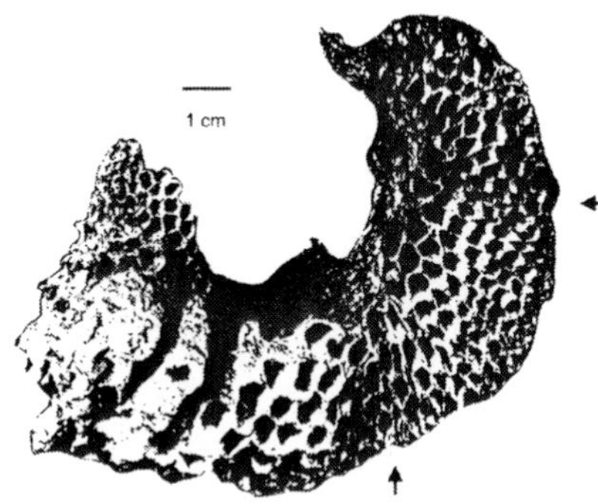

Figure 5.4
A large, mature nest of *Belonogaster griseus*. The pedicel is at the top. Cells were added sequentially from there in a clockwise array. The most recently used cells, at lower left, are conspicuously larger than those that preceded them. Arrows indicate regions of cell compression caused by weight of larvae in cells at the expanding margin of the nest. From Marino Piccioli and Pardi (1978). Reprinted with permission.

Another component of cell size difference, brought to my attention by my colleague István Karsai, is that the shape of cocoons spun by *Polistes* changes over the course of the colony cycle. Early cocoons are flat, and many are actually recessed within the cell. Later cocoons are attached at the cell rim and may be slightly domed, and still later ones stand well above the cell rim as a tall dome. Such cocoons can extend the total cell length by one-fourth to one-third (Rau 1930a). Karsai and I collected data on cocoon shape from a population of *Polistes metricus* in nest boxes in Franklin County, Missouri. I made a contour gauge by

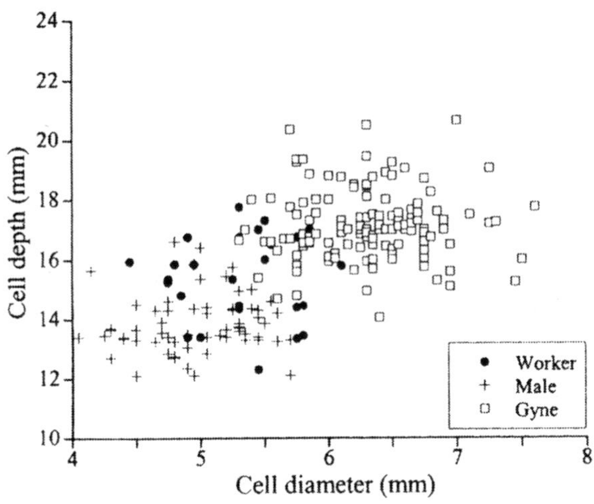

Figure 5.5
Distribution of workers, males, and gynes according to cell diameter and cell depth in three nests of *Polistes snelleni* from Hokkaido, Japan. Originally published as Figure 2 in K. Inagawa, J. Kojima, K. Sayama, and K. Tsuchida, 2001, "Colony productivity of the paper wasp *Polistes snelleni*: comparison between cold-temperate and warm-temperate populations," *Insectes Sociaux* 48:259–265. Reprinted with permission.

using paintbrush bristles pinioned in parallel between two layers of velvet glued to light cardboard. Pressing the tips of the bristles against a flat surface brought the tips into a straight line (figure 5.6, left), which when gently pressed against a cocoon (figure 5.6, center) caused bristles to slide differentially through the velvet and conform to the contour of the cocoon (figure 5.6, right). Cocoon height was then measured indirectly from the contour gauge by means of a machinist's rule scaled in 64ths of an inch and then converted to millimeters. The data (figure 5.7) are published for the first time here with Dr. Karsai's kind permission. Clearly, the spinning larvae affect cell size in a way that reflects and accommodates the known pattern of difference in adult size (Haggard and Gamboa 1980). And, as noted by Rau (1930a, p. 143), "the larvae then assist to a very great extent in building the nest, yet seldom get credit for doing anything useful."

Dimorphism between workers and queens is concomitant with evolution of swarm founding in Polistinae. *Polybioides* (Turillazzi et al. 1994), swarm-founding *Ropalidia* (Sô. Yamane et al. 1983), and the basal epiponine genera *Apoica* (Jeanne et al. 1995b) and *Agelaia* (Jeanne and Fagen 1974; Hunt et al. 2001a) all are dimorphic. The caste difference in *Polybioides* involves hairlike structures on the eyes of queens in addition to size differences, and caste difference incorporating these hairlike structures seems to be unique among polistines. The difference in swarm-founding *Ropalidia* involves both size (queens are larger) and shape. *Apoica* queens and workers differ in proportional relations among

Figure 5.6
The cocoon gauge is made of velvet glued to two pieces of light cardboard that are taped at the ends with velvet facing velvet. Bristles from a paint brush are pinioned in parallel within the velvet sandwich, and these slide lengthwise when pressed gently from one end. Tapping the gauge on a flat surface brings one end of the bristles into parallel (left), and when that end is then gently pressed against a cocoon (center), the resulting displacement (right) can be measured. A metal machinist's rule scaled in 1/64ths of an inch gave reasonable precision when used in the field.

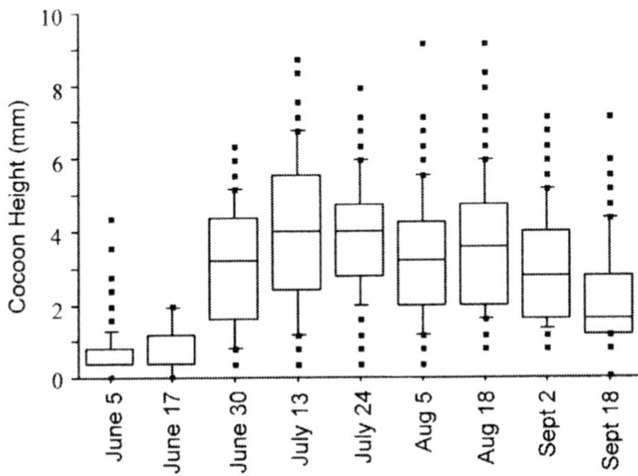

Figure 5.7
Polistes metricus cocoon heights (above the rim of the paper cell) at Shaw Nature Reserve, Gray Summit, Franklin County, Missouri, in the summer of 1998. Cocoons of foundress-reared larvae (June 5 and 17) are flat or nearly so; indeed, cocoons in late May can be recessed 1–3 mm within the rim of the cell. Cocoons of worker-reared larvae (June 30 and beyond) are higher, excepting a decrease in the mean height at the end of the season and occasional outliers throughout the season. Bars within boxes are means; boxes are ± 1 SD; error bars are ± 2 SD; dots are outliers. Sample sizes, beginning on June 5, were 133, 26, 72, 120, 134, 136, 238, 197, and 75. Figure courtesy of Andrew V. Suarez, University of Illinois at Urbana-Champaign.

body dimensions (Jeanne et al. 1995a), although they are the same overall size, or queens are slightly smaller (Shima et al. 1994). *Agelaia* and a number of other epiponine genera (e.g., *Epipona*: Hunt et al. 1996) have larger queens, and queens and workers also differ in shape. Other epiponines, however, differ little or not at all in either size or shape (Noll et al. 2004). The basis for size dimorphism in swarm-founding polistines is unknown. I once carefully examined the nest comb of *Agelaia yepocapa* for possible cell size differences that might represent the two female castes. Visual inspection revealed no cell difference, but some cocoons were somewhat domed and stood a bit above the majority of cocoons, which were only very slightly convex.

In Vespinae, cell size affects adult size. There are two expressions of this phenomenon in females—one among workers and one between workers and gynes. In large-colony species, worker size increases from small initial workers to large workers emerging later, with perhaps a slight drop in worker size near colony dissolution (figure 5.8). The size increase reflects increasing cell diameters in successively built combs (figure 5.9). Concomitant with increasing worker size, larger workers live longer, the number of workers in the colony increases through both emergence and retention, and the larva-to-worker ratio decreases

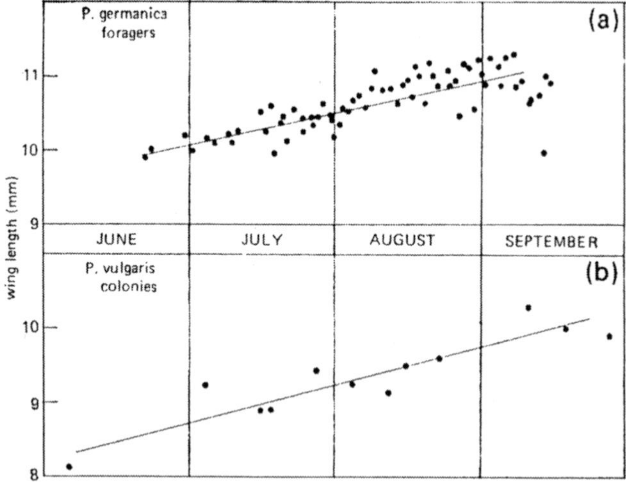

Figure 5.8
Size, measured as wing length, of workers of the large-colony yellowjackets *Vespula germanica* (collected as foragers) and *Vespula vulgaris* (taken from collected colonies) across a season in Great Britain. Worker size increases throughout the season, excepting a slight drop at season's end. (The genus initial "*P.*" signifies *Paravespula*, which has been synonymized with *Vespula*.) From Spradbery (1973). © J. Philip Spradbery. Reprinted with permission of the author.

(Spradbery 1971, 1972, 1973). Thus, the later, larger larvae are better nourished. Although worker cell size increases continuously in successive combs, there is a gap between the largest worker cells and the gyne cells. Discretely larger gyne cells are produced when worker numbers are large, and they are situated closer than worker cells to the nest entrance through which foragers return, a circumstance that generally results in more abundant feeding (Archer 1972). Gyne larvae, therefore, receive the most abundant nourishment of all. The difference between workers and gynes is more than quantitative, however. Just as there are discrete morphological differences, there are discrete developmental differences.

Figure 5.10 shows fresh weights of worker and gyne *Vespula maculifrons* larvae. I collected the data in collaboration with my colleague Neil Chernoff, and they are published for the first time here with Dr. Chernoff's kind permission. Vespine worker and gyne larvae are similar in weight through the first three instars. Size divergence begins in the fourth instar and becomes pronounced in the fifth instar. This reflects longer times spent in later larval instars by gyne larvae (Wafa and Sharkawi 1972), which would lead to cumulatively greater provisioning of gyne larvae. Worker and gyne eggs are equipotent (Spradbery 1973). First-instar through third-instar larvae from gyne cells, when transferred into worker

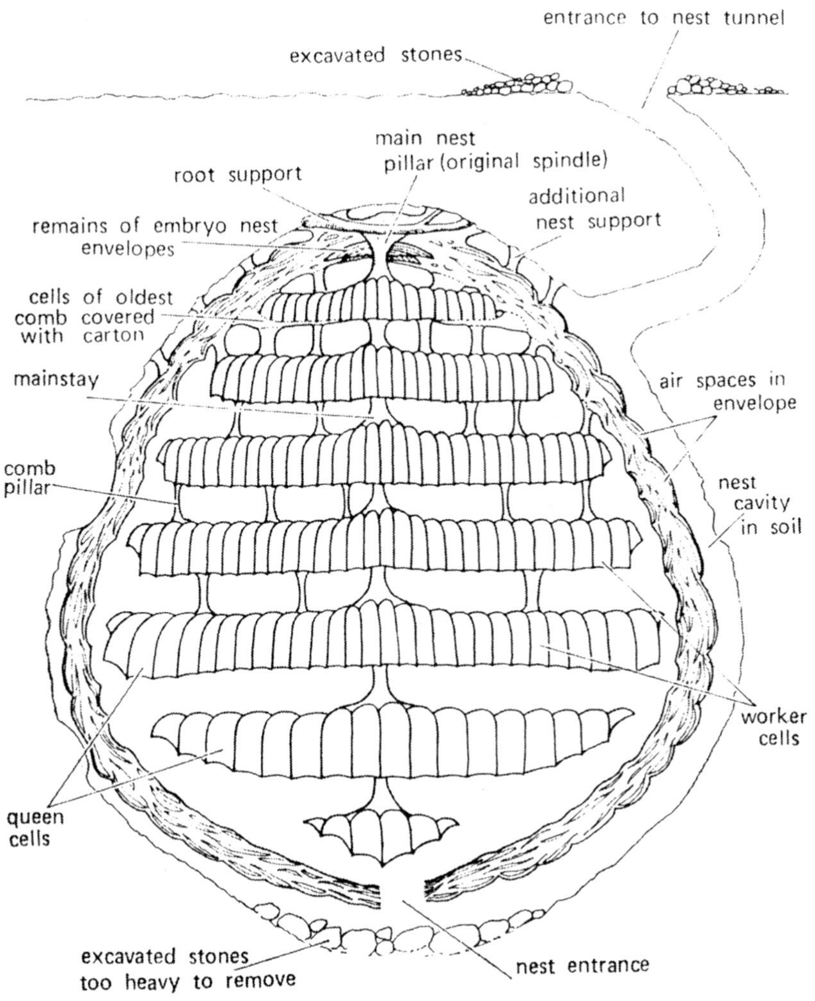

Figure 5.9

Cross-section of a subterranean nest of *Vespula vulgaris*. Nest cell size increases in successively constructed combs (from the top down) and is the basis of increasing worker size. Gyne larvae are reared in discretely larger cells ("queen cells") in the lowermost combs. From Spradbery (1973). © J. Philip Spradbery. Reprinted with permission of the author.

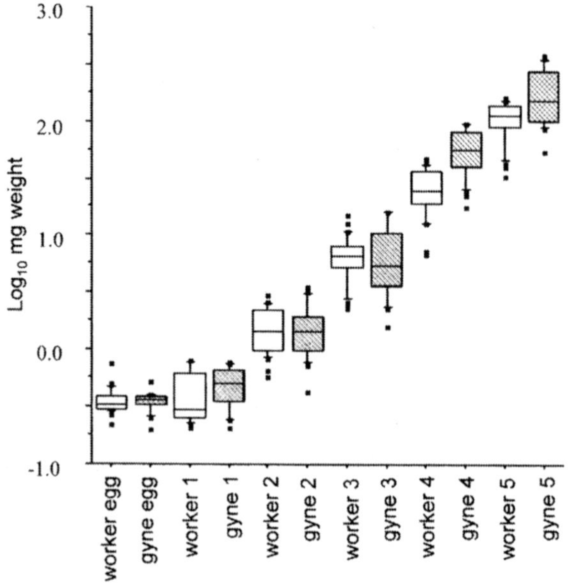

Figure 5.10
Weight of eggs and larvae of workers and gynes of *Vespula maculifrons*, plotted as $\log_{10}$ of the wet weight. Larvae were assigned to instar (numbers 1–5) by head capsule width and to caste (worker/gyne) by the size of the nest cell from which they were taken. Male brood was present in the nest but was easily discerned by nest cell size and location and thus was excluded. Larval sizes of female castes are similar through three instars but diverge thereafter. No gyne larvae had yet pupated, so final weights of the gynes would have been greater. Figure courtesy of Andrew V. Suarez, University of Illinois at Urbana-Champaign.

cells, develop as morphological workers, and first- through third-instar larvae transferred from worker cells into gyne cells develop as gynes (Ishay 1975b). Ishay (1975b) found, however, that fourth- and fifth-instar larvae transferred from worker to gyne cells developed as workers despite the larger cell size. Fourth- and fifth-instar gyne larvae were too large to fit into worker cells.

The dichotomy in developmental pathways revealed by such exchanges suggests that a developmental switch occurs between the third and fourth instars and is a function of provisioning differences concomitant with nest cell size. That is, the differentiation of worker and gyne larvae may have similarities to that of the honey bee *Apis mellifera* (Wirtz 1973). In honey bees, cell size difference is the proximate cue for differential provisioning, both quantitatively and qualitatively, by worker bees (Winston 1987). The provisioning difference, in turn, prompts the differential expression of several genes (Corona et al. 1999; Evans and Wheeler 1999; Hepperle and Hartfelder 2001). The gene expressed most strongly in honey bee queen larvae but not in worker larvae is one with high sequence similarity to a family of genes that encode storage proteins. A greater number of uniquely expressed genes occur in larvae of honey bee workers than in larvae of queens. Proteins encoded by such caste-specific genes may be regulators of development (Nijhout 1999). One piece of evidence strongly suggests that a similar mechanism of dichotomous caste determination based on differential gene expression (although not involving storage proteins: Hunt et al. 2003) also exists in vespines.

The large-colony yellowjacket *Vespula germanica*, which is invasive in the temperate zones of the Southern Hemisphere, has an unusual feature of its life cycle in Australia. There, under benign environmental conditions and apparently high food availability, gynes are sometimes reared from worker-sized nest cells (Spradbery 1993). Although worker nest cells are only about half the volume of gyne cells (Spradbery 1973; Ishay 1975b), tall, domed silk cocoons enclose the gyne pupae in worker cells and accommodate their much larger size. The larger gyne cells are scattered throughout the worker comb, and gynes may be reared adjacent to workers or males. Even so, the difference between gynes and workers is as distinctly dichotomous as when gynes are reared exclusively in gyne cells. Thus, worker and gyne development show a switchlike discontinuity between larvae reared in adjacent, similar nest cells, rather than continuous gradation between the two castes.

Diapause and Quiescence

Solitary vespids in seasonal environments do something that social vespids never do: they pass the unfavorable season in prepupal diapause. Even in populations of such solitary vespids, however, there is interindividual variation. Many solitary vespids, as well as other solitary wasps and bees, in seasonal (especially temperate) environments are "bivoltine"—they have two generations per year. A population begins the favorable season when individuals in diapause become physiologically active, undergo metamorphosis, and emerge from their natal nests to mate and reproduce. The first offspring that they produce have uninterrupted development and emerge in mid-season. Later offspring (if females live long enough) plus offspring of the first-emerged generation feed and grow in their nest cells but then enter developmental arrest when larval growth has ended, the meconium has been voided, and storage proteins have been synthesized. The developmental arrest, called diapause, is not linked to immediate environmental conditions. There must, however, be some cue that trips the developmental switch. That switch has not yet been studied in vespids, but diapause in a fly was shown to be controlled by genes differentially expressed in diapausing and nondiapausing immatures (Flannagan et al. 1998). The conclusion drawn from the fly data was that diapause is not a mere slowing down of physiology but is instead a specific developmental pathway (Denlinger 2002). A similar mechanism of differential gene expression almost certainly must exist in diapausing/

nondiapausing solitary vespids. All social Vespidae, however, undergo uninterrupted development, and so none diapause as larvae.

Social vespids in seasonal environments face the same problem as solitary species, but they have a different solution. They pass the unfavorable season in quiescence, which is a behavioral state characterized by minimal activity and is responsive to immediate environmental cues. Wasps in quiescence may become active on warm days in either temperate winters (Strassmann 1979) or in cool tropical montane cloud forests (Hunt et al. 1999). Quiescent wasps thus face two problems. Those in temperate regions must resist freezing, and all wasps in quiescence must have adequate nourishment reserves to survive until the next feeding opportunity.

The absence of prepupal diapause in social Vespidae may be more than coincidentally related to the construction of paper nests. Wasps are freezing intolerant (Gibo 1976). Even so, they are able to mobilize cryoprotectants such as sugars and glycerol and thereby survive subfreezing temperatures (Gibo 1972, 1976; Strassmann et al. 1984a). This can be expensive: gynes of the vespine *Dolichovespula maculata* lose 70% of lipid reserves, 79% of sugar reserves, and more than 80% of their glycogen reserves during winter quiescence (Stein and Fell 1992). *Polistes* categorized as workers or gynes have been subjected to tests of cold tolerance, and putative gynes survived significantly longer (Strassmann et al. 1984a; Solís and Strassmann 1990). The basis for this difference begins during larval development. Newly emerged adult female *Polistes metricus* that had been hand supplemented above ad libitum feeding levels survived significantly longer in the cold (Karsai and Hunt 2002).

Gynes feed as adults. Indeed, feeding on provisions brought to the nest by foragers, feeding on larval saliva, and feeding on cannibalized larval tissue are behaviors that often contribute to the dissolution of a colony (Mead and Pratte 2002). In temperate zones, gynes also extensively visit fall flowers for nectar (Hunt, personal observation), and they may derive nourishment from pollen in the process. The nutritional aspects of such feeding have not been examined. Successful quiescence almost certainly reflects the combined effects of larval nutrition and additional feeding as an adult. Workers and males lack the nutrient reserves that enable successful quiescence.

Questions Arising

How many ants other than Attini lack endogenous proteases? Many of the liquid foods consumed by ants contain free amino acids, and worker ants may have little need for protein.

Do bees have endogenous proteases? Pollen is so enzyme-rich that selection might not act strongly against the loss of endogenous production. The flip side of the question, then, is to ask if the enzyme richness of pollens has been selected in part for efficacy in facilitating protein nourishment of bees that act as pollen vectors.

What are the quantities and role of ingested proteases in social wasps? Do *Polistes* that malaxate caterpillars derive physiologically useful quantities of enzymes from them? Maybe one way to tackle some of these questions would be to look at midgut proteases in male wasps, bees, and ants.

What variation occurs in the nutrient contents of larval saliva as a function of nourishment level, colony cycle, or larval gender? I have proposed that poorly-fed larvae early in a *Polistes* colony cycle produce copious amounts of saliva. Is the amount more than is produced by later, better-fed larvae, and do the earlier larvae produce a more nutrient-rich saliva? Do males produce less (or more)? The labial gland produces three different proteinaceous products: larval saliva, silk for the pupal cocoon, and, apparently, the proteinaceous glazing that is applied to nests (see chapter 6). The physiology of this gland merits attention.

Do gynes of tropical *Polistes* have storage protein residual in their fat body after metamorphosis? Or, more generally, are there physiological correlates that distinguish workers and gynes of tropical *Polistes*, as seems to be the case with temperate-zone *Polistes*?

What mediates caste dimorphism in swarm-founding Polistinae? No one has yet attempted an experimental approach. For size-dimorphic species, larval nourishment must certainly play a role.

Do swarm-founding polistine wasps incorporate storage protein as part of their caste differentiation? A first step to answering this and many other questions is to collect a few specimens and analyze them in the laboratory. I think of this as laboratory natural history, and I think of Hunt et al. (2003) as a favorite example. We didn't do a controlled experiment; we did natural history at the lab bench. Our findings crystallized new thinking and opened the door to new hypotheses and new research.

Colonies

6

When an inseminated female *Polistes* founds a nest alone, she re-creates in miniature her own evolution. She is, de facto, a solitary wasp. Her solitariness is not some degenerate state derived from sociality but is instead the ancestral state. She, and she alone, progressively provisions her larvae in a nest that she alone has constructed. She rears the first of her offspring all the way to their emergence as adults, and she does this by herself. She has all of the morphology, behaviors, and physiology of the solitary wasp that she is. Her status and role as a solitary wasp constitute one of the most profoundly important perspectives on wasp sociality to have and to keep in mind. Her status and role as a solitary wasp shape everything that follows the emergence of her first offspring. However, *Polistes*, the "city founder," wasn't so-named without reason. In time, a metropolis emerges. Like any city, it has constructed components and living occupants. And, as in any city, construction costs aren't cheap, and the occupants aren't all alike. Webs of interaction shape the occupants' diversity, which is the core of wasp sociality. All of this gets underway before nest construction begins.

Colony Founding and Nest Construction

Except in asynchronous tropical populations, *Polistes* gynes emerge from quiescence in which they have passed the unfavorable season. It has been a long time since the last meal, and nutrient reserves are depleted. In particular, ovaries are undeveloped. Before a gyne can lay eggs, therefore, she must feed. She does so at flowers (Hunt, personal observations of multiple species in Missouri) and perhaps on the hemolymph of prey.

In time, as the ovaries develop, the wasp selects a site and initiates a nest. The gyne becomes a foundress. Nest cells are added serially, and an egg is laid in each cell until no new cells are added as the foundress begins to provision the larvae that sequentially eclose from the eggs (Morimoto 1954). In *Polistes metricus* in Missouri, this plateau in nest size increase occurs at about 20 nest cells. The slowing or pause in new cell addition could be caused by the foundress cannibalizing newly laid eggs (Miyano 1980), taking them from marginal nest cells and using them to feed newly eclosed larvae in central nest cells (Mead et al. 1994). Presence of empty nest cells into which to oviposit obviates the foundress's need to initiate additional cells (Deleurance 1956). When the first larvae pupate, the foundress resumes cell initiation. The plateau in nest size increase may more fundamentally reflect the physiology of the foundress. *Polistes metricus* foundresses have large ovaries when nests are small and contain only eggs, but when foundress nests contain larvae the foundress's ovaries become smaller (Haggard and Gamboa 1980), and egg laying pauses (Bohm 1972a). The energetic costs of brood provisioning apparently diminish a foundress's ability to mature eggs in her ovaries, which could lead to the plateau in new nest cell construction. Pupation of the first larvae and, especially, foraging by the first offspring to emerge would lead to restoration of the foundress's reproductive capacity, and nest growth would resume.

The first-emerged offspring are females that forage for provisions, nectar, water, and pulp. The foundress, now a queen, greatly reduces her foraging activities, although she remains very active on the nest. She intercepts provision loads and uses them to provision larvae, takes incoming pulp loads and initiates new nest cells, and nourishes herself not only with the hemolymph of larval provisions but also by means of larval trophallactic saliva and the crop contents of nestmate adults via interadult trophallaxis. Her daughter workers do most of the foraging, most of the nest construction, and most of the larval provisioning. The nest grows to reach the size it will have at the end of the season.

The dynamics of nest growth have been modeled by Karsai et al. (1996). The central feature is a cell cycle diagram (figure 6.1) that shows the occupancy of a single nest cell. An empty nest cell stimulates oviposition by the queen. That egg then passes through embryogenesis, larval growth, and pupation and emerges as an adult. The consequent empty cell stimulates the queen to oviposit yet again. Based on the data of Mead et al. (1994), who found that newly laid eggs in *Polistes dominulus* were cannibalized to feed newly hatched larvae, a loop from "egg" to "empty" is included. Pupae are an additional (although unexplained) stimulus to the queen's oviposition. Parameter values for a mathematical model were taken from various field studies, and the fit of model output to field data is remarkably good (figure 6.2).

With continued colony growth, gynes and males become so numerous that drinking larval saliva and intercepting incoming provision loads do not adequately meet their nutritional needs. The gynes and males then begin to cannibalize the brood, leading to numerous empty nest cells, and nest growth stops

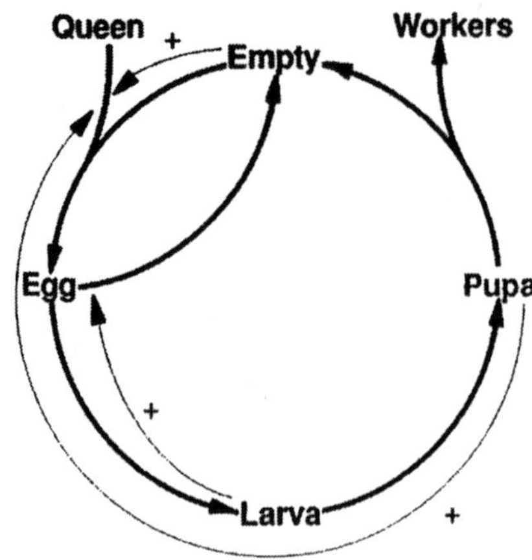

Figure 6.1
Cell cycle, larval development, and positive feedback in *Polistes* wasps. Empty cells and pupae stimulate egg laying by the queen (+). Eggs are fed to new larvae resulting in empty cells (+). Originally published as Figure 1 in I. Karsai, Z. Penzes, and J. W. Wenzel, 1996, "Dynamics of colony development in *Polistes dominulus*: a modeling approach," *Behavioral Ecology and Sociobiology* 39:97–105. © Springer-Verlag 1996. With kind permission of Springer Science and Business Media.

(Duncan 1939; Pardi 1951; Deleurance 1955; West Eberhard 1969; Strassmann 1989b; Mead and Pratte 2002).

Alloparental Offspring

The transfer of much of the brood-rearing and nest-building activities from the foundress to the first of her offspring is the defining moment of polistine sociality. One moment the foundress is a solitary wasp rearing her brood alone; the next moment she is queen of a small society. The key question in the evolution of sociality concerns those first offspring. What factors cause them to stay at their natal nest and perform alloparental care behaviors? Each female has the ability to become reproductive. Even so, idiobiont females lack the nutritional reserves to reproduce immediately (chapter 5); therefore foraging for protein is the appropriate behavior to elevate nutrient levels and foster their own reproduction. Because protein foraging in *Polistes* plays out in the framework of brood provisioning, however, yet another layer of complexity comes into play.

Paul Marchal (1896, 1897; figure 6.3) drew a distinction between two modes of caste determination. In species well beyond the sociality threshold, such as ants and honey bees, Carlo Emery and Herbert Spencer had ascribed worker/queen differences to nutritional conditions during larval development, with low nourishment leading to "trophic castration" of workers. In paper wasps such as *Polistes*, however, Marchal could discern no developmental predisposition to caste and instead ascribed worker/queen differentiation to the physiological costs of brood care by workers, which he called *castration nutriciale*. "Nutritional castration," as used by Wilson (1971), is a misleading translation of the French (Starr 1982). I propose using the phrase "nursing castration," drawn from the Latin *nutrix* (to nurse or foster-mother), to more accurately capture Marchal's meaning while distinguishing it from trophic castration (Starr 1982).

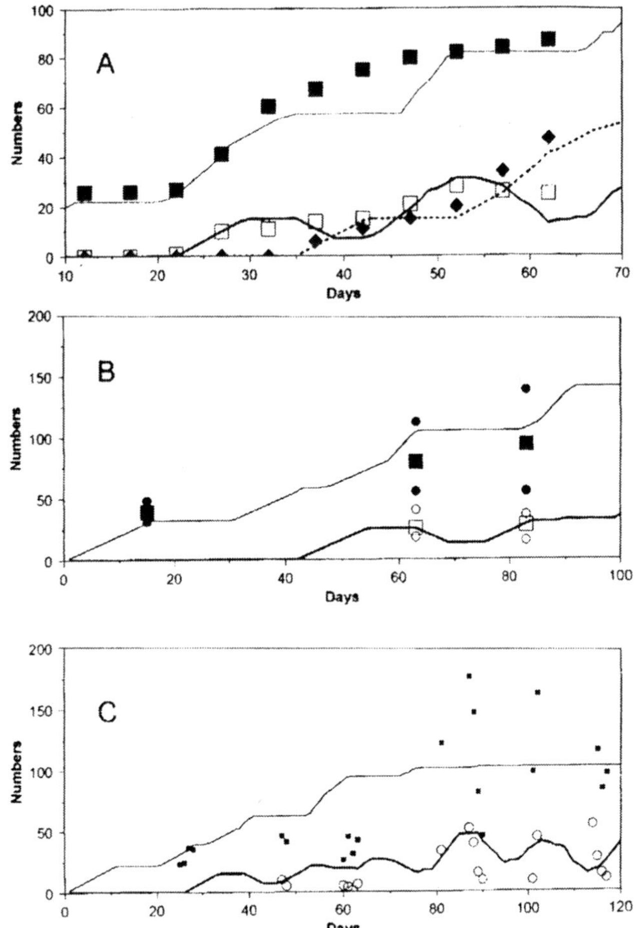

Figure 6.2
Natural data (symbols) and model predictions (lines). Nest size and adult number are cumulative (thin/broken lines; filled symbols); number of pupae is instantaneous (thick lines; open symbols). (A) model parameters and data on *Polistes dominulus* from Mead et al. (1994). (B) model parameters and data on *P. dominulus* from Röseler and Röseler (1989). (C) model parameters and data on *P. dominulus* from Turillazzi (1980). Originally published as Figure 8 in I. Karsai, Z. Penzes, and J. W. Wenzel, 1996, "Dynamics of colony development in *Polistes dominulus*: a modeling approach," *Behavioral Ecology and Sociobiology* 39:97–105. © Springer-Verlag 1996. With kind permission of Springer Science and Business Media.

Figure 6.3
Paul Marchal in his laboratory in the Agronomical National Institute, Paris, 1923. © Boyer/Roger-Viollet. Published with permission.

Emile Roubaud (1916) combined Marchal's two modes of caste determination when he suggested that in independent-founding wasps (Roubaud primarily studied *Belonogaster*) the costs of worker behavior can amplify the disadvantage of poor larval nourishment. I have previously endorsed Roubaud's position (Hunt 1991), and I continue to do so, although in modified form (chapter 8). At the minimum, it has come to be known that the propensity, at least, for worker/queen differentiation during larval development is widespread in social wasps, including species that have no, or very subtle, morphological differences between castes (Gadagkar et al. 1988, 1990;1991; O'Donnell 1998; Hunt et al. 2003).

A young adult female paper wasp that experiences low larval nourishment may be susceptible to exacerbation of any developmental nutritional deficiency as soon as it emerges from pupation. This can occur by means of a behavior that, to my knowledge, is found among wasps only in social forms. Newly emerged social wasps do not leave the nest, or at least they do so very little, during the first few days of adult life (Jeanne 1972; Dew and Michener 1981; Strassmann et al. 1984b; Post et al. 1988; Matsuura 1991). Indeed, they may not be able to fly when newly emerged (Rau and Rau 1918; Hunt, personal observation of *Polistes metricus*). Therefore young wasps are integrated into the social milieu, including dominance behaviors of the queen and interactions with other nestmates, before they forage for the first time. A trajectory toward subdominance can be established before they ever leave the nest.

Offspring that emerge with low or no food reserves from their own development, as is characteristic of the first brood of *Polistes*, may be disposed to remain in the social milieu of their natal nest in order to obtain protein nourishment for their own reproduction. When female Stenogastrinae undertake alloparental care at their natal nest, they gain access to nutrition sources that enable them to reach independent reproductive capacity (chapters 2 and 4). In *Polistes*, that same scenario also may play out en route to sociality, although this has not previously been suggested. In most first-brood female *Polistes*, low nourishment levels remaining from larval development are exacerbated when those females undertake alloparental care, and under these constraints first-brood females rarely reproduce. Instead, the alloparental caregivers become workers. It is noteworthy, however, that at least some first-brood females do reproduce. Replacement queens (Strassmann and Meyer 1983; Suzuki 1985, 1997, 1998), laying workers (Miyano 1980, 1986), foundresses of satellite nests (Strassmann 1981; Page et al. 1989), foundresses of new mid-season nests (Rau 1941), and foundresses of replacement nests following nest destruction (Dani and Cervo 1992; Mead et al. 1995) all come from the cohort that contains alloparental caregivers. In a number of these cases, wasps become egg layers after having first been foragers (O'Donnell 1996).

A first-generation *Polistes* female will forage for protein in the form of caterpillars, but most of the protein that she forages is provisioned to nestmate larvae. Although a provisioning wasp can retain some of the hemolymph of the malaxated caterpillar for her own nourishment (Hunt 1984; Greenstone and Hunt 1993), even that protein may not become available for her reproduction. Recall that the basic nest construction material used by *Polistes* is fibers stripped from dead wood, although, rarely, other pulp sources may be used (Duncan 1928). The fibers, softened with water at collection, are mixed with oral secretions, and additional oral secretions may be applied to existing construction (Réaumur 1742b; West Eberhard 1969; Downing and Jeanne 1983, 1987; Singer et al. 1992). Possible sources of the added secretions are the maxillary glands or even crop regurgitate, but the most likely source is the labial glands (Janet 1903)—the same glands that, in larvae, produce trophallactic saliva and silk. Whatever its glandular source, the material is proteinaceous, and it is expensive. Solitary foundresses temporally partition their protein foraging toward either nest secretions or larval provisioning (Kudô 2000, 2002). In nests with alloparental offspring, secretions in or on nest paper can represent from 10% to nearly 20% of foraged protein (Kudô et al. 1998). Stenogastrinae, in contrast, add little secretion to nests, and the stenogastrine secretion is low in protein content compared to that of polistines (Kudô et al. 1996). Alloparental caregivers in *Polistes* and similar independent-founding polistines therefore not only regurgitate proteinaceous food from their crops to feed larvae or nestmate adults, they also lose assimilated protein by way of glandular secretion as they construct and reinforce the nest. Shaped by their idiobiont heritage to seek protein to initiate their own independent reproduction, they experience an ongoing, context-dependent protein deficiency (with regard to their own reproduction) by virtue of the behaviors they pursue as colony members engaged in brood care. That is, they remain in a state of nursing castration.

Supplementation Experiments

The nutritional basis of caste determination is conventional wisdom: "Remember, any worker can be a queen if she gets enough of the right food when she's young" (Jukes 1990, p. 1359). If this is true and worker behavior in *Polistes* is context dependent, then worker behavior should be malleable or perhaps could be eliminated altogether. All one has to do is manipulate the context that affects individual development. Anthony Rossi and I showed that honey supplementation to preemergence nests of *Polistes metricus* led to large fat bodies in first offspring (Rossi and Hunt 1988), and it now seems plausible that the fat body of offspring from supplemented nests should also contain storage protein. Such well-endowed females shouldn't work. Margaret Dove and I undertook to find out if this speculation might be correct. We replicated the Rossi and Hunt experimental protocol, but rather than collect and assay individuals, we instead monitored colony-level response variables (Hunt and Dove 2002). We anticipated that well-endowed first-brood offspring of supplemented colonies would not perform alloparental care behaviors, and that the supplemented colonies therefore might not persist or, at least, would achieve lower success than controls. We were wrong. Supplemented colonies survived at the same level as controls (figure 6.4), reached a larger size (figure 6.5), produced more offspring (figure 6.6), and had a higher frequency of reproductives among total offspring (figure 6.7).

The larger nest size with supplementation suggests a possible explanation for the larger number of offspring. Recall that in *Polistes dominulus* newly laid eggs are cannibalized to provision newly eclosed larvae (Mead et al. 1994). In the case of our experiment, the supplemental honey could have been used instead of eggs to provision new larvae. Because we did not monitor eggs during the study, this is speculation. If this occurred, however, the plateau in nest construction that is apparent in figure 6.5 would reflect a physiological pause in egg production by the foundress. Upon resumption of oviposition after pupation of the first larvae, there would have been few vacant nest cells, and therefore the foundress would have begun adding new nest cells sooner than would have occurred in nests where egg eating left empty cells. The larger nest size and the dramatic early August increase (figure 6.5) in number of pupal cocoons both could reflect this scenario. However, why the initial offspring of supplemented colonies performed alloparental care rather than behave as gynes remains an open question.

Rossi and I had interpreted the higher fat content in first-brood offspring of honey-supplemented colonies as evidence that colonies in the initial (preemer-

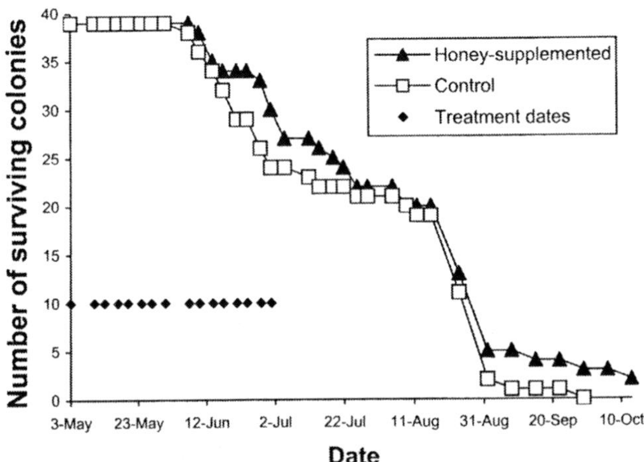

Figure 6.4
Number of surviving colonies over time for a cohort that began with 39 honey-supplemented and 39 control colonies of *Polistes metricus*. Treatment colonies received honey droplets as food supplementation twice weekly from May 3 to July 1 on dates shown by the symbols at $Y = 10$. From Hunt and Dove (2002). Reprinted with permission of The Royal Entomological Society, London.

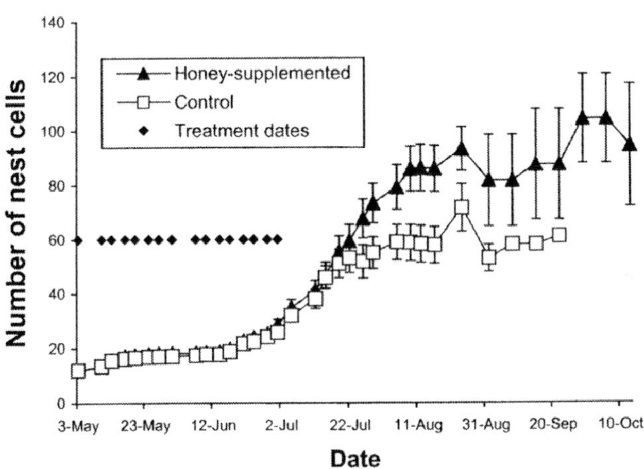

Figure 6.5
Mean ± standard error of number of cells per nest in honey-supplemented and control colonies of *P. metricus*. Sample sizes at each data point are the values of figure 6.4. Decreases in mean cell number reflect termination of nests with high cell numbers as the sample size diminished sharply in mid-August. From Hunt and Dove (2002). Reprinted with permission of The Royal Entomological Society, London.

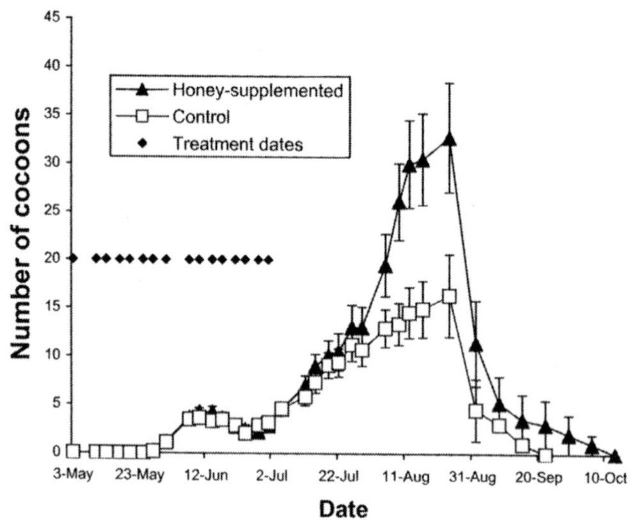

Figure 6.6
Mean ± standard error of the number of pupae (cocoons) per colony in honey-supplemented and control colonies of *P. metricus*. From Hunt and Dove (2002). Reprinted with permission of The Royal Entomological Society, London.

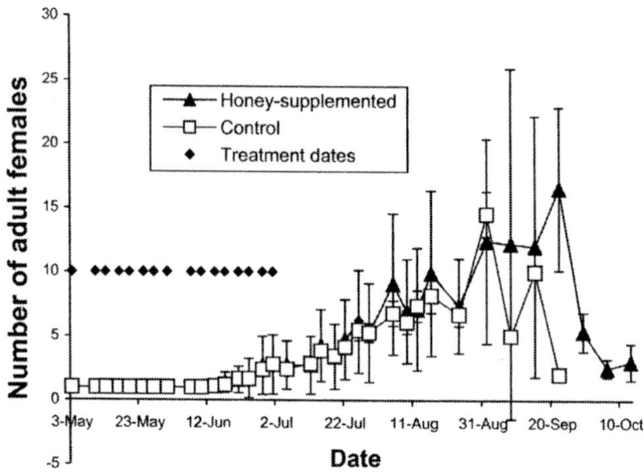

Figure 6.7
Mean ± standard error of the number of adult females present at the nest for honey-supplemented and control colonies of *P. metricus*. Because there was no difference in the number of adults present, despite the larger number of pupae in supplemented nests (figure 6.6), it was inferred that a higher number and frequency of offspring in supplemented colonies departed the nest—an activity characteristic of gynes and males. From Hunt and Dove (2002). Reprinted with permission of The Royal Entomological Society, London.

gence) phase in natural conditions are food limited. Klahn (1981), who performed supplementation experiments on *Polistes fuscatus*, reached the same conclusion. Klahn used a mix of honey and proteinaceous foods and made them available to single foundresses at their nests from eclosion of the first larva until emergence of the first adult. The consequence of supplementation was stronger in a year in which supplementation was offered by a tube at the nest than in another year in which supplementation was offered in dishes near the nest. Sites also differed between years. Supplementation during the preemergence period led to the production of more workers and to the initiation of more nest cells by the queen during the supplementation period. Colonies that had been supplemented in the first year of Klahn's study produced 16 times more gynes and males (numbers of each not specified) than did unsupplemented colonies (Klahn 1981).

Jon Seal studied effects of supplementation of *Polistes metricus* beyond offspring emergence. Seal replicated the honey supplementation protocol of Rossi and Hunt (1988), but he continued it until the experiment was terminated during gyne emergence. Additionally, he provided caterpillars by placing them on top of the nest. A caterpillar placed on top of a nest had been provisioned to larvae in an early feeding experiment (Rau 1929), but Seal could not be certain that the caterpillars he provided were similarly provisioned to larvae. In any event, in this experiment as in those that preceded it, supplemented colonies produced more gynes (Seal and Hunt 2004). The number of male offspring was unaffected by supplementation.

Mead and Pratte (2002) supplemented *Polistes dominulus* colonies beginning at the emergence of the first adult offspring and obtained the same result as other supplementation experiments: more offspring, in this case including more males, were produced in the latter part of the colony cycle. The result, therefore, is robust. Supplementation at any phase of the colony cycle leads to greater production of gynes and sometimes also of males. This demonstrates that natural colonies typically exist in conditions of limited nourishment.

Œcotrophobiosis and Diminishment Experiments

Emile Roubaud (figure 6.8) was a microbiologist and a distinguished researcher on insect vector-borne tropical diseases (http://www.pasteur.fr/recherche/socpatex/pages/Roubaud.html). While in the Congo, he watched wasps as a diversion from his primary research. His careful observations of *Belonogaster* led him to focus on a phenomenon that had been foreshadowed in observations by Réaumur (1742b) and described by du Buysson (1903) and Janet (1903): larvae of polistine and vespine wasps produce saliva that the adults drink. Roubaud placed drinking of saliva together with direct provisioning of larvae by adults as the two components of a reciprocal food exchange that he called *œcotrophobiosis* (Roubaud 1916). William Morton Wheeler (1918) renamed the reciprocal exchange *trophallaxis*—the term we use today—and simultaneously started the long slide of the use of the term into confusion (Wilson 1971). Focus on trophallaxis became

Figure 6.8
Emile Roubaud (1882–1962), bacteriologist at the Pasteur Institute, member of the Science Academy (France), 1938. © Boyer/Roger-Viollet. Published with permission.

a distraction from Roubaud's main point—that adult attraction to larval saliva was the very basis for wasp sociality. A key passage from Roubaud (1916) is translated in Wilson (1971, pp. 281–282):

> The retention of the young females in the nest, the associations between isolated females, and the cooperative rearing of a great number of larvae are all rationally explained, in our opinion, by the attachment of the wasps to the larval secretion. The name *œcotrophobiosis* (from σιχοσ, family) may be given to this peculiar family symbiosis which is characterized by reciprocal exchanges of nutriment between larvae and parents, and is the raison d'être of the colonies of the social wasps. The associations of the higher vespids has, in our opinion, as its first cause the trophic exploitation of the larvae by the adults. This is, however, merely a particular case of the *trophobiosis* of which the social insects, particularly the ants that cultivate aphids and coccids, furnish so many examples.

Wilson (1971, p. 283–284) reviewed the ensuing half century of history of thought and research on trophallaxis and concluded:

> The principal significance of the findings on the vespine wasps is, in my view, that they demonstrate for the first time that larvae can

> behave in an altruistic manner toward adults and that they thus contribute, by virtue of their behavioral patterns, to the homeostatic machinery of the colony. . . . At the same time, trophallaxis is seen to be not at all the driving force behind social evolution, as envisaged by Roubaud and [William Morton] Wheeler, but only one of a number of forms of communication and nutritive exchange that have been built, with variations, in the course of social evolution.

West-Eberhard (1978a) also argued that Roubaud's hypothesis should be discarded, a proposition with which I was once in partial agreement: "The focus of West-Eberhard's discontent was Roubaud's proposition that attractiveness of larval saliva to adults was the main cause of social cohesion among them and, indeed, that idea should be dismissed" (Hunt 1991, p. 431). Wilson, West-Eberhard, and I all were wrong.

The season after the supplementation experiment conducted with Margaret Dove, I undertook a diminishment experiment. Assisted by students Margaret Williams and Mary Rosenthal, I structured a field experiment in which I took larval saliva from fifth-instar larvae in field nests of *Polistes metricus*. The results (Hunt and Dove 2002) were nothing short of stunning. Diminished colonies had significantly lower survivorship than controls (figure 6.9), had smaller final nest sizes (figure 6.10), and produced fewer offspring (figures 6.11, 6.12). Indeed, the diminished colonies probably produced no gynes at all. The onset of these dire consequences was apparent as early as the third diminishment treatment,

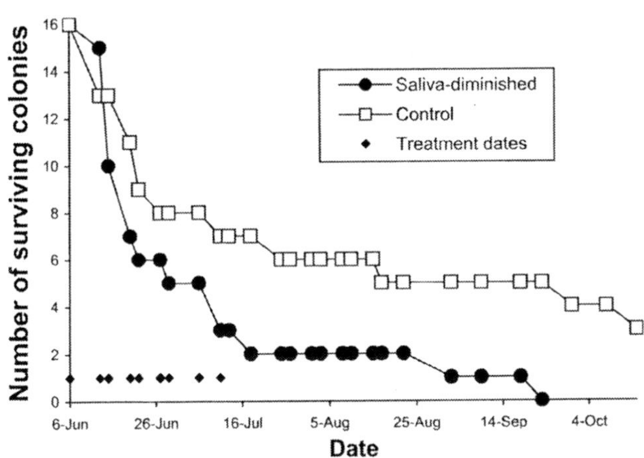

Figure 6.9
Number of surviving colonies over time for a cohort that began with 16 saliva-diminished and 16 control colonies of *Polistes metricus*. Treatment colonies had saliva taken from fifth-instar larvae on dates from June 6 to July 11, shown by the symbols at $Y = 1$. From Hunt and Dove (2002). Reprinted with permission of The Royal Entomological Society, London.

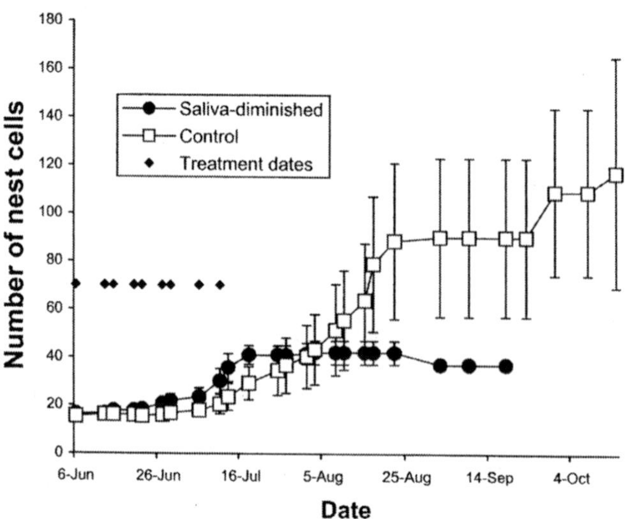

Figure 6.10.
Mean ± standard error of the number of cells per nest in saliva-diminished and control colonies of *P. metricus*. Sample sizes are the data of figure 6.9. Decrease in mean cell number for saliva-diminished colonies reflects sample size decrease in mid-August. From Hunt and Dove (2002). Reprinted with permission of The Royal Entomological Society, London.

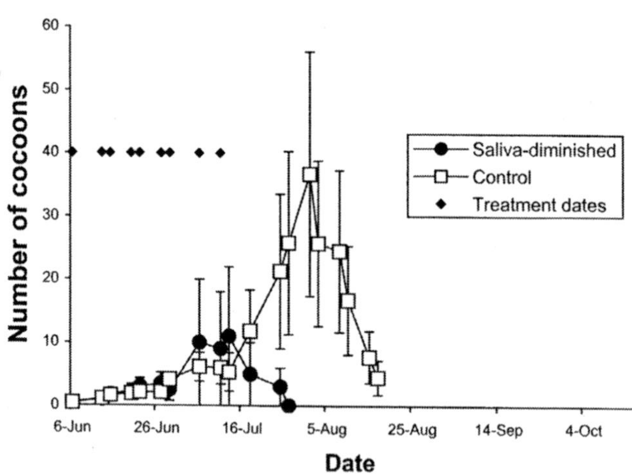

Figure 6.11
Mean ± standard error of number of pupae (cocoons) per colony for saliva-diminished and control colonies of *P. metricus*. From Hunt and Dove (2002). Reprinted with permission of The Royal Entomological Society, London.

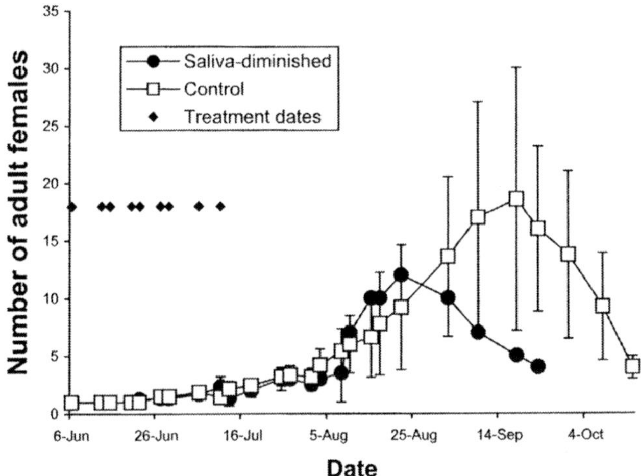

Figure 6.12
Mean ± standard error of number of adult females present at the nest for saliva-diminished and control colonies of *P. metricus*. From Hunt and Dove (2002). Reprinted with permission of The Royal Entomological Society, London.

and after the seventh diminishment I scaled back from two treatments per week to one. Taking saliva from larvae on only a few occasions spaced days apart caused foundresses to disappear from some colonies and the remaining colonies to lapse into deeply diminished productivity. It was a dramatic result.

Kumano and Kasuya (2001) performed field experiments in which they divided nests of *Polistes chinensis antennalis* and allowed workers to have access to both halves. Workers remained only with nest fragments that contained larvae. Analogous results have been obtained in studies of Vespinae. Ishay and Ikan (1968a) removed larvae from nests of *Vespa orientalis* whose workers were confined to foraging enclosures, and despite continuing oviposition by the queen, the workers did not care for new, small larvae, and they abandoned the nests. When larvae were removed from colonies that had freely foraging workers, queens died and workers abandoned the nests. When larvae were covered with a thin, transparent nylon sheet, the queen died and workers abandoned the nest. When larvae were covered with mosquito netting, newly emerged adults beneath the netting transferred foraged provisions from workers outside the netting to larvae within it and transferred saliva from larvae to the adults outside. New comb was constructed outside the netting, and these colonies continued development (Ishay and Ikan 1968a). These experiments show that when adults lack access to larval saliva they do not maintain a colony. For reasons I elucidate in chapter 8, I now substantially agree with Roubaud (1916) that adult attraction to larval saliva is central to the origin of colony life in Polistinae and Vespinae, although the story is more subtle and complex than Roubaud envisioned. Before telling that story, however, there is a colony-level phenomenon I have not yet addressed, and a chapter on populations is needed to put the story into context.

Dominance and Cofoundress Interactions

In 1948 Leo Pardi startled the English-speaking world of science with the republication (Pardi 1948) of research previously published in Italian (Pardi 1942, 1946) in which he reported that cofoundresses and workers of *Polistes dominulus* establish dominance hierarchies among themselves. Behavior of such sophistication, dependent on individual recognition, had never before been reported in an insect. This dramatic finding established the conceptual framework for much of our thinking about *Polistes* since then. Pardi's characterization of the scenario is worth quoting at length:

> The associated founding females coexist in the nest until the eclosion of the workers, and soon a work distribution is established among them (Heldmann 1936; Pardi 1942)—one, the leader female, remains on the nest, lays eggs intensively, does less building than her associates, and, as will appear later, dominates the associates (Pardi 1942). The other females (auxiliary females) go out more frequently, build, bring solid and liquid food, lay eggs less, and are dominated. Soon the auxiliary females finish laying, and their ovaries gradually become reduced. After the eclosion of the first workers, the auxiliary females are eliminated sooner or later from the association, either because the leader paralyzes them by a sting or because the continuous hostility of the leader reduces them to exhaustion (Pardi 1942). The leader (which now can be considered the veritable queen of the nest) remains alone with the workers, goes on with laying eggs, and from this moment leaves the nest rarely.
>
> As is known, all the other social functions, except laying eggs, are undertaken by the workers, which build, feed the larvae, regulate the temperature by fanning their wings, defend the nest, and so on. From the eggs of the queen there continue to be born other females with the character of workers and also some males, but the males make their appearance in great number only in August. On the other hand, during this period of time, as the colony is growing and trophic conditions are improving, the females that are born are always larger and have more fat reserves; these females do not maintain the typical behavior of the workers and stay much on the nest. These females, unlike their elder sisters, tolerate copulation and, after fecundation, they live over the winter and found the nests in the next spring. (Pardi 1948, p. 2; reprinted with permission of The University of Chicago Press)

Pardi thus described formation of a dominance hierarchy among cofoundresses. He went on to describe a second dominance hierarchy among worker offspring. The queen is alpha in both. Pardi described that individuals higher in the dominance hierarchy spend more time on the anterior face of the nest comb, and they receive regurgitated liquid foods from adult nestmates much more often than they give it. Dominant individuals lay more eggs, inspect more cells (which would include drinking larval saliva), and rest more. Less dominant individuals forage more and build more.

Dominance among cofoundresses affects their reproduction: "the trophic advantage, which is connected with a very frequent dominance function, influences the ovarian development; as the α-individual gets more and more developed ovaries, the other associated females have a tendency toward regression. At the end of the polygynic association, for instance, α possesses very well-developed ovaries; those of the auxiliary females are more regressed, the lower their position in the dominance scale" (Pardi 1948, p. 9). Pardi noted two probable causes of ovarian regression among the subordinate cofoundresses. For individuals near alpha in social rank, "the continuous subtraction of nutritive liquids among the more frequently dominated wasps causes a state of trophic deficiency to which, in part at least, the ovarian regression might be attributed" (p. 10). However,

> There is also another way through which dominance can influence the fertility and the sterility of the components of the colony: the work distribution. This point is supported by the observation that the auxiliary females of lower position in the polygynic association, which work hard and are compelled to give regurgitated liquid less frequently than the auxiliaries nearer to the leader in social status, show at the end of the polygynic association a much more conspicuous ovarian regression. In my opinion, the fact cannot be explained alone by the initial small differences in size of the ovaries in the associated females, but it should be interpreted as a proper 'castration for work,' caused by energy consumption dependent on intense work. (Pardi 1948, p. 10)

In other words, the subordinate cofoundresses that engage in brood care experience nursing castration.

Pardi reported that if the alpha should be lost during the cofoundress phase, she will be replaced as egg layer by the beta cofoundress. If the alpha should be lost after daughters have emerged and cofoundresses are no longer present, she will be replaced as egg layer by the highest ranking worker, who is beta in the hierarchy at that time. Successive losses of alpha individuals result in successive replacements in the same manner.

In closing, Pardi made a famous pronouncement: "In *P. gallicus* [as *P. dominulus* was known at the time] the production of sterile and fertile forms is an indirect consequence of the dominance system, and it is independent of the nutrition received during the larval life" (Pardi 1948, p. 13). This assertion has greatly influenced vespid biologists (O'Donnell 1998), but that influence has been misguided due to misinterpretation of Pardi's meaning. In the long passage quoted above, Pardi said: "as the colony is growing and trophic conditions are improving, the females that are born are always larger and have more fat reserves; these females do not maintain the typical behavior of the worker and stay much on the nest. These females, unlike their elder sisters, tolerate copulation and, after fecundation, they live over the winter and found the nest in the next spring" (1948, p. 2). In other words, there are two types of female offspring: those that undertake "the typical behavior of the workers" and those that do not, and there is a clear correlation between improved trophic conditions and the latter. The latter are "always larger," which can only be based on greater larval nourishment, and they have "reserves" of fat. Reserved from where and when? Larval development is the only logical conclusion. Pardi clearly recognized differentiation among female offspring as a function of larval nourishment. Individuals less well nourished become workers, and later, better nourished individuals become nonworkers. Thus, Pardi's famous pronouncement refers not to the female brood in toto but instead to partitioning of reproduction within each of two cohorts, first among the cofoundresses and then the among worker daughters of the current generation. In each cohort, reproduction is sorted out by dominance, and there is no indication that larval nutrition plays a role within a cohort. Assortment among individuals of the nonworker cohort of the current generation (the gynes) occurs only when they become cofoundresses the following season. However, nutritionally based partitioning of the female offspring into two categories, workers and gynes, is an integral part of Pardi's view of sociality in *Polistes dominulus*.

Even when Pardi is interpreted correctly, however, there is a chance that his proposition that assortment of reproduction among cofoundresses does not have a larval nutrition component may not be correct. Hunt et al. (2003) suggested that quantitative differences in storage protein reserves from the preceding year could be a factor in ordering the dominance hierarchy that mediates differential reproductive success among cofoundresses, and those differences could include a component of nourishment retained from larval growth.

Questions Arising

Does a factor other than or in addition to egg cannibalism cause the plateau in nest size when the first larvae eclose? It would be important to know whether the egg-laying foundress is temporarily incapable of oviposition due to nutritional depletion caused by producing and provisioning her initial brood. Physiological depletion of the foundress could, for example, be a significant factor in explaining why usurping and inquiline wasps (chapter 7) are able to be dominant and successful in colony takeovers at this stage of the colony cycle.

What causes nest growth to resume when the first larvae pupate? Perhaps there is a lowered demand for foraging by the foundress(es), enabling them to resume oviposition. Or perhaps reduced access to larval saliva induces the foundress(es) to leave the nest and forage. Or is it that there are no empty nest cells for oviposition? Experimental tests of behavioral cues could be a gateway to elucidating mechanisms.

Are first-emerged offspring of *Polistes* in fact set on the road to subdominance before they ever leave the nest? The inability of newly emerged social wasps to fly has been described infrequently, and there have been no specific studies of the phenomenon and its consequences. The activities of paper wasps in the first 3 days of their adulthood may be one of the most significant unstudied aspects of social wasp natural history.

Do adult female wasps in the first brood of supplemented colonies retain storage protein from their larval development? That is, can supplementation during larval growth induce physiological characteristics of gynes?

Do adult females vary in their responsiveness to larval saliva? Foundresses, workers, and gynes all drink saliva; do they do so at similar frequencies or in similar amounts? Does the nutrient content of saliva vary among larvae, and, if so, are there patterns of variation according to stage of the colony cycle, sex of the larvae, or environmental variables?

Does dominance rank among a group of cofoundresses correspond to levels of nutrient reserves at the time of cofounding? If so, do nutrient levels in foundresses include a component carried forward from larval development?

7 Populations

The terms "solitary" and "social" divert attention from a perspective essential to understanding insect social evolution. Whether a wasp lives alone or in a colony, all wasps are members of populations. Members of a population make up the gene pool, interbreed among themselves, and exhibit differential reproductive success. Thus, populations are the primary context in which natural selection operates. Failure to think at the population level is a shortcoming of much social insect evolutionary research. Even when thought is directed at populations, however, technical problems come to the fore. Demographic studies are daunting. They generally require one or several seasons of field work. Data can be difficult to collect, and some phases of the life history may be nearly impossible to document. Missing data from one life cycle phase can cast data from other phases into uncertainty. It is no surprise that demographic studies of wasps are few and far between (although life tables exist for parasitoid hymenopterans), and none of them encompasses a full life cycle. This doesn't mean, however, that the fragmentary studies that are available can be dismissed or ignored. In fact, they have much to tell us.

Because social life histories evolved from solitary antecedents, it is necessary to know the demography of both life history types in order to identify and assess consequences of the differences. Therefore, an objective of this chapter is to establish realistic images of the demography of exemplary solitary and social wasps in order to envision demographic aspects of the solitary-to-social life history transition. I have long sought to document the demography of a nest-constructing solitary eumenine wasp, but I have never encountered a suitable study population such as that of the African wasp *Synagris cornuta* tantalizingly illustrated in Wheeler (1923, p. 64). Camillo (1999) gives a life table for immatures of the eumenine *Brachymenes dyscherus* in Brazil, and he reports nests with up to 62 cells, presumably the production of a single female. In the absence of demographic data on adult female eumenines, I have had to rely on a surrogate.

When I made my first trip to the tropics to study wasps, my goal was to get to know swarm-founding epiponines. I was near the end of my stay when my attention was directed to a roughly made wooden cabinet in a workshop. It contained bee-keeping equipment that had been abandoned when Africanized bees appeared 7 or 8 years earlier at the site in northwestern Costa Rica. The cabinet had been unopened for that span until only a day or two before I was directed to it. I was stunned when I swung open the door. Within the cabinet were more than 800 mud nests of a single species of solitary wasp, *Sceliphron assimile*. *Sceliphron* species, the common black and yellow mud daubers well known to naturalists, are members of the Sphecidae, and so they do not exemplify ancestors of social Vespidae. They do, however, very nicely exemplify nest-constructing solitary wasps in general. I did not hesitate; I dove right in. *Sceliphron assimile* had already been the subject of excellent studies on developmental mortality and emergence success (Freeman 1973, 1977) and a mark–recapture study of female longevity (Freeman 1980). However, the key demographic variable, female reproductive success, was undocumented. The nests in the cabinet were ideal for such an analysis (Hunt 1993).

An Exemplary Solitary Wasp

Female *Sceliphron* species are hunting wasps that construct with mud, provision with spiders, and do so one nest cell at a time (Mitchell and Hunt 1984; Ferguson and Hunt 1989). The number of nest cells a wasp constructs in her lifetime, therefore, is a measure of her reproductive longevity. For the 866 *S. assimile* nests in the cabinet, I assumed that all cells constructed by a single female were represented in each nest (discussed in Hunt 1993), and so the number of cells per nest constituted data on age at death, measured as number of nest cells. I treated these data to survivorship analysis (table 7.1), and the resulting survivorship curve (figure 7.1) was stunningly log-linear. Females that initiated nesting in the cabinet experienced a constant probability of death (Type II survivorship), which was in accord with the mark–recapture study by Freeman (1980). In Type II survivorship, causes of death do not vary as a function of the wasp's age. Such deaths would be primarily ecological (being caught in a storm; being taken by a predator) and not physiological ("old age"). I once encountered an aged wasp of another mud dauber species struggling to crawl along the ground (figure 7.2). For several years I showed the photograph to seminar audiences and called it "the rarest wasp in the world," because the probability of living to such an advanced state of wear and tear is miniscule.

The number of nest cells is not a direct measure of reproduction for a mud dauber, because not all immatures reach adulthood. I quantified the frequency of immature deaths by counting cells with uneaten provisions, with mummified larval remains, or with evidence of parasitoid presence, and I also found that

Table 7.1
Life table for 866 female *Sceliphron assimile* that initiated nesting over a span of 7 or 8 years inside a wooden cabinet.

X	l_x	m_x	L_x	M_x	L_xM_x
1	1.000	0.182	0.239	0.182	0.043
2	0.761	0.225	0.152	0.407	0.062
3	0.609	0.267	0.122	0.647	0.079
4	0.487	0.310	0.099	0.984	0.097
5	0.388	0.353	0.083	1.337	0.111
6	0.305	0.395	0.052	1.732	0.090
7	0.253	0.483	0.054	2.170	0.117
8	0.199	0.480	0.042	3.650	0.111
9	0.157	0.523	0.030	3.173	0.095
10	0.127	0.565	0.020	3.738	0.075
11	0.107	0.608	0.027	4.346	0.117
12	0.080	0.650	0.020	4.996	0.100
13	0.060	0.693	0.012	5.689	0.068
14	0.048	0.721	0.011	6.410	0.070
15	0.039	0.721	0.008	7.131	0.057
16	0.031	0.721	0.008	7.852	0.062
17	0.023	0.721	0.007	8.573	0.060
18	0.016	0.721	0.002	9.294	0.019
19	0.014	0.721	0.002	10.015	0.020
20	0.012	0.721	0.003	10.736	0.032
21	0.009	0.721	0.000	11.475	0.000
22	0.009	0.721	0.000	12.178	0.000
23	0.009	0.721	0.002	12.899	0.026
24	0.007	0.721	0.005	13.620	0.068
25	0.002	0.721	0.000	14.341	0.000
26	0.002	0.721	0.002	15.062	0.030

From Hunt (1993).

the probability of adult emergence for any cell was independent of the number of cells per nest. The estimated probability of emergence was not equal to the probability of producing a female, of course, and to make that estimate I had to use other data. Freeman and Johnston (1978) reported that the probability of producing a male *S. assimile* is highest in the first-built nest cell and that it declines in successively built cells, until the 13th cell and beyond, at which point only females are produced. The probabilities of producing a female thus were lowest at the first cell and increased to 1.0 at cell 13. I multiplied the per-cell probability of producing a female (from the Freeman and Johnson study) by the probability of successful emergence (from my study) to generate estimates of age-specific female fecundity (table 7.1).

Figure 7.1
Log_{10} of age-specific survivorship, measured as number of cells per nest, for a cohort of 866 female *Sceliphron assimile* that initiated nesting in the wooden cabinet. From Hunt (1993); figure courtesy of Andrew V. Suarez, University of Illinois at Urbana-Champaign.

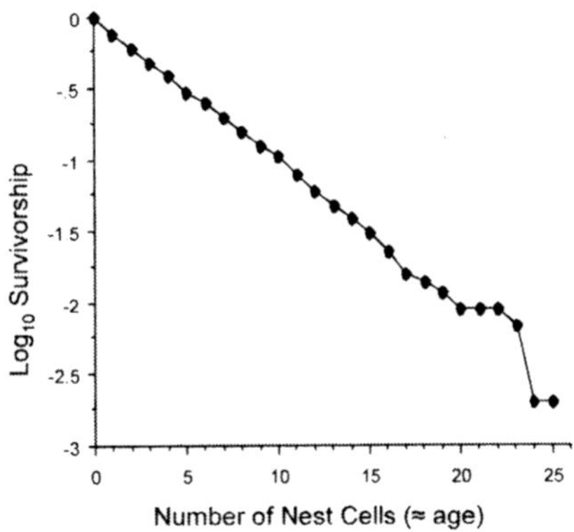

In traditional life table analyses, age-specific survivorship (l_x) is multiplied by age-specific female fecundity (m_x), and the sum of those products equals the net reproductive rate (R_0). A net reproductive rate of 1 means that each female produces, on average, one female offspring. I could calculate R_0 from the estimates I had in hand, but to do so was meaningless. I had no data on dispersal, immigration, or on survivorship between the time that a female emerged from pupation until the time she initiated nesting. My data addressed only the fraction of the population that initiated nesting. Faced with a dilemma, I recast the demographic analysis to address only the portion of the life span from nest initiation to death.

In traditional life table analyses, age-specific survivorship, l_x, is the proportion of the initial population alive at the beginning of age interval x. Values of l_x begin at 1.0 at $x = 1$ (the start of the life table) and decline to 0 at the next age interval beyond the one in which the last individual has died. I reformulated

Figure 7.2
This aged female blue mud dauber, *Chalybion californicum*, was found crawling feebly along the ground. The ragged wingtips are indicative of a long lifetime of foraging. Geriatric wasps such as this, of any species, are almost never encountered. *Chalybion* species are spider-hunting mud daubers (Sphecidae) that rent empty nests of other mud daubers (Landes et al. 1987; Hunt 1993). Although solitary, they exhibit the very interesting behavior of aggregating, sometimes in large numbers, each night (Landes and Hunt 1988).

survivorship as the proportion of the population living to but not beyond any age interval, x, as:

$$L_x = l_x - l_{x+1}\,.$$

The values of L_x (table 7.1) showed that an estimated 23.9% of the population did not survive beyond the construction of one nest cell; another 15.2% did not survive beyond the second nest cell; another 12.2% did not survive beyond the third, and so on in steady decline. I then reformulated fecundity as the estimated total production of female offspring by a female living to but not beyond age x as:

$$M_x = \sum_{0}^{x} m_x\,.$$

The product of these two estimates, L_xM_x, is the age-specific recruitment of female offspring produced by an average female that lives to age x. The sum of those products yields a net recruitment rate, R_n, that differs from R_0 only by rounding error but is biologically interpretable, because it addresses only the span of adult life from the initiation of nesting until death. An average female *S. assimile* that nested in the wooden cabinet produced 1.61 female offspring. Thus the population could sustain 38% mortality between emergence and initiation of nesting and still remain stationary at $R_0 = 1.0$.

Calculation of L_xM_x values has an added and very valuable benefit. The value for each age, X, is the portion of the net recruitment rate that is contributed by females that live to but not beyond age X. A plot of those values (figure 7.3) makes it easy to see that the bulk of female recruitment in the population is attributable to females that live to make from 4 to 12 nest cells. The few females that lived to produce more than 20 nest cells also had high relative recruitment. Nests of up to 54 cells of *Sceliphron fistularium*, each apparently the work of a single female (Camillo 2002), would embody very high recruitment for the few females that lived long enough to produce them, assuming that other demographic variables are similar to those of *S. assimile*.

Polistes

One spring in Missouri I was following *Polistes* wasps as they visited flowers. I had been following a *P. fuscatus* for a few minutes when she dropped onto low grass, crawled beneath the blades, and disappeared. After waiting a few minutes without her reappearing, I kneeled and looked for her. I found her under a tuft of grass, on a nest of six cells only 2 centimeters above the ground! I had never seen such a nesting site before. Immediately my focus shifted to nests rather than foraging, and over the course of the next 2 weeks two students and I found 73 nests of *P. fuscatus* and *P. metricus* in an area of about one-third hectare. One was beneath a dilapidated bench; three were on low branches of shrubs, and the remainder were under vegetation within 2 or 3 centimeters of the ground. Disappointingly, I was unable to monitor the nests after that, but I have no doubt

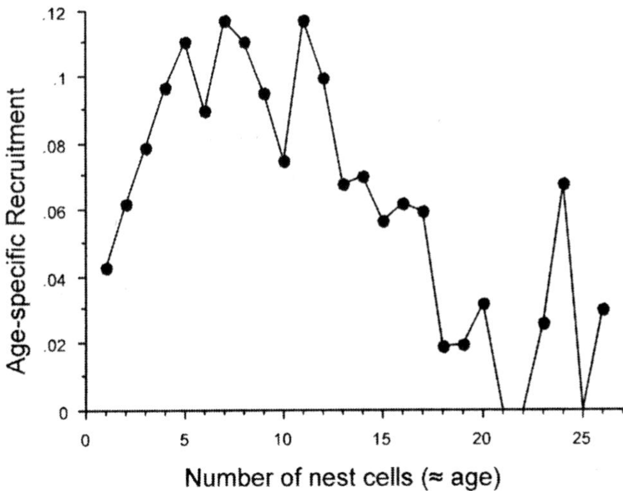

Figure 7.3
Age-specific recruitment values for the cabinet population of *S. assimile*. Each value is the portion of the net recruitment (R_n = 1.61) that is generated by females that live to but not beyond each age, *X*. From Hunt (1993); figure courtesy of Andrew V. Suarez, University of Illinois at Urbana-Champaign.

of what would have happened to them. In nearly three decades of field work at the same site and in that region, I have seen scarcely a handful of mid-season nests on shrub branches and none whatsoever at ground level. There is little doubt in my mind that, excepting perhaps the nest under the bench, all nests were doomed to fail long before producing reproductive offspring.

The fate I envision for those nests is borne out by data of published studies. Investigators who have sought foundress nests and followed them through the nesting season have reported consistent results. There is substantial failure of nests before reproductive offspring emerge (table 7.2). I have not encountered a full life table analysis for any *Polistes* population, and an attempt to generate life table data in my lab (DeMarco 1982) failed when half the study population was vandalized. There are, however, a handful of nest survivorship curves or, at least, data from which such curves can be generated (figure 7.4).

No study has documented the pattern and variability of fecundity for a population of paper wasps. The possible pattern of age-specific relative recruitment for a *Polistes* population in a seasonal environment is shown in Figure 7.5. Only gyne offspring are considered to be fitness units, and as in nearly all life table studies, males are excluded from consideration. Figure 7.5 nonetheless encapsulates a central feature of *Polistes* populations. The figure as drawn has $x = 0$ at the initiation of nesting. Now envision the superimposition of any of the survivorship curves of figure 7.4, and it is an inescapable conclusion that the

Table 7.2

Failure rates of single-foundress and multiple-foundress colonies of *Polistes*.[a]

Species	% Failure to reach worker stage		% Colonies multiply founded	References
	Single foundress	Multiple foundress		
P. annularis	80 (72)	20 (326)	86 (637)	Strassmann (1989a)
P. biglumis	58 (24)		0 (24)	Lorenzi and Turillazzi (1986)
P. canadensis	97 (143)	84 (393)	77 (532)	Pickering (1980; tabulated in Hughes 1987)
P. chinenesis	60 (421)		0 (421)	Matsuura (1977, cited in Miyano 1980)
P. chinenesis	64 (125)		34 (185)	Nonacs and Reeve (1993, personal communication to D. C. Queller)
P. fuscatus	93 (45)	83 (36)	44 (81)	Gibo (1978)
P. fuscatus	40 (277)		37 (632)	Klahn (1981, personal communication to C. R. Hughes, cited in Hughes 1987)
P. fuscatus	47 (55)		52 (288)	Noonan (1979)
P. jadwigae	53 (90)		0 (90)	Matsuura (1977; cited in Miyano 1980)
P. nimpha	68 (37)		0 (37)	
P. riparius (= *biglumis*)	71 (63)		0 (63)	Sô. Yamane and Kawamichi (1975)
P. snelleni	84 (38)		1 (100)	Sô.Yamane (1969)

[a]Sample sizes in parentheses. Table from Queller (1996).

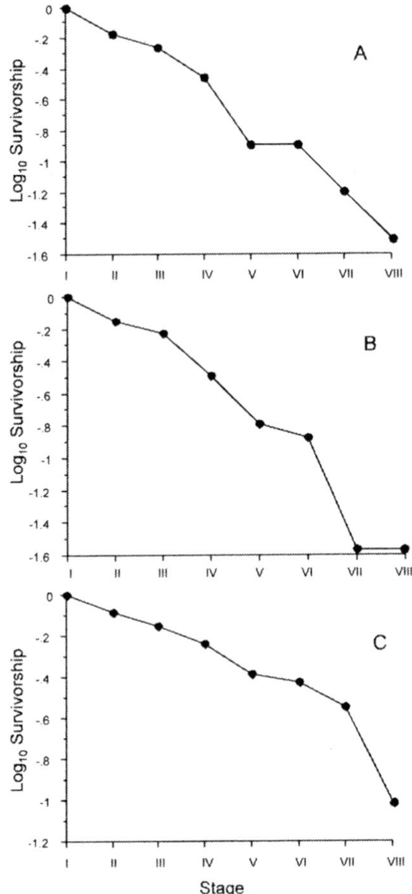

Figure 7.4
Survivorship curves for colonies of *Polistes chinenesis antennalis* from three Japanese populations as reported by Myiano (1980): (A) Funabashi, 1975 (n = 31); (B) Kanazawa, 1976 (n = 37); (C) Itako, 1977 (n = 162). "Phases" are I: from nest founding by the queen to first hatching of larva; II: from first hatching to first cocoon spinning; III: from first spinning to emergence of the first worker; IV: from first worker emergence to first reproductive emergence; V–VIII: time-sequence periods of the social stage of Yoshikawa (1962). Nests were built on the stems of withered plants near the ground or inside holes of small concrete blocks. I am grateful to Professor Shinya Miyano for providing the original data for the Funabashi and Kanazawa populations. Figure courtesy of Andrew V. Suarez, University of Illinois at Urbana-Champaign.

majority of foundresses in one generation have been recruited from a minority of foundresses of the preceding generation, and the foundresses are only a fraction of the number of gynes produced in the preceding generation. This raises the question of how selection could have shaped this demography. An answer can be drawn from annual plants.

Paper Wasps as Annuals

Conventional wisdom on annual plants once held that they are classic r-selected organisms (Pianka 1970) and, as such, they should grow rapidly and disseminate seeds quickly to minimize the chance of pre-reproductive mortality and maximize their contribution to the gene pool. Although this generalization may be applicable to ephemeral plants or to annuals growing in

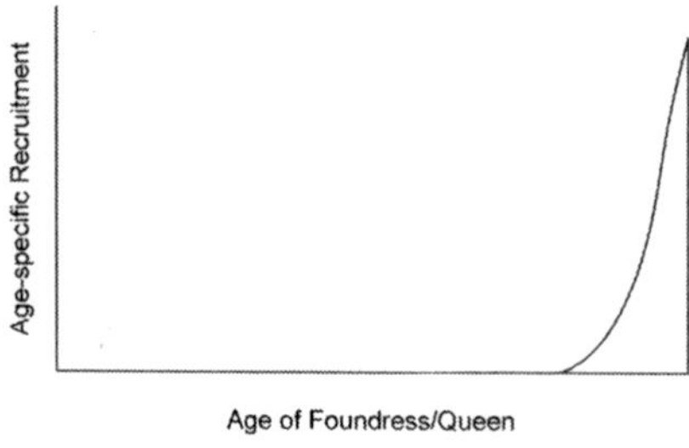

Figure 7.5
Conjectural age-specific recruitment values for a population of *Polistes* in a seasonal environment.

marginal or unpredictable environments, a very different model may apply to most annual plants. Schaal and Leverich (1981) proposed that selection should, in fact, favor annual plants that disseminate their seeds as late as possible in the growing season when several criteria are met: (1) there is a two-stage life cycle—active (the growing plants) and dormant (the seeds); (2) there is an age-structured dormancy pool (not all seeds are the same age); (3) risk of mortality during the dormant phase is a function of time spent in the dormant phase (due to fungi, seed predators, disturbance, etc.); and (4) there is a synchronous onset of the active phase (germination). Schaal and Leverich's model is presented in figure 7.6. The key feature is that the later the deposition of seeds into the dormancy pool, the higher the survivorship at germination after dormancy.

Polistes wasps in seasonal environments are directly analogous to annual plants of the Schaal-Leverich model. There is a two-stage life cycle—nesting and quiescence. Not all quiescent individuals are the same age, having been produced over a span of weeks or months of the active phase. Initiation of nesting often occurs over a very short span of days (West Eberhard 1969) and sometimes even on a single day (Rossi and Hunt 1988). But what about the fourth criterion—mortality as a function of time spent in the dormancy pool? None of the demographic studies cited in the preceding section address survivorship during non-nesting portions of the life cycle. "Enormous mortality" of quiescent *P. annularis* gynes was observed in Missouri (Rau and Rau 1918; Rau 1930c), whereas there was no apparent mortality of *P. instabilis* quiescent in tropical Costa Rican cloud forests during the dry season (Hunt, personal observation; D. H. Janzen, personal communication). In temperate species, at least, survivorship from emergence to initiation of nesting probably is more like that of *P. annularis* than *P. instabilis*. The estimated female mortality of dispersing (= gyne) female *P. jadwigae* in Japan was 52% and 78% in 2 years of study (Hirose and Yamasaki 1984). Of 500 to 2000 female *P. fuscatus* marked in each of several years in Iowa, 10–20% were relocated the following spring (Klahn 1988). Of more than 600 female offspring *P. metricus* paint-marked in one year in Missouri, only 2 were observed the following year (Hunt and Dove 2002), although many of the marked offspring may have been workers.

Figure 7.7 is a conjectural survivorship curve for *Polistes metricus* in Missouri. Survivorship for an annual cohort of gynes would decline from the time

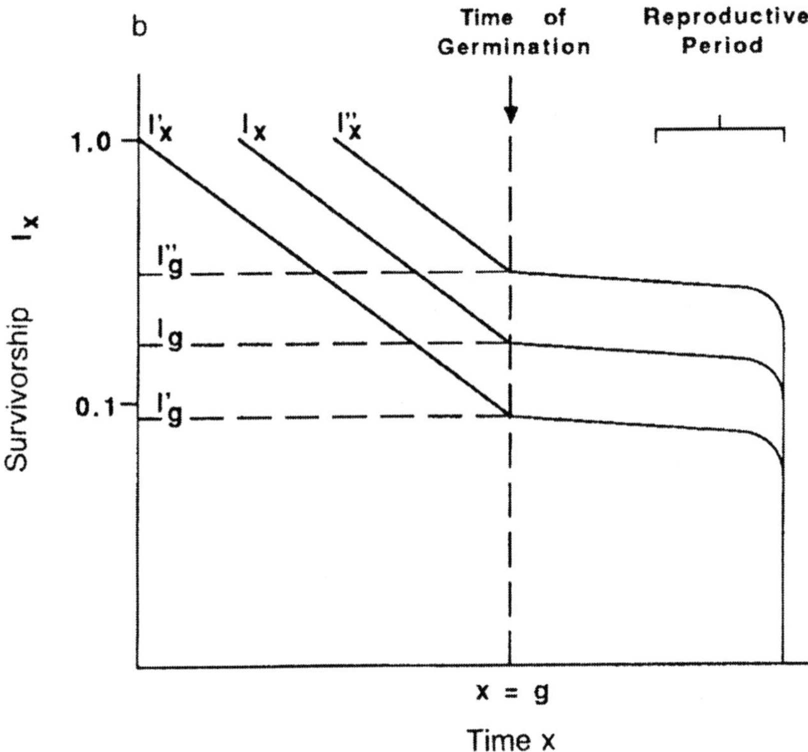

Figure 7.6
The Schaal-Leverich model for delayed seed set in annual plants. Idealized survivorship curves show the effects of different times of entry into a seed pool on the time of germination ($x = g$). Early seeds have curve l'_x, while late seeds have curve l''_x. Clearly, $l'_g < l_g < l''_g$. Substitute "dormancy pool" for "seed pool" and "nest initiation" for "germination," and the figure would apply directly to *Polistes* populations in seasonal environments. Originally published as Figure 1 in B. A. Schaal and W. J. Leverich, 1981, "The demographic consequences of two-stage life cycles: survivorship and the time of reproduction," *The American Naturalist* 118:135–138, published by the University of Chicago Press. © 1981 by The University of Chicago. Reprinted with permission.

the first gyne emerges in one season until the last member of the cohort dies in the following season. Earliest gyne emergence is in mid-July (line A in figure 7.7), and dissolution of that year's colonies occurs at the end of August (line B). Gynes forage at fall flowers until they enter quiescence (line C), from which they may emerge intermittently on warm winter days before final emergence in March (line D). Nests are initiated around April 15 (line E), and the first offspring, which are alloparental females, emerge in late June (line F). When alloparental offspring undertake foraging and construction, queens

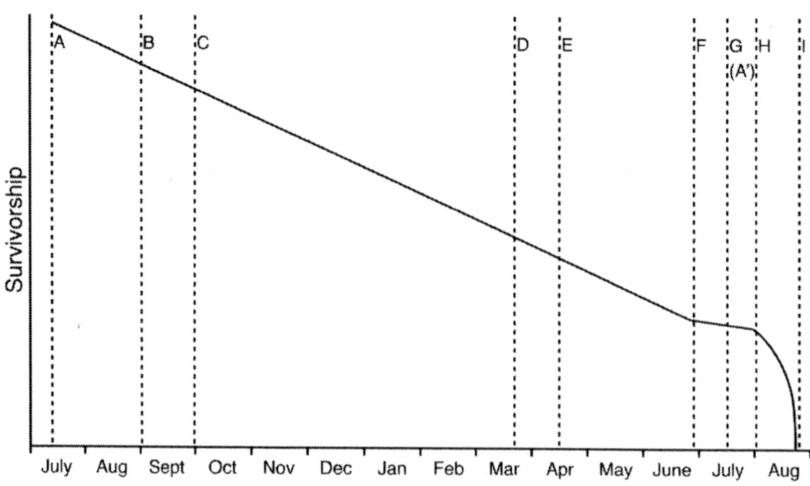

Figure 7.7
A conjectural survivorship curve for gynes/foundresses/queens of *Polistes metricus* in Missouri. Survivorship declines with age across the maximum possible life span of a single female, from first emergence of a gyne in mid-July of one year (A) until death of the last queen at the end of the nesting season the following year (H). Milestones along the way are B, final departure from nest and foraging at fall flowers; C, entry into quiescence; D, emergence from quiescence and foraging at spring flowers; E, nest founding; F, emergence of first alloparental offspring; G, emergence of first gyne offspring; H, first dissolution of gyne-producing colony; I, last dissolution of gyne-producing colony.

remain at the nest, and the rate of Type II deaths diminishes (see Reeve 1991) until the earliest colony dissolution (line H), at which time the rate becomes increasingly precipitous until the final colony dissolution (line I). Note that the production of the first gyne offspring (line G) would correspond with first gyne emergence (line A) for the ensuing generation.

The time span in figure 7.7 from line A to line B (lines G to I of the preceding generation) encompasses production of fitness units (gynes) and is therefore the most fundamentally important phase of the life cycle. It is the portion of the life cycle encompassed by the Schaal-Leverich model for delayed reproduction. Selection will favor gyne production close to line B of figure 7.7. By looking at the G to I portion of the figure, it is easy to visualize that highest fitness accrues to the very few colonies that continue gyne production until final colony dissolution. A small number of highly successful colonies will have very high representation in the next generation. Selection for delayed gyne production, presented here from the perspective of annual plant demography, is also accommodated by the model of bang-bang switching from production of workers to reproductives articulated by Oster and Wilson (1978).

Given the critical importance of the G to I phase of the life cycle, selection should shape adaptations that enhance the probability of a gyne in one season becoming an egg layer at time G the following season. The model as I have drawn it in figure 7.7, however, shows Type II survivorship from gyne production in one season until emergence of first offspring in the next. Type II survivorship is the epitome of stochastic selective factors at work. For many years I assumed that natural selection acts on traits such as saliva production and saliva solicitation that may affect the probability that a foundress gains the alloparental aid of her first offspring and enhances her chance to survive long enough to produce gyne offspring. The perspective presented in figure 7.7, however, led me to think that although selection may shape the variables of individual development, stochastic components of *Polistes* demography may select as strongly or more strongly on life history variables other than variables of individual development. That is, there is variability on which selection can act to modify potential demographic outcomes in a stochastic environment, but the components of such variability are not reflected in figure 7.7. Instead, such variability is embodied in alternative reproductive strategies.

Usurpation and Inquilinism

A newly founded *Polistes* nest isn't valuable to a predator—it's a bit of paper and a few milligrams of eggs. In time, however, it becomes a worthwhile target. The brood of a nest just before first offspring emerge can contain several grams of larvae, and it is defended by only a single foundress. Such nests are often taken by predators (Rau 1930a; Klahn 1988; Strassmann et al. 1988; Makino and Sayama 1991). The foundress is usually not killed, and suddenly she is faced with reproductive options. To suspend further nesting is the zero-fitness option, so most nestless females either initiate a new nest or try to commandeer one. Renesting by a solitary foundress is less likely to yield reproductive offspring than is commandeering (Klahn 1988; Makino and Sayama 1991), although if offspring females are already present, renesting with them is often undertaken (Makino 1989). For solitary foundresses that lose their nest, selection favors usurpers. They attempt, often successfully, to displace another foundress from her nest and to take over her immature brood. That is, they become social parasites. Usurpers often destroy (presumably eat) eggs and small larvae in the nest they have taken over (Makino 1989), particularly in nests with many large larvae and/or pupae (Klahn 1988). They lay eggs, and offspring of the initial foundress provide alloparental care for the brood of the usurper, although this does not always occur without dissention (Klahn 1988; Makino 1989). If the destroyed brood is eaten by the usurping female or fed to other larvae, then the behavior has a trophic benefit (Klahn and Gamboa 1983). At the same time, there is adaptive benefit for the usurper in terms of the nature of the host brood being reared. Pupae and large larvae of the host that are spared by the usurper will emerge as alloparental caregivers, whereas host eggs and small larvae, if reared to adulthood, could emerge as nonworking gynes and males (Klahn and Gamboa 1983; Makino 1989). Thus, even though usurped nests are less productive on average than nonusurped nests, usurpers can contribute reproductive offspring to the gene pool despite the loss of their original nest.

The success rate for usurping is sufficiently high that it could be favored as a primary, rather than secondary, nesting behavior. To delay reproduction until near the emergence of the first conspecific offspring (line F in figure 7.7) might in some way lower the risk of mortality during the preemergence nesting period (between lines E–F in Figure 7.7), and it might simultaneously conserve the usurper's nutrient reserves. Usurpation as a primary reproductive behavior has been proposed to occur in populations of the European *P. dominulus* that now are established in northeastern North America. In an intensive study of a population of preemergence colonies in which all nesting females were marked, the continuous appearance of unmarked wasps joining, adopting, or usurping nests was interpreted as evidence of a large population of non-nesting gynes (Nonacs and Reeve 1993, 1995). In an experimental study, Starks (1998) found that not all females in a captive *P. dominulus* population initiated nests, and orphan nests were adopted by females that had not previously nested. The inescapable artificialities of working with an enclosed population make this result tentative. Nonetheless, there is a clear indication that usurpation and/or adoption as a primary reproductive strategy can occur, at least in the adventive population of *P. dominulus* in North America.

Usurping is the exclusive reproductive behavior of three *Polistes* species in southern Europe (Cervo and Dani 1996). These social parasites are inquilines that usurp the nest of another *Polistes* species and utilize the alloparental behavior of the host offspring to rear reproductive offspring of their own. Specialized traits of the inquilines include altitudinal migration to cold, high elevation dormancy sites, which means that the inquilines emerge from quiescence only when the nesting cycle of host species in the lowlands is well underway. Inquilines' eggs develop more rapidly than those of hosts, and larvae of the inquilines are provisioned more frequently than are host larvae and consequently develop more rapidly (Cervo et al. 2004). In these species, nest initiation and alloparental care have been deleted from the life history, and there are specific adaptations to inquilinism.

Multiple Foundresses

The foregoing discussion has addressed only single-foundress colonies. However, much, and perhaps most, of the literature on *Polistes* foundress nests and colony success addresses two or more foundresses on a single nest. Many temperate and most tropical *Polistes* have multiple foundresses. Sometimes multiple foundresses become multiple queens in a colony producing reproductive offspring (Hunt et al. 2003), but in most instances a single female lays most or all of the reproductive-destined eggs. If all but one of the cofoundresses have no reproductive offspring, what were the selective agencies and adaptations that shaped the evolution of multiple foundresses? There are two variations of co-founding: one in which an initial solitary foundress is joined by one or more additional foundresses, exemplified by *P. fuscatus* and *P. erythrocephalus* (West Eberhard 1969), and one in which the nests may be established by a group of foundresses, exemplified by *P. annularis* (Strassmann 1983). In the first case foundress groups are generally small, and in the second case the foundress groups can number into the teens or twenties.

A small foundress group of *P. fuscatus* is not inherently superior to solitary founding in terms of development time for immatures, nest size, or total production of offspring (Gibo 1974). *Polistes metricus* in Kansas has both single-foundress and multiple-foundress colonies, which form primarily by joining (Gamboa 1978) or occasionally by retention of the now-subordinate foundress when a usurper invades (Gamboa 1978; Gamboa and Dropkin 1979). Multiple-foundress *P. metricus* colonies produce their first workers earlier than do single-foundress colonies, but only because joining wasps tend to associate with nests founded early (Gamboa 1980). When nest densities are low, productivity of single- and multiple-foundress colonies is similar (Gamboa 1978). At higher nest densities, however, multiple-foundress colonies are more resistant to conspecific usurpation and have higher productivity (Gamboa 1978). Lower rates of usurpation in multiple-foundress than in single-foundress nests have also been recorded in *P. fuscatus* (Klahn 1988) and in *P. dominulus* (Nonacs and Reeve 1995).

In a field study of *P. fuscatus* in Ontario, multiple-foundress colonies were more likely to refound after nest predation, and if refounded they were more productive than refounded single-foundress colonies (Gibo 1978). A corresponding result was found in an experimental study of *P. bellicosus* in Texas (Strassmann et al. 1988). Natural predation on nests did not vary as a function of number of females on a nest, either in the preemergence or postemergence phases of the nesting cycle. However, when nests were removed, simulating predation, nests with more females on them were significantly more likely to build a new nest, and rebuilt nests with more females on them were more likely to be successful.

In *P. annularis* in Texas, foundress number is a major determinant of nest success. Nests attended by fewer foundresses were more likely to fail, with single-foundress nests particularly vulnerable, whereas larger foundress groups built larger nests that contained more pupae at the time of first emergence and that were more likely to survive to produce reproductives (Strassmann 1989a). This finding was not directly linked to predation, and usurpation was not correlated with foundress number (Queller and Strassmann 1988). In *P. fuscatus* in Michigan, multiple-foundress nests were more likely than single-foundress nests to produce reproductives, and among colonies that produced reproductives there was a significant positive correlation between foundress number and productivity (Noonan 1981). These cases exemplify a pattern that has been often documented and that can be related to the survivorship pattern conceptualized in figure 7.7. Wasps die or lose their colony for diverse reasons, including usurpation and predation, and the demographic pattern of such mortality or failure dictates the commonly observed pattern: colonies with more than one foundress are less likely to lose all foundresses before achieving reproductive success (Strassmann and Queller 1989; Reeve 1991; Nonacs and Reeve 1995; Shreeves et al. 2003).

The membership of multiple-foundress groups is often ascribed to a well-known component of *Polistes* behavior—*Polistes* foundresses are philopatric. That is, they tend to establish their nests near the nest site where they were reared (West Eberhard 1969). This should facilitate cofounding as a kinship-based behavior because most foundress groups consist of nestmates from the previous year (Klahn 1979; Strassmann 1996). Despite this, foundress groups are often not sisters (Strassmann 1996) and can even be unrelated (Queller et al. 2000). Therefore, the enhanced demographic success of multiple foundress nests must be sufficiently advantageous for joining behavior to be adaptative even if cofoundresses have low relatedness or are unrelated. A tendency to cofound nests in environments where the chance of colony success is low would initiate a sweepstakes dynamic: although the odds would be heavily against winning anything at all, the (fitness) payoff for the lucky winner could be huge. If ecological and demographic factors make the probability of success as a single foundress low relative to multiple-foundress groups, colony queens whose reproductive offspring have a stronger propensity to cofound than do offspring of other queens in the population will elevate the probability that one of their offspring will be a successful colony queen in the next generation. When cofoundresses are kin, kin selection might enhance selection favoring cofounding behavior, but kin selection is not necessary to explain either the origin or maintenance of cofounding.

Opportunity for Selection

In a population that is not increasing or decreasing in size, a net reproductive rate (R_0) of 1.0 means that an average female that enters the population produces one female offspring. The variance in reproductive success around the population mean constitutes the opportunity for selection (Arnold and Wade 1984). If all members of a population reproduced with equal success, there would be no variance and, therefore, no opportunity for selection. As variance increases, the opportunity for selection becomes greater. From this perspective it is clear that the opportunity for selection differs between solitary and social wasps and among the life history variations of social wasps.

Solitary wasps have high probability of low fitness. In the example of *Sceliphron assimile*, which could be typical for solitary wasps in general, it seems plausible that half, or perhaps more, of the females that initiated nesting produced at least one female offspring. Many would have produced several female offspring, but few would have produced more than 10. The variance in reproductive success in solitary wasp populations is low, and so the opportunity for selection to increase longevity and/or fecundity is also low. And, given the mass provisioning behavior of most solitary wasps, there is no possibility of interactions among nestmates that might provide variability on which selection could operate to enhance longevity or fecundity.

Social wasps have very low probability of very high fitness. *Polistes* nests that have produced tens to hundreds of gynes can be found at the end of every nesting season. However, if the population is stationary and has a net reproductive rate of 1.0, then the average female that entered that generation will have produced one female offspring. Clearly, most females produced no gynes; a modest number produced some, and a very few produced a large number. The opportunity for selection is great. Simultaneous progressive provisioning, with direct interaction between the foundress and her multiple offspring, provides the context in which selection could act on variable traits to enhance the longevity and reproductive output of the foundress. The web of interactions among the foundress, immatures, and emerged offspring constitute sociality, and sociality evolved as a framework that enhances the probability of a foundress's reproductive success in a population that has high variance in reproductive success among foundresses. Zero reproduction by early emerging daughters is a small price for a foundress to pay if the social behavior of those first daughters leads to hundreds of gynes. The intensity of selection for traits that enhance the probability of such an outcome will be strong. Sociality in *Polistes* thus reflects the evolution of adaptations that enhance foundress reproductive success in a demographic context that continually militates, and often strongly so, against the likelihood of success.

With matrifilial sociality seen as the fundamental life history pattern of *Polistes*, variations on that theme fall into perspective. Usurpation and adoption can be seen as behaviors that place the usurper/adopter into a foundress role closer to the emergence of her alloparentally reared gynes than if she had initiated the nest. These behaviors may have evolved as adaptive responses to loss of a foundress nest, and selection clearly can favor these behaviors as a primary reproductive strategy with nest founding deleted from the life history. The extreme form of this social parasitism, inquilinism, deletes both nest founding and alloparental offspring from the life history. Usurping, adopting, and inquilinism are adaptations that enhance the demographic probability of reproductive success for the individual wasps that express these behaviors.

Joining and multiple founding constitute another means to the same end. The presence of multiple adult females on preemergence nests is favored by selection in situations where preemergence nest failure is particularly high and, consequently, variance in reproductive success is greater than in single-foundress populations. Reproductive contests among the cofoundresses certainly occur, and they can be of interest at the colony level. However, the behaviors of joining, nestmate tolerance, dominance interactions, and their concomitant behaviors evolved and persist as a route to reproductive success of the colony queen. For a queen to have most of her gyne offspring become subordinates in cofoundress associations in the next generation is a small price to pay if just one of her daughters becomes the queen of a colony that produces hundreds of gynes (that themselves will have the propensity to join or cofound). A demographic perspective thus is key to understanding foundress associations. The strength of selection in a population with very high variance in fitness will maintain cofoundress behavior despite low relatedness among cofoundresses and low or zero fitness of subordinates. From a population demographic perspective, the dominant–subordinate interactions that have been the focus of much of the *Polistes* literature are nothing more than the fine-tuning of behaviors that result in very high reproductive success of a very small percentage of the females that constitute a generation.

Stenogastrines, Vespines, and Swarm Founders

A demographic perspective can also shed some light on other social wasps. Stenogastrines probably have greater variance in reproductive success than do solitary wasps but less than that of *Polistes*. The intensity of selection on social traits will therefore be less than in *Polistes*. Progressive provisioning and nestmate interactions provide the context for behaviors on which selection could operate, raising the question, to be addressed in the following chapter, of why stenogastrine colonies remain small and, consequently, variance in reproductive success remains relatively low.

Large-colony vespines represent the extreme among wasps for variance in reproductive success. In a population with a net reproductive rate of 1.0, each queen of one generation will produce, on average, one queen in the next generation, yet the numbers of gynes from successful nests of large-colony yellowjackets commonly run into the thousands. That means that there are thousands of pre-reproductive deaths and colony failures before the next generation of successful colonies. It is no surprise that caste difference between vespine workers and gynes is as distinct and well regulated as it is. Neither is it a surprise that usurping may be more common than nest initiation in some Vespines. The opportunity for selection on variable traits that affect yellowjacket reproductive success is greater than in any other independent-founding wasps.

Swarm-founding polistine wasps represent a novel demographic pattern. If new queens can replace old ones in a colony, and if a colony can construct a new nest to replace an old one, then colonies are theoretically capable of open-ended existence. What, then, is their demography? The closest parallel to the *Polistes* model would be to look at the number of daughter queens produced by each queen in a colony. It seems possible that such numbers would be low, and so variance among queens would also be relatively low. A different perspective would be to look at the number of daughter colonies produced by reproductive swarms emanating from a colony. This view addresses the fact that solitary existence is impossible for an individual swarm-founding wasp. The colony has become the functional organismal unit—a polycorporal entity analogous to a multicellular organism, albeit one with ever-changing components. Few data address the demography of swarm-founders in a manner applicable to this perspective. A large data set on *Parachartergus colobopterus* (Strassmann et al. 1997) reveals a Type II pattern of colony survivorship (figure 7.8). A similar pattern

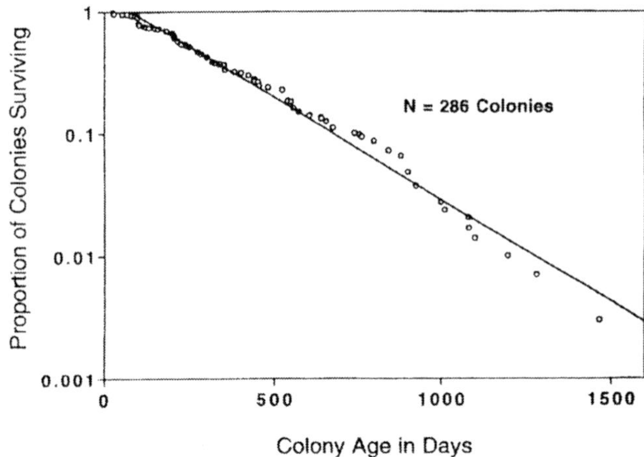

Figure 7.8
Survivorship of 286 colonies of the swarm-founding epiponine wasp *Parachartergus colobopterus* at Maracay, Venezuela. Originally published as Figure 3 in J. E. Strassmann, C. R. Solís, C. R. Hughes, K. F. Goodnight, and D. C. Queller, 1997, "Colony life history and demography of a swarm-founding social wasp," *Behavioral Ecology and Sociobiology* 40:71–77. © Springer-Verlag 1997. With kind permission of Springer Science and Business Media.

was generated from nest size data on a population of *Polybia occidentalis* colonies (Kaufman et al. 1995). It is difficult, however, to know how to interpret these data. How many new colonies are founded by reproductive swarms versus absconding swarms? How many nest terminations are colony failures versus absconding colonies? I am brought back to an assertion made at the beginning of this chapter. Demographic studies are daunting.

Questions Arising

What is the demography of a population of solitary, nest-constructing eumenines? If Carpenter (1982, 1991) is correct that Eumeninae is monophyletic, then any findings on demography of a eumenine can shed no more light on the origin of sociality in Vespidae than does the example of the sphecid wasp given in this chapter. I reject that notion as implausible, and I still hope to find a suitable eumenine study population.

What is the pattern of survivorship across the full life span for *Polistes* gynes in a single-foundress population? What is the shape of the age-specific recruitment curve for a single-foundress *Polistes* population? To what extent do usurpation and adoption change the probability of successful reproduction for single-foundress *Polistes*? To robustly document the demography over the full life cycle of a naturally nesting population of *Polistes* would be a lastingly important contribution to our understanding of *Polistes* biology and evolution. The same demographic questions also could be asked for a species of Stenogastrinae.

Is there firm evidence of usurpation or adoption without prior nesting as a primary strategy in *Polistes* species other than the invasive U.S. population of *P. dominulus*?

How can multiple foundresses of swarm-founding Epiponini be treated to demographic (life-table) analysis? Demographic studies are daunting, but they are essential to a complete understanding of all levels of social evolution.

8 *The Dynamic Scenario of Social Evolution*

Armed with knowledge of individual and colony development in a population context, it is possible to construct a dynamic scenario of caste differentiation (which is the essence of sociality) and to hypothesize how natural selection fosters gyne production late in the colony cycle. An earlier model of vespid social evolution, the "ovarian ground plan" hypothesis, sheds light on several central issues. An additional key insight for building the dynamic scenario comes not from vespid wasps, but from worker honey bees examined in a framework called the "reproductive ground plan" hypothesis. A central feature of the dynamic scenario, presented here as the "diapause ground plan hypothesis," builds on the reproductive ground plan idea and adds a new and significant component to the mix of ideas. These ground plan models have common elements, and other elements are complementary. When the components are synthesized appropriately, a picture emerges in which many mysteries of social vespid evolution can be seen, perhaps for the first time, in proper perspective. The preceding three chapters put most of the pieces into place. Descriptions are given first of the ovarian ground plan and reproductive ground plan ideas. These are followed by a short but necessary excursion into life cycles of solitary wasps. This is followed by a discourse on the diapause ground plan. The dynamic scenario of polistine social evolution then emerges as a conceptual synthesis.

The Ovarian Ground Plan Hypothesis

A model for dichotomy of reproductive and nonreproductive castes in social wasps has long been proposed by West-Eberhard (1987b, c, 1988, 1992a, 1996).

She makes reference to a "set of associations—among ovarian development, hormone activity, and behaviour" (West-Eberhard 1996, p. 293), which she calls the ovarian ground plan. Her model, the ovarian ground plan hypothesis, rests on a three-part foundation. One part is a proposed cycle of ovarian activity in a solitary ancestor of social wasps. A second part is the context-dependent expression of alternative behaviors from a single genotype (phenotypic plasticity) that can occur among wasps, each with the solitary ovarian ground plan, that nest together. The third part is the evolution of a switchlike mechanism that regulates the phenotypic expression into two castes, queens and workers. Emphases on intragenerational nest sharing by adults as the context for caste evolution and on dominance interactions as its driving force appear in her earliest writing on social evolution (West 1967), and they are made explicit as a proposed explanatory framework for caste evolution in her "polygynous family hypothesis" (West-Eberhard 1978a).

Observations of a group-living eumenine wasp, *Zethus miniatus* (West-Eberhard 1987c), contributed significantly to West-Eberhard's vision of an ovarian cycle. She later emphasized the importance of a switch mechanism (West-Eberhard 1992a). These observations and emphases converge in the full ovarian ground plan hypothesis, in which she proposes (West-Eberhard 1996, p. 293):

> The origin of reproductive "castes" (queens and workers) in wasps might be classified as a "reciprocal deletion," for it probably involved the decoupling of [an] ancestral reproductive cycle into two parts with one expressed in workers and the other in queens (West-Eberhard 1987[c]). Each caste lacks the "deleted" set of traits expressed in the other, and the two alternatives are mutually "dependent" or complementary morphs (see West-Eberhard 1979 and Gadagkar [1996]): they stay together and co-operate (or parasitize each other!) in the same colony, compensating each other's deficiencies.

This same model was presented in West-Eberhard (1987c), where it was accompanied by a figure, shown here as figure 8.1.

To clearly envision the model, consider that a nest-building solitary wasp (chapter 2) exhibits cycles of behaviors: nest cell construction, ovipositioning, foraging and provisioning, then construction again. These behaviors may be underlain by cyclical ovarian activity: activation leading to oviposition, followed by decreased activity during foraging and provisioning, followed by reactivation. West-Eberhard envisions that reproductive behaviors expressed cyclically in a solitary wasp could become decoupled among nest-sharing adults. Some individuals would express the ovipositional role and others express the foraging and construction roles, leading to reproductive and worker castes. As a potential underlying mechanism, West-Eberhard (1996, p. 301) proposes that "a decoupling of the ovarian and behavioural influences of JH [juvenile hormone] in workers and queens could initially have been achieved by incidental and/or socially imposed nutritional differences between the two castes that differenti-

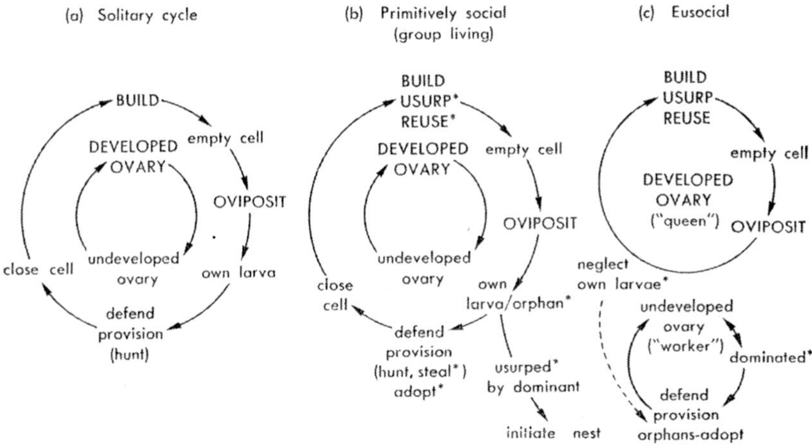

Figure 8.1
West-Eberhard's ovarian ground plan hypothesis. The solitary cycle (a) represents envisioned coupling of ovarian and behavior cycles of a progressively provisioning solitary wasp. The primitively social (group living) cycle (b) is based on observations of the communal eumenine *Zethus miniatus*. West-Eberhard proposes that (b) could have evolved from (a) via contextual change with the advent of group life due to selection for remaining at the natal nest. The eusocial cycle (c) represents *Polistes*, in which West-Eberhard envisions that selection for success in social competition has led to effective mechanisms of control by dominant group members over reproduction by subordinate group members, leading in turn to increasing numbers of orphan larvae and ovary-suppressed females inclined to adopt them. Asterisks mark novel components that could reflect selection producing group living and effective dominance. From West-Eberhard (1987c), which should be consulted for additional details. Reprinted with permission from Scientific Societies Press, Ltd.

ated their responses to JH, producing two classes of females." Some time after that, a switch mechanism could have come to be associated with the caste difference. "In social insects . . . the switch regulating caste is immediately 'conditional' on dominance rank (relative aggressiveness) and associated differences in ovary size and hormone titer. . . . Thus, the 'worker' caste of social insects likely originated as a condition- (rank-) and hormone-sensitive alternative, subject to the evolution of thresholds for escape . . . and manipulation" (West-Eberhard 1989, p. 256).

In an examination of some components of the ovarian ground plan hypothesis, Giray et al. (2005) studied *Polistes canadensis* in Panama. They examined a particular workerlike behavior: nest-guarding against a conspecific intruder. They also performed a manipulation in which treatment wasps received an application of a juvenile hormone (JH) analog, and they quantified JH levels in some wasps. Application of the JH analog led to higher frequencies and earlier expression of workerlike behaviors, and reproductives had higher JH levels than workers.

In the text of Giray et al. (2005, p. 3333), reproductives are "queens from postemergence nests," "queens of preemergence single foundress nests," and "queens of multiple foundress nests." In their figure 4 (p. 3333), these are all lumped together as "gynes." The term "gyne" is not used in their text, and in their figure 5 (p. 3333) it is used interchangeably with "queen." Thus Giray et al. fail to discriminate between two widely-recognized behavioral categories of social wasps. Queens lay eggs, are usually inseminated, are behaviorally dominant, and are usually nonforaging wasps in full reproductive mode. In this book, the term "gyne" has been used to characterize females that are worker-fed as larvae and emerge in the latter part of a nesting cycle in reproductive diapause. Gynes do not work, and they do not reproduce in the season of their emergence. Instead, they pass an unfavorable season in quiescence and then become foundresses (and queens if they are fortunate) in the next favorable season. The use of the term "gyne" by Giray et al. does not fit or incorporate this concept. Giray et al. did, however, note the possibility that "a portion of newly emerging females in this population of *P. canadensis* during the periods (late wet season) of our observations and experiments were in reproductive diapause" (p. 3334). Indeed, they were. They were gynes as the term is used in this book.

The Reproductive Ground Plan Hypothesis

A body of work being put forward in the framework of the reproductive ground plan hypothesis is showing how "variation in maternal reproductive traits gives rise to complex social behaviour in non-reproductive helpers" (Amdam et al. 2006, p. 76). In particular, the work is synthesizing variables of ovarian development, hormones, and storage proteins as they affect or correlate with suites of behaviors in worker honey bees, *Apis mellifera*, that could be characterized as more versus less reproductive, even though the workers are not reproducing. Honey bee workers can lay eggs, although the majority of them never do so (Winston 1987). Thus honey bee workers, which are not reproductives, possess elements of reproductive physiology, and these elements play roles not related to egg production. The initial discovery came from investigations of honey bee workers as a model system to study aging.

Honey bee workers in temperate zones have a bimodal distribution of longevity. Bees emerging in spring and mid-summer are called summer bees and have a mean life span of 25 to 35 days, whereas bees emerging in late summer are called winter bees and have a mean life span of 6 to 8 months (Maurizio 1950; Free and Spencer-Booth 1959). Protein levels in the hemolymph seem to be a major determinant of honey bee life span (Maurizio 1950; Schatton-Gadelmayer and Engels 1988). The most abundant hemolymph protein in honey bees is vitellogenin (Engels and Fahrenhorst 1974; Cremonez et al. 1998), which is synthesized in the fat body and is the primary egg yolk protein (Chapman 1998). Modeling had supported the hypothesis that high levels of brood care by workers can result in short potential life span, whereas low levels of brood care can result in long potential life span (Omholt 1988). When vitellogenin dynamics were integrated, a clear picture began to emerge. Amdam and Omholt (2002) reviewed the role of vitellogenin in honey bee workers, and they presented a model that supports a role for vitellogenin in the age difference. Honey bee workers called nurse (or hive) bees secrete brood food from their mandibular and hypopharyngeal glands and feed it to larvae (Winston 1987). Vitellogenin in the hemolymph is converted by hive bees' hypopharyngeal glands into the brood food component called royal jelly (Amdam et al. 2003). Thus a protein that plays a widespread role in insects as an egg protein has been used in a novel reproductive role.

Modeling also showed that vitellogenin dynamics affect the longevity differences in honey bees (Amdam and Omholt 2002). When hive bees become foragers in the well-known sequence of age-related behavior changes in workers (age polyethism: Winston 1987), vitellogenin levels drop. Life spans of foragers are short. However, in bees that face low demands for brood care at the end of the season, vitellogenin levels remain high and can be augmented by feeding. High vitellogenin levels sustain the bees through a long life span. In worker honey bees, vitellogenin—normally a reproductive protein—functions as a storage protein.

Continuing investigations of vitellogenin dynamics in worker honey bees, using a novel experimental framework, suggest even broader relationships between reproductive physiology and worker behaviors. Strains of honey bees have been artificially selected for a single behavior difference—high levels of pollen hoarding in the hive versus low levels of pollen hoarding (Page and Fondrk 1995). Although selection was based on only the single, colony-level trait of pollen hoarding, the strains show individual differences in the age at which they begin to forage, locomotory activity at emergence, associated learning performance, and sensory perception, as well as physiological differences (see Amdam et al. 2004). Bees of the high pollen-hoarding strain have higher levels and synthesis rates of hemolymph vitellogenin than do low pollen-hoarding bees (Amdam et al. 2004). High-hoarding bees with high levels of vitellogenin behave in ways that suggest that their ground plan reproductive physiology is switched on, and low-hoarding bees with low levels of vitellogenin behave in ways that suggest that their ground plan reproductive physiology is switched off. Amdam et al. (2004, p. 11350) proposed that "associations between behavior, physiology and sensory tuning in workers with different foraging strategies indicate that the underlying genetic architectures were designed to control a reproductive cycle." Thus the variability in behaviors of worker honey bees reflects the reproductive physiology inherited from the solitary ancestor of honey bees. Amdam et al. (2004, 2006) further suggested that exploitation of the reproductive ground plan plays a fundamental role in the evolution of social insect societies, although none of the reproductive ground plan investigations to date has tackled the question of queen/worker divergence in honey bees.

Life Cycles in Solitary Wasps

To analyze evolution of nest architecture in social vespid wasps, Wenzel (1993) adopted a holistic approach and looked at nest construction as an entire, complex behavior in which the series of component behaviors was treated as developmental steps toward the larger unit. Component behaviors then were compared across taxa according to their distinctive characteristics and position in the sequence. Nest building sequences were compared in terms of the embellishment, acceleration, or deletion of steps. Using a similar conceptual approach, Truman and Riddiford (1999) envisioned the evolution of insect metamorphosis as based on extension and deletion of stages of individual development. Stimulated and inspired by the work of Wenzel and of Truman and Riddiford, I attempted a similar holistic approach for an analysis of life cycles in wasps.

In the simplest possible life cycle for a solitary wasp, generations succeed one another in unbroken progression (figure 8.2). Such a simple life cycle is unlikely to occur in nature, because in aseasonal tropics where individual development is direct and resources are continuously available, the generations of a population almost certainly would be asynchronous and overlapping rather than a series of distinct generations as shown in the figure. Asynchronous, overlapping generations may occur in some tropical Eumeninae (chapter 3). Seasonality, whether wet/dry or warm/cold, brings generations into synchrony. The simplest seasonal life history would be annual generations that correspond to yearly cycles of favorable and unfavorable conditions (figure 8.3). Adult life span could be less than the full length of the favorable climate conditions if brood rearing is tied to a specialized resource that has limited temporal availability. Many solitary bees specialize on the pollen of only a single plant or narrow range of plants, for example, and have only brief seasons of adult activity. Among temperate-zone insects, some pass the unfavorable (winter) season as eggs or as

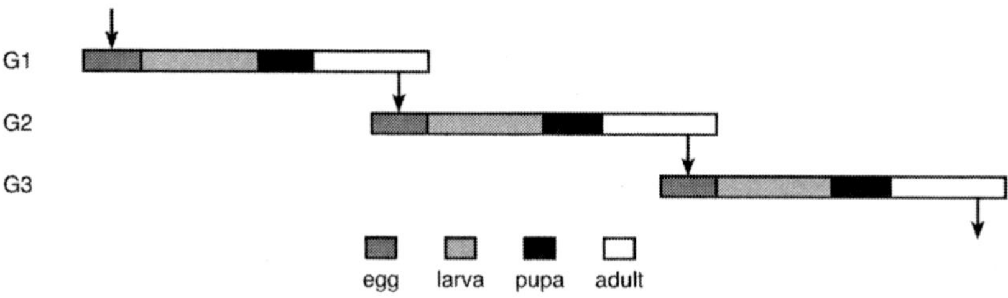

Figure 8.2
Life cycles of a multivoltine solitary wasp in an aseasonal environment. Each generation passes through its life stages (egg, larva, pupa, adult) without diapause or quiescence. Generations, designated G1, G2, G3, and so on, are continuous. Arrows between generations indicate oviposition by adults of one generation to initiate the next generation. In natural populations, generations would more likely be overlapping and asynchronous rather than the serial progression of generations as shown here.

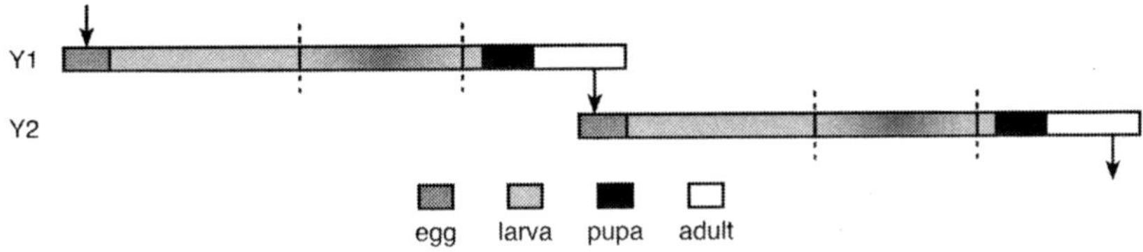

Figure 8.3
Life cycles of univoltine wasps with one annual generation per year in seasonal environments. Years are designated Y1, Y2, and the unfavorable season is denoted by the shaded area. Wasps pass the unfavorable season either by extension of larval life as prepupae as shown or by extension of adult life incorporating quiescence.

incompletely developed larvae. Wasps, however, pass the unfavorable season either as postdefecation last instar larvae (prepupae) in diapause (figure 8.3) or, rarely, as adults in quiescence. Prepupal diapause characterizes the majority of temperate wasps (O'Neill 2001), but a few species of solitary Pompilidae (Richards and Hamm 1939; Evans 1970), Sphecidae (Maneval 1939; Grandi 1961), and some bees (Kemp et al. 2004) overwinter as adults.

Two-generation (bivoltine) life cycles have one generation that completes its life span in a single season and one that extends its life span from one season into the next (figure 8.4). Among solitary Vespidae, for example, the eumenine *Monobia quadridens* has two generations per year in Missouri (Hunt, personal observation): one that emerges in mid-summer and one that overwinters as a prepupa and emerges the following spring. Many solitary bees and wasps in temperate zones have bivoltine life cycles (Seger 1983), and meager data sug-

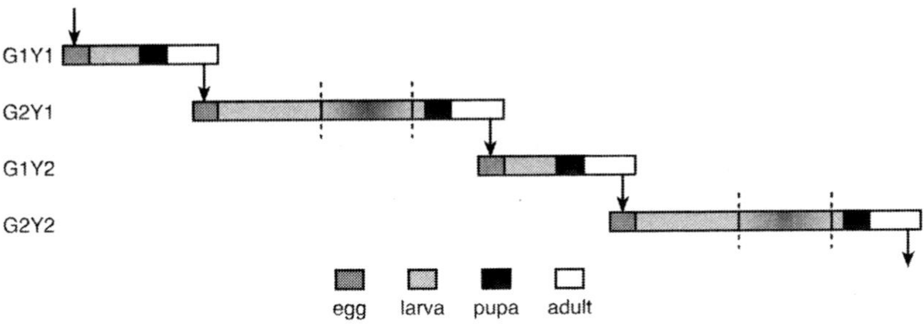

Figure 8.4
Life cycles of bivoltine wasps in seasonal environments with two annual generations, designated G1 and G2, per year, designated Y1, Y2. The unfavorable season is denoted by shading. One generation passes through its life stages, as in figure 8.1, entirely during the favorable season, whereas a second generation passes the unfavorable season either by extension of larval life as prepupae as shown or by extension of adult life incorporating quiescence.

gest that some tropical species may be bivoltine (e.g., *Ammophila centralis* [Sphecidae]: Menke and Parker 1996). At least two bivoltine species of Sphecidae overwinter as adults (O'Brien and Kurczewski 1982a, b).

Some, and perhaps many, two-generational wasp species are partially bivoltine (Seger 1983), which means that they have overlapping generations during the active season (figure 8.5). In partial bivoltinism, adults from the previous year's second generation contribute to both of the generations in the current year. Life cycles of bivoltine and partially bivoltine species differ in one fundamental aspect that depends on whether the unfavorable season is passed in prepupal diapause or adult quiescence. In the case of prepupal diapause, all adults enter into reproduction shortly after they emerge from pupation. In the case of adult quiescence, in contrast, adults of the first generation enter into reproduction shortly after they emerge from pupation, but adults of the second generation are reproductively inactive until the following nesting season. The second generation has longer adult life spans than the first generation.

The Diapause Ground Plan Hypothesis

The life cycle of *Polistes* (figure 8.6) is like that of a partially bivoltine wasp with adult quiescence (Hunt and Amdam 2005). The first female offspring have non-diapause physiology and can engage in reproduction soon after emergence, whereas later emerging females have diapause physiology and do not enter into

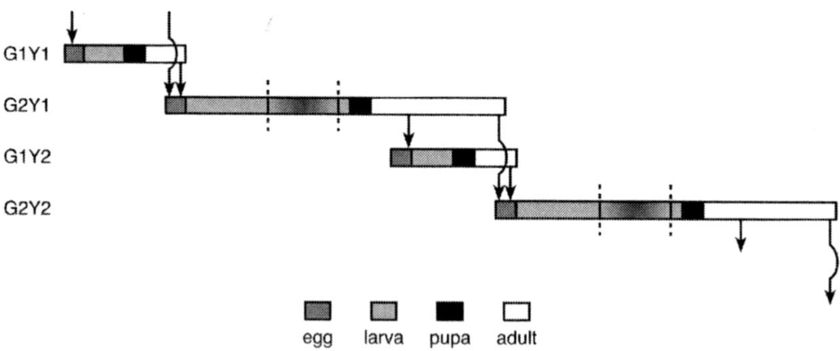

Figure 8.5
Life cycles of partially bivoltine wasps in seasonal environments with two generations, designated G1 and G2 per year, designated Y1, Y2. The unfavorable season is denoted by shading. The pattern differs from that of bivoltine species in figure 8.3 in oviposition pattern. G1 receives eggs only from G2, whereas G2 receives eggs from both G1 and from G2 preceding it. One generation passes through its life stages, as in figure 8.1, entirely during the favorable season, whereas a second generation passes the unfavorable season either by extension of larval life as prepupae as shown or by extension of adult life incorporating quiescence.

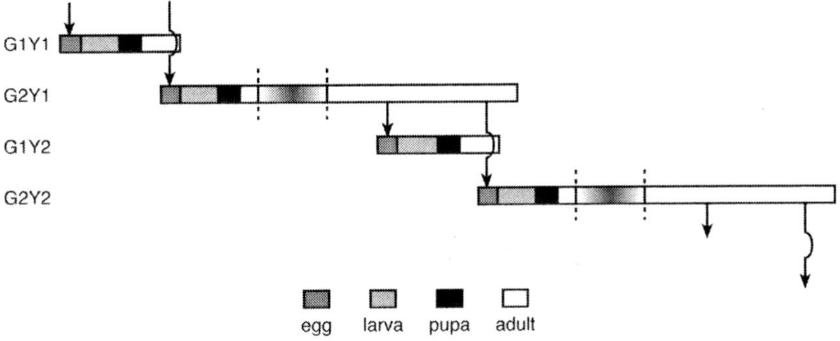

Figure 8.6
Life cycle of *Polistes* in a seasonal environment. The pattern is that of a partially bivoltine species that passes the unfavorable season (shaded) in adult quiescence. G2 females from year Y1, after quiescence, lay the eggs for both G1 and G2 in year Y2. G1 females are alloparental caregivers that lay no eggs; therefore oviposition, signified by an arrow from adults of G1 to eggs of G2, has been deleted from the life cycle. The life span of adult G2 females is extended.

reproduction but instead enter quiescence and reproduce the following year. In a major variation on that basic pattern, oviposition by females that engage in alloparental care (the workers) has been deleted from the life cycle. However, recognizing that the first-emerged females possess the underlying ground plan of wasps physiologically primed to reproduce provides a powerful framework for understanding the origin and nature of worker behavior.

Traits that characterize workers and gynes are identical to suites of traits that would be present in a bivoltine solitary wasp with adult diapause (figure 8.7). Larvae respond during development to a food cue and diverge onto one of two developmental pathways. Low food availability leads to the G1 pathway, which is signaled by slow larval development (due to the slow nutrient inflow) coupled with short pupation time (Karsai and Hunt 2002) and no storage protein residuum in emerging adults (Hunt et al. 2003). High food availability leads to more rapid larval development coupled with longer pupation time (Karsai and Hunt 2002) and residual storage protein in emerging G2 pathway adults (Hunt et al. 2003). G1 females are in a "reproduce now" physiological state, and they forage for protein, care for brood, and construct nests. The expression of these behaviors depends on colony conditions, as indicated by two branching points in the G1 sequence. If a queen is living and few workers are present, a G1 female will express these maternal behaviors allomaternally at her natal nest. This is worker behavior. However, if the queen should die, a G1 female can develop her ovaries, mate if males are present, and become a replacement queen. If the queen is living and many workers are present, a G1 may depart the natal nest and found a satellite nest in mid-season. Because G1 females

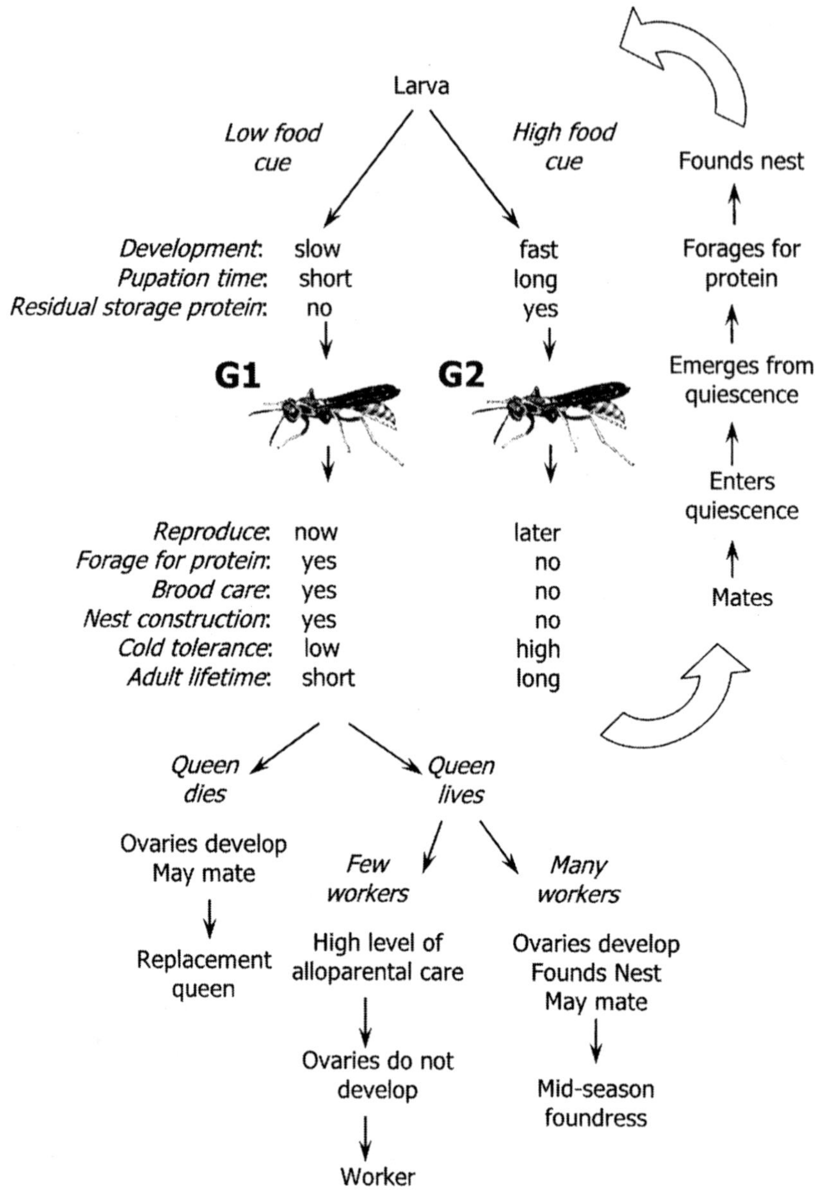

Figure 8.7
The *Polistes* life cycle incorporates fundamental elements of the diapause ground plan. G1 female offspring exhibit characteristics of non-diapause, and G2 female offspring have characteristics of diapause. Behavior pathways that are open to each type of offspring are shaped by the underlying ground plan. From Hunt and Amdam (2005).

have low cold tolerance, they do not survive quiescence, and lifetimes are short, especially for workers. G2 females, in contrast, are in a "reproduce later" physiological state. They are nonreproductive and express no maternal behaviors. Their cold tolerance is high, and they successfully pass quiescence during a life span that can reach nearly 1 year. They mate before quiescence, and after emerging from quiescence they break reproductive diapause, enter a reproduce-now physiological state, and exhibit maternal behaviors.

Worker behavior in wasps is maternal care behavior. First-emerged *Polistes* females correspond to G1 females of figures 8.4 and 8.5, and they are physiologically primed to reproduce. They readily engage in maternal care, but the context of the colony into which they emerge as young adults shapes the maternal behavior they express. Newly emerged *Polistes* and other paper wasps do not leave the natal nest, or do so only briefly, for a period of up to several days after their emergence (chapter 6), a phenomenon unknown in solitary wasps. By the time of their first foraging flight, newly emerged female paper wasps will have fed on larval saliva rather than on flower nectar as would newly emerged solitary wasps. Indeed, in describing the paper wasp *Mischocyttarus drewseni*, Jeanne (1972, p. 85) wrote, "when an adult emerged from its cocoon its first act was to visit one or two larvae, where it apparently obtained this secretion. These contacts were quite long, frequently lasting 30–60 seconds." As was noted in chapter 5, the larval saliva is a significantly richer source of amino acids than is any flower nectar. Because wasps emerging early in the colony cycle are G1 wasps that are physiologically predisposed to seek protein for self-nourishment, the amino-acid rich saliva of larvae in their natal nest becomes a powerful attraction that induces them to remain. Roubaud's (1916) suggestion that attractiveness of larval saliva is what retains adults at the nest has been borne out by experimental studies (Kumano and Kasuya 2001), and in the context of the diapause ground plan this attraction now appears to be deeply fundamental to the initiation of sociality. When wasps disposed to reproductive behavior subsequently undertake maternal behaviors, they do so at the place where they feed on proteinaceous saliva rather than at a nest of their own making that would contain no saliva-producing larvae. This marks the threshold of sociality: they remain at their natal nest, and there they perform alloparental care. This is worker behavior. Behavioral dominance by the queen combined with energetic costs of nest construction and foraging constrain or prevent egg production by these G1 females, even though they are physiologically primed to reproduce. The physiological disposition to reproduce in G1 females is nonetheless borne out by the exceptions to worker behavior. Some early emerging female *Polistes* become replacement queens on their natal nest, or they found and lay eggs in satellite nests (chapters 3 and 6; figure 8.7). Wasps that undertake these reproductive activities are G1 females, and many of the wasps that undertake reproduction in these ways have previously been foragers (O'Donnell 1996).

Workers have heretofore been thought of as "sterile workers" and gynes as "reproductives." This traditional terminology obscures the true nature of caste in the offspring of social vespid wasps. In terms of the diapause ground plan and differences in reproductive physiology and behaviors during the season in which the wasps develop and emerge as adults, workers are reproductive and gynes are not.

I have long argued (first in Hunt 1975; best in Hunt 1991), as have many others (reviewed in O'Donnell 1998), that differential nourishment during larval development underlies the caste dichotomy in *Polistes*. Nonetheless, a specific mechanism whereby differential nourishment could be translated into worker and gyne castes was not proposed until the work of Hunt and Amdam (2005). The crucial step toward identifying that mechanism was the finding that *Polistes* offspring in early July contain no storage protein as adults, whereas late July female offspring do (Hunt et al. 2003). The similarity of this pattern to the pattern being elucidated in worker honey bees by Amdam and Omholt (2002) and by Amdam et al. (2003, 2004) was the gateway to synthesis, insight, and understanding.

The storage protein in *Polistes* is hexamerin, not vitellogenin, but its roles are probably similar to those of vitellogenin in worker honey bees. G2 wasps, like winter bees, have high levels of storage protein in their fat body, survive the winter, and have long adult lifetimes. They also forage little or not at all for protein in the fall but do so in the spring, and they are reproductively inactive in the fall but reproductively active (the worker bees make brood food) in the spring. G1 females (the workers) are physiologically cued to seek protein as a reflection of their idiobiont heritage of feeding to support ovarian activation. This almost certainly plays a role in the attractiveness of the larval saliva, which then plays the central role in retaining G1 females at their natal nest. The physiological effects of interactions between retained wasps and their adult and larval nestmates may result in heterochronic early expression of genes controlling maternal care behavior (Linksvayer and Wade 2006), leading to alloparental care. G2 females also will feed on larval saliva and intercept food brought to the nest, but they are physiologically primed to store protein rather than to activate their ovaries, and they do not express maternal care genes or behaviors until after diapause has been passed.

Prior Evidence

In papers published in 1949 and 1952, Deleurance (cited in West Eberhard 1969) described the seasonal separation of the brood of *Polistes dominulus* into two kinds of females, workers and *fondatrices-filles* ("girl foundresses"). The latter, which in this book are called gynes, Deleurance characterized as lethargic nonforagers, and he considered them to be in reproductive diapause. West-Eberhard, who distinguished worker and nonworker females in *Polistes fuscatus*, concurred: "behavioral differences between workers and non-workers are indeed of the kind which might be associated with differences in reproductive (ovarian or hormonal) physiology" (West Eberhard 1969, p. 47). She noted that workers are aggressive, that they participate in building and brood care, and that they are "more like queens than are non-workers" (p. 47), whereas the nonworkers are passive, subordinate, and do not build or forage. West-Eberhard went on to note that although workers have reduced ovaries that are devoid of oocytes, "it is possible that they have active neurosecretory systems which affect their behavior even though ovarian development (yolk deposition) is repressed through work" (p. 47).

In the early 1970s, M. K. Bohm (1972b) discovered a pattern that reflects the diapause reproductive ground plan. She collected nests of *Polistes metricus* near Lawrence, Kansas, at three times: early June, early July, and early August. Wasps emerging from pupal cocoons of the collected nests were isolated and maintained in the laboratory under either summer (long day, warm) or fall (shorter day, not as warm) conditions, and then they were dissected and their ovarian activity quantified. Most of the wasps that showed ovarian development under these conditions were collected in June and maintained on summer conditions. In terms of the diapause ground plan, the June wasps maintained on summer conditions were G1 wasps in a favorable environment. The ovaries of wasps collected in July and August did not become active even under summer environmental conditions, except for some of the July wasps maintained on summer conditions that had been treated with a topical application of JH. In terminology used here, the July and August wasps were G2 wasps. Because all experimental animals were pupae when collected and maintenance conditions were identical, the basis of the difference, excepting the hormone application treatment, can only have occurred during larval development.

A related result was obtained in an experiment by Mead et al. (1995), who took offspring early in mid-season in France from nests of *Polistes dominulus* and tested nest founding ability of those offspring at three times in the ensuing months. Immediately after removal from the nest, 34 of 37 wasps that founded nests were workers. Two and four months after removal, few wasps founded nests, but seven of the ten that did so had been workers and eight of the ten were among the first-emerged wasps from their natal nests. Workers survived winter quiescence at about half the rate of later emerging, nonworking offspring, and after winter quiescence most of the nest foundings were by nonworking wasps that had emerged relatively late from their natal nest. In terms of the model just introduced, the workers were G1 females primed to reproduce in the season of their rearing but not to survive winter quiescence, and the later nonworkers were G2 females with their reproduction turned off until after winter quiescence.

Queen replacement by workers on the natal nest and mid-season nest founding (chapter 3) can be interpreted in terms of G1 females primed to reproduce now (figures 8.6, 8.7). Worker mating and worker egg laying, heretofore seen as a paradox (Suzuki 1997, 1998), are not paradoxical if workers are G1 females. Ovarian development in workers but not in gynes (Haggard and Gamboa 1980) is what one would expect if workers are G1 and gynes are G2. The fit of a vast amount of natural history literature to the diapause ground plan hypothesis is a strong argument in its favor. However, fit of observations to the hypothesis doesn't constitute a direct test.

Simulation Studies

To test whether a food-activated switch during larval development was sufficient to generate the caste patterns observed in *Polistes*, my colleague Gro Amdam created an individual-based model that incorporates basic features of *Polistes* development (Hunt and Amdam 2005). Individual-based models can translate life history information into demographic data if the basis for demographic pattern-formation is known or hypothesized. Such models are conceptually easy to understand, because the unit of the model, the individual, is also the biological unit. The digital wasps are governed by a set of rules that translate life history information on each individual into demographic data for a colony. Parameter values for life history information were set to simulate *Polistes* in nature (see appendix at the end of this chapter). The only input variable that changed between runs of the model was the amount of food available to the digital foragers. As a rule of the model, well-nourished individuals became gynes, and individuals that were more poorly nourished became workers.

Results from four runs of the model are shown in figure 8.8, and each of the patterns can be matched to examples from nature. Moderate food levels produce the dynamics of figure 8.8A, which is the typical pattern reported in every study of *Polistes* in the temperate zone: a peak of workers precedes a peak of gynes. Because the model incorporates random day-to-day fluctuations in food supply, moderate food levels can also produce two distinct phenomena that occur in *Polistes*: a minor peak of early gynes (Reeve et al. 1998) and a minor peak of late workers (Dapporto et al. 2005; figure 8.8B). Higher food levels, shown in figure 8.8C, lead to early termination of brood rearing (Strassmann 1989b), whereas very low food levels (figure 8.8D), lead to few workers and almost no gynes (Hunt and Dove 2002).

Summary data from repeated runs of the model illustrate the effect of nutrient level on the model and demonstrate the consistency of model results (figure 8.9). At the lowest food level at which larvae develop, the development times of large brood (instars 4–5) are the longest and have the highest standard errors (figure 8.9A). As the amount of available nutrients increase, development times decline toward a minimum with no variability. The time from nest initiation until the peak of gyne production (figure 8.9B) is highly variable at low food levels, reaches a maximum at an intermediate food level, and declines to a constant at saturation. The number of gynes (figure 8.9C) increases from very few to a constant as food increases.

Values of response variables at the food level indicated by arrows in figure 8.9 are of particular interest. Here, the number of gynes has the largest variance and spans the full range for the observations (0–60 gynes). At the same time, the number of days until the colony reaches maximum gyne production is the largest. These results describe colonies that vary greatly in gyne production, with

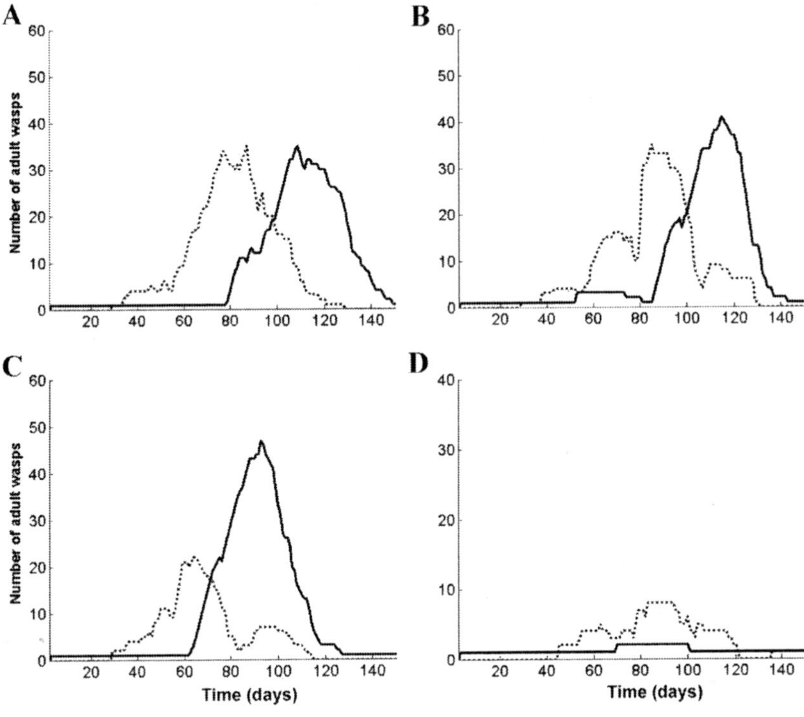

Figure 8.8
The simulation model generates numbers of workers (dotted lines) and gynes (solid lines) present on a nest. The appendix of chapter 8 gives details of the model. Four runs are shown, and the amount of food available to individuals that forage is the only input variable changed between runs. Moderate food levels always generate a peak of workers, followed by a peak of gynes (A). Day-to-day random fluctuations in food generated by the model can result in early gynes, late workers, or both in the same run (B). More food leads to an earlier worker peak and to earlier and more gynes (C); food demands of those gynes caused a late peak of workers and early termination of brood rearing. Very low food conditions result in production of few offspring, almost all of which are workers, and early termination of brood rearing (D). From Hunt and Amdam (2005).

many colonies producing few gynes, and a few colonies producing many gynes. High variability among colonies, with a few colonies placing large numbers of gynes into quiescence at the termination of the nesting season, exactly characterizes *Polistes* populations in a temperate seasonal environment (Hunt 1991; chapter 7).

The fit of the model output to actual observations in nature, both in general pattern and in particular variations, demonstrates that a nutrient-dependent switch mechanism is sufficient to explain the fundamental caste differentiation

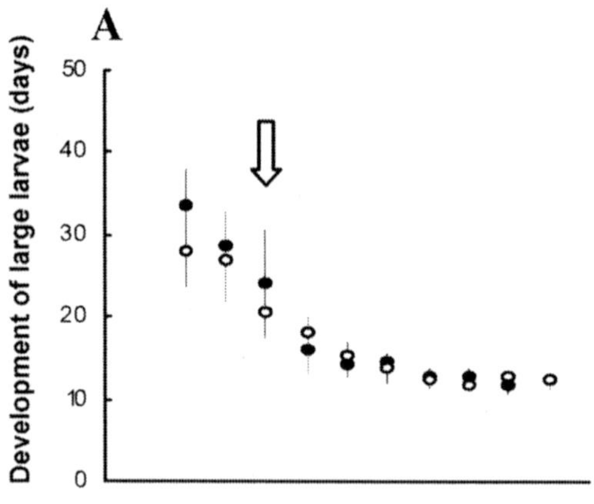

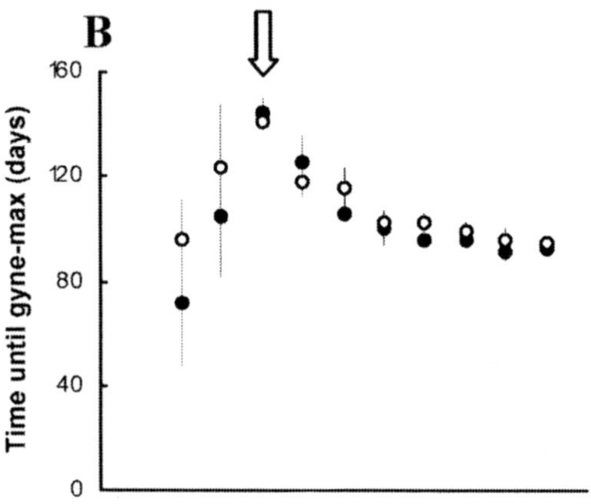

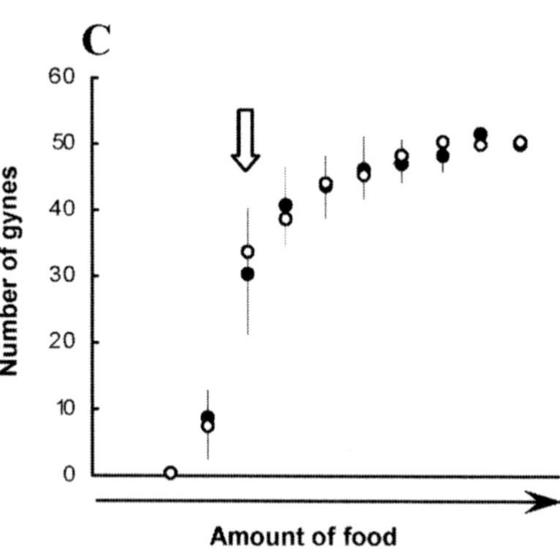

Figure 8.9
Summary statistics from two sets (n = 15 each) of simulations illustrate both the consistency of results and the strong effect of food on model output. Open and filled circles are the set means; vertical lines are standard errors. Development time for large larvae is long and has a high standard error at low food levels, and both development time and standard error decrease with increasing food (A). Time from the start of the simulation until the peak of gynes present on the nest has high variance at low food levels and reaches a maximum at an intermediate food level (B). Number of gynes produced increases with increasing food level (C). Arrows denote simultaneous occurrence of the longest time until peak gyne production and highest standard error in number of gynes produced. Natural *Polistes* populations are characterized by late gyne production and high variance among colonies in numbers of gynes produced (Hunt 1991). From Hunt and Amdam (2005).

in *Polistes*. A correlation between feeding and caste determination had long been suspected (Hunt 1991; O'Donnell 1998), but a framework that combines a nutritional switch with a specific hypothesis on its regulatory origin had been missing. Within the explanatory framework put forward by Hunt and Amdam (2005), however, the nutritional cue is secondary to the putative pathways on which it can act. We identified a regulatory ground plan, and it is the ground plan that in the solitary condition produces a developmental bifurcation in diapause pathways and the reproductive behavior and physiology associated with each pathway. Therefore, we not only provided a specific hypothesis about the developmental circuits that have been co-opted in social evolution, but we also provided an explanation for the suite of traits associated with each resulting adult phenotype (figure 8.7).

Inquilinism

In inquilinism, offspring of the founding queen rear offspring of the inquiline, whose offspring do not work (chapter 7). From the perspective of the diapause ground plan model, offspring of the founding queen are G1 females, and offspring of the inquiline queen are G2 females. Inquiline gynes migrate in fall to high, cold elevations, where they remain quiescent longer than do other species, and they do not return until the nesting season in the lowlands is well underway. Once an inquiline invades a colony and lays eggs, its larvae are provisioned before the larvae of the host, even though the workers and host larvae are the same species and the inquilines are a different species (Cervo et al. 2004). As a consequence, the inquiline larvae develop more rapidly (Cervo et al. 2004). The inquilines are species that have deleted G1 offspring from their life cycle. The physiological and/or behavioral cues that mediate the higher levels of brood care for inquiline larvae by host workers (Cervo et al. 2004) could be a gateway to learning cues and mechanisms that regulate the G1/G2 switch in noninquiline species.

Parasitism

Twisted-wing parasites, which comprise the insect order Strepsiptera, spend a major portion of their life cycle as endoparasites of other insects (Kathirithamby 1991). *Polistes dominulus* females in Italy, when infected by strepsipteran parasites, depart the colony and gather in aggregations away from any nest, a phenomenon discovered by W. D. Hamilton (Hughes et al. 2004a). At maturity, a female strepsipteran in a host wasp situates herself to protrude between segments of the wasp's gaster, and there she is mated by winged males that have emerged from other host wasps and live for only a few hours. Hughes et al. (2004a) interpreted the wasp aggregations as a parasite-induced modification of behavior to facilitate parasite mating. Infected females that harbor inseminated strepsipterans then pass the winter in quiescence together with noninfected females, and they are the source of strepsipteran larvae that attack new hosts the following spring

(Hughes et al. 2004b). Wasp larvae are the host stage that becomes newly infected (Hughes et al. 2003), and parasitism changes the behavior of infected females after they emerge from pupation, as confirmed by artificial infestation of larval wasps and monitoring adult behavior (Hughes et al. 2004a). Parasitized wasps in both the field and lab were "nominal workers" (Hughes et al. 2004, p. 1039), but the parasitism changed their adult behavior to resemble that of gynes—they did not work at their natal nest but instead departed and joined aggregations that contained gynes. In terms of the diapause ground plan model, the parasites change the infected wasps from G1 females into G2 females. Discovering the chemical signal or other mechanism whereby the parasites accomplish this physiological change in their host could be a gateway to learning mechanisms that regulate the G1/G2 switch in nonparasitized wasps.

Regulation

What regulates the G1/G2 pathway bifurcation? A first-level answer is that hormones mediate the switch, with JH and ecdysteroids almost certainly being major players. In the experiment described earlier in this chapter, Bohm (1972b) was able to induce ovarian activity in some of the wasps emerging in July that were maintained on summer conditions by the topical application of JH. In nature, JH plays a central role in caste differentiation. At a deeper level, there will be differential expression of regulatory genes. The regulatory cascade remains to be discovered. The driver, however, is almost certainly larval nutrition or another variable linked to larval nutrition. Saliva solicitation could be the external cue. Early larvae could be solicited more frequently and/or more aggressively for their saliva, and they may surrender more of it. Later larvae could be solicited less frequently or aggressively, perhaps simply as a consequence of a changing larva-to-adult ratio as the colony size increases . A different cue could be increased quantity or rate of provisioning of later larvae due to the presence of workers. Either of these variables, or a combination of them, or another factor correlated with them, could stimulate a hormonal response that then bifurcates the storage protein regulatory pathway that characterizes the two caste phenotypes and underlies their dichotomous behaviors.

There is an inverse relation between larval development time and pupation time in *Polistes fuscatus* (West Eberhard 1969). *Polistes metricus* larvae raised under restricted nourishment conditions had long larval development times but short pupation times, whereas larvae with nourishment supplementation had shorter larval development times and longer pupation times (Karsai and Hunt 2002). From the perspective of the bivoltine reproductive ground plan model, the short pupation time of G1 larvae could reflect the developmental trajectory of wasps selected to feed on protein at emergence and develop their ovaries. G2 larvae will not develop their ovaries after emergence, and the longer pupation time could reflect selection to synthesize and sequester storage protein, thereby prolonging the pupation period. Pupation in wasps is measured from the spinning of the cocoon until emergence, and so it encompasses multiple phases of development:

the silk-spinning fifth-instar larva, the postdefecation quiescent fifth larval instar (= the prepupa), the pharate pupa (within the larval cuticle), the pupa, the pharate adult (within the pupal cuticle), and the callow preemergence adult. Any or all of these stages could be lengthened in G2 larvae, but if storage protein synthesis causes the extended development, it is likely that extension occurs in the prepupal stage (Hunt and Amdam 2005).

It is important to note that G1 and G2 phenotypes reflect environmental conditions and developmental trajectories, not time periods. The non-diapause/diapause generations of partially bivoltine solitary wasps are sequential, just as are worker/gyne castes of *Polistes*. Both cases reflect environmental influence. However, in solitary wasps the principal environmental cues come from the ambient environment, whereas in *Polistes* the colony social environment translates or supersedes ambient variables.

Ancestry

In the initial presentation of the diapause ground plan hypothesis (Hunt and Amdam 2005), bivoltinism was proposed to characterize the ancestor of *Polistes*. This proposition immediately brought forth challenges from systematists, because the current phylogeny gives no indication of a bivoltine ancestor nor of a sister group characterized by bivoltine members (chapter 4). Because I had let go of the current phylogeny of Vespidae several months before the insight that now is the diapause ground plan hypothesis, I was not particularly concerned by these challenges. I think that new phylogenetic data will reveal relationships other than those of the current phylogeny. However, to ask whether these relationships do or do not show bivoltine ancestry to Polistinae will not be a test of the ground plan hypothesis, because an additional insight has changed my perspective on the nature of polistine ancestry.

In a conversation addressing the Hunt and Amdam (2005) hypothesis, Gene E. Robinson mused, "what if G1 wasps were invented by *Polistes*?" It was immediately clear that Robinson was on to something, although I now think he was right in concept but wrong in detail. Gene expression differences underlie diapause expression in blow flies (Flannagan et al. 1998; chapter 5), and gene expression differences specifically associated with diapause/non-diapause now have been reported in a diversity of insects, including beetles (de Kort and Koopmanschap 1994; Yocum 2001, 2003), moths (Yamashita 1996), and flies (Chen et al. 2005). In a number of cases a hexameric storage protein has been specifically associated with the diapause phenotype (Šula et al. 1995; Kłudkiewicz et al. 1996). In the boll weevil, for example, diapause occurs in the adult stage, diapausing adults are characterized by expression of hexamerin, and the difference in non-diapause and diapause expression can be induced by differenital feeding of newly eclosed adults (Lewis et al. 2002). It is also important to note that some taxa exhibit intraspecific variation in diapause expression. Populations of a solitary wasp, *Trypoxylon politum* in the family Sphecidae, in the southeastern United States, are partially bivoltine, whereas populations north of central

Virginia are univoltine (Brockmann 2004). Variation among partially bivoltine populations suggest a latitudinal gradient of responsiveness to environmental cues affecting diapause, and transfer of individuals between populations reveals plasticity of development as well as traces of local adaptation (Brockmann 2004).

If diapause is latent in the ground plan of many holometabolous insects, and if diapause can be differentially expressed intraspecifically in response to environmental context (and both of these propositions are supported by living examples), then it is possible that diapause expression first appeared in the context of the proto-*Polistes* life history in the absence of a diapausing ancestor. That is, the ancestor of the proto-*Polistes* could have been multivoltine in a tropical environment. Given the architecture of its nest, with simultaneously open cells containing progressively provisioned larvae, the stage could have been set for emerged offspring to remain at their natal nest to obtain protein nourishment for their own reproduction, as seems to be the case in living Stenogastrinae (chapter 2). If nourishment provisioned by alloparental first-emerged wasps to later-developing larvae was sufficient to trigger a nutritionally mediated diapause pathway, and if the wasps had been nesting in a seasonal environment in which diapause could have been demographically advantageous (chapter 7), the *Polistes* life cycle could have been established de novo. The novel invention would have been the G2 developmental pathway leading to the gyne phenotype. This scenario is plausible, and it is fully concordant with all the patterns and implications that have been presented and discussed in this volume. I therefore think it is appropriate to characterize the hypothesis for the ground plan that underlies caste differentiation in *Polistes* as the diapause ground plan hypothesis rather than the bivoltine ground plan hypothesis.

The Dynamic Scenario

Sociality as seen in *Polistes* paper wasps (chapter 3) evolved from an ancestral solitary wasp characterized by prey hunting, nest-building, and idiobiont reproductive physiology (chapter 1). Selection for traits that would reduce brood loss to parasitoids led to the nest architecture and progressive provisioning seen in *Polistes* today. The nest architecture and the simultaneous rearing of an uneven-aged brood led to the emergence of first offspring in a context in which they could have interactions with both their mother and their nestmate larvae. Contact with saliva of their nestmate larvae (with larva–adult trophallaxis having first appeared in the context of matrifilial solitary nesting) intersected with the idiobiont predilection for feeding on protein sources and led to retention of emerged wasps, at least for a short time, at their natal nest. In this framework, dominance interactions between mother and daughter could lead, via the flexibility inherent in idiobiont reproductive physiology, to the modulation of ovarian development in the female offspring. Without doubt, the daughter wasps all were destined to undertake independent nesting and reproduction, and selection may have favored life history variations such as those seen in

Stenogastrinae (chapter 2). Indeed, polistine sociality probably evolved via a stenogastrine-like stage.

Longer-term residency at the natal nest by early-emerged females led to the expression of maternal behaviors. Although expressed alloparentally as care of their sisters rather than of their own offspring, the wasps remained destined for independent reproduction. The brood provisioning by retained early-emerged offspring fostered a qualitative change in subsequently developing offspring. The higher nourishment levels during larval development effected by multiple foragers and provisioners activated a latent ground plan diapause mechanism. Later emerging wasps were consequently in reproductive diapause. If this occurred in a benign, relatively aseasonal tropical environment, there would have been little, if any, adaptive advantage to reproductive diapause. However, in a seasonal environment, which could have been either tropical wet/dry or temperate warm/cool, the gyne phenotype would have been an adaptation for initiating the next generation in the ensuing favorable season. When this adaptation occurred, selection began to favor specific components of the system that could lead to increased gyne production late in the favorable season. Sociality as seen in *Polistes* today emerged.

Synthesis

Three variables—ovarian development, hormones, and storage proteins—are the common threads that link the ovarian, reproductive, and diapause ground plans. All three ground plan hypotheses are appropriately focused on ancestral traits of reproductive physiology and development, and all ask how these traits have been adaptively modified to yield the origin and elaboration of sociality. Careful focus on the same physiological traits from this same perspective will probably shed light on sociality not only in *Polistes* and *Apis* but also in taxa that have heretofore not been examined in this way.

There are several points of difference between West-Eberhard's ovarian ground plan hypothesis and the perspectives on polistine social evolution presented in this book. In the ovarian ground plan hypothesis the worker caste is envisioned as evolving first among nest-sharing females of a single generation, as exemplified by *Zethus miniatus*, whereas here I have argued that the context for worker evolution is solitary nesting that incorporates mother–daughter interactions, as in a single-foundress *Polistes* colony. To resurrect some terminology from the 1970s, the ovarian ground plan hypothesis is a semisocial (within-generation) model, and the diapause ground plan hypothesis is a subsocial (between-generation) model. Second, in the ovarian ground plan hypothesis the underlying physiology is proposed to be a continuous ovarian cycle in which queens independently reflect the ovary-activated, egg laying phase of the cycle, and workers independently reflect the ovary-inactive, foraging phase of the cycle. It seems unlikely that the ovarian cycle, as envisioned by West-Eberhard (1987c, 1996), is truly cyclical, as is the ovarian cycle of *Homo sapiens*, but the elements of the envisioned cycle clearly reflect the reproductive physiology and dynamics of idiobiont ancestral forms (chapter 1). Third, in the ovarian ground plan hypothesis it is proposed that separation of the ovarian cycle into two phases would have been followed by the adaptive evolution of a regulatory switch, whereas in the diapause ground plan hypothesis, the underlying physiology of caste difference is proposed to be the bifurcation of ontogenetic pathways expressed as non-diapause and diapause in many insect taxa. West-Eberhard has appropriately focused attention on the importance of a regulatory switch, but she has not identified a specific mechanism. In her ovarian ground plan hypothesis, a switchlike mechanism is proposed to evolve de novo with or after caste divergence, whereas the diapause ground plan hypothesis of caste divergence is based on context-dependent co-option of a preexisting, switchlike physiological mechanism.

The heart of the difficulty with West-Eberhard's ovarian ground plan hypothesis is that it includes no pathway of gyne production. It is important to stress, however, that there is broad compatibility between the ovarian and diapause ground plan hypotheses in one key aspect. All *Polistes* females do, indeed, have the capacity to reproduce, and behavior interactions among nestmates can and do differentiate members of a cohort into reproductives and workers. These interactions occur in two different contexts: at the divergent ends of G1 pathway (figure 8.7) and among G2 females cofounding a colony. In both of these cases, behavior interactions assort individuals into reproductive or worker roles. Observations of queen replacement in *Mischocyttarus drewseni* (Jeanne 1972) strongly suggest a role for reproductive competition in serial queen replacements on a single colony, and reproductive competition among cofoundresses leading to a single egg-layer has been much studied since the pioneering observations by Pardi (1948). In a similar manner, dominance assortment of reproductive and worker roles may describe the situation under which Sakagami and Maeta (1987a, b) were able to artificially induce sociality by forcing nest sharing by two bees of the genus *Ceratina*, and dominance assortment may describe sociality that arises among nest-sharing adults of a single generation in many bees with small-colony sociality (Michener 1969, 1974; Lin and Michener 1972). Thus, West-Eberhard's focus on differential nourishment of adults affecting JH-mediated differential expression of reproductive and worker behaviors is highly appropriate and probably applicable to several taxa, including some parts of the *Polistes* colony cycle. The ovarian ground plan hypothesis does not, however, describe ontogenetic partitioning of *Polistes* offspring into worker (non-diapause, G1) versus gyne (reproductive diapause, G2) phenotypes.

The reproductive, ovarian, and diapause ground plan hypotheses all have a common thread: nutritional differences can be the basis for phenotypic differentiation. The reproductive ground plan hypothesis, however, focuses on worker honey bees and does not tackle the question of queen/worker divergence, and it does not currently include a switchlike mechanism as a component of differentially expressed worker phenotypes in honey bees. The diapause ground plan hypothesis presents a plausible model for the origin of a switch mechanism that differentiates nondiapause and diapause pathways, but the original presentation of the idea (Hunt and Amdam 2005) did not include a detailed scenario for the retention of early-emerged offspring at their natal nest, and it did not address the physiology that would underlie the expression of allomaternal behaviors by these wasps. The ovarian ground plan hypothesis more closely approaches that component of the scenario. In sum, all three hypotheses have a common focus of investigation, and a synthesis of selected parts of the three hypotheses leads to the dynamic scenario.

Other Social Wasps

If the dynamic scenario describes sociality in *Polistes* in seasonal environments, does it accurately describe paper wasps in aseasonal tropics? One consideration to keep in mind is that virtually no tropical sites are truly aseasonal; almost all have wet and dry (or wetter and less wet) seasons to which many organisms are attuned. But, the tropical wasps themselves are the best evidence that the dynamics described for *Polistes* in seasonal environments also hold for tropical wasps. Almost all independent-founding polistine wasps in the tropics have determinate, but asynchronous, colony cycles (chapter 3), as exemplified by *Mischocyttarus drewseni* (Jeanne 1972). That is, although colonies may be founded at any time of the year, individual colonies show similar trajectories of growth and decline. That colonies of paper wasps in aseasonal tropics have determinate trajectories despite benign environments strongly suggests that the G1/G2 demographic dynamics remain in place in these tropical wasps even in the absence of seasonal patterning.

Ropalidia marginata is the only independent-founding paper wasp studied to date that has asynchronous and indeterminate colony cycles and single colonies with potentially very long lifetimes (Gadagkar 2001). Queen replacement on long-lived *R. marginata* colonies occurs as a normal course of events, and new colonies are founded throughout the year (Gadagkar 2001). A worker/gyne dichotomy is not apparent. Even so, when newly emerged females are isolated and provided nourishment, about half found nests and about half do not (Gadagkar et al. 1988). This striking result was confirmed in a second experiment (Gadagkar et al. 1990). In a third replication of the isolation experiment, larval nourishment was strongly implicated as playing a role in the nest founding difference, with wasps better nourished as larvae being more likely to found nests (Gadagkar et al. 1991). This is the opposite of the situation hypothesized for *Polistes* by Hunt and Amdam (2005), in which better larval nourishment leads to reproductive diapause rather than to reproductive behavior. Nonetheless, it seems promising that a search for G1/G2 dynamics in *R. marginata* or in any other tropical paper wasp will reveal a switchlike bifurcation of two developmental pathways regulated by nourishment as a social/environmental variable.

Swarm-founding polistine wasps, all of which are tropical, are an open question. The morphological castes that occur in many species (chapter 3) could coincide with components of nondiapause/diapause phenotypes. However, caste totipotency has been proposed in epiponines that lack morphological castes (Strassmann et al. 2002). Perhaps the diapause developmental pathway has been selectively lost from the life history of some swarm founders. Recruitment of replacement queens from among newly emerged females is consonant with the notion that all females are nondiapause G1s with their reproduction typically suppressed (pheromonally?), and queen loss enables ovarian development in recently emerged individuals with a nondiapause phenotype in circumstances when ovarian development is not suppressed. Components of nondiapause/ diapause phenotypes have not yet been sought in swarm-founding wasps, and the search should provide some interesting tests of the diapause ground plan hypothesis.

If vespine wasps have a caste difference physiology that does not involve storage proteins (chapter 5), then, as mentioned in chapter 4, evolution of sociality in Vespinae may be more interesting than heretofore was apparent. Certainly that seems likely to be so with regard to Stenogastrinae. In chapter 4 I argued that Stenogastrinae evolved sociality independently from that in Polistinae. Indeed, if Stenogastrinae evolved from a solitary ancestor in tropical rainforests, the underlying reproductive ground plan of Stenogastrinae would be that of a solitary wasp with a life cycle like that shown in figure 8.1. All female offspring would be nondiapause phenotypes, and they would be physiologically primed to enter reproduction after emergence as adults. This model is not only consistent with all the natural history information on Stenogastrinae presented in chapter 3, it also offers a powerful argument for why stenogastrine colonies never become large.

Stenogastrinae never evolved large colony size because there are no diapause females. Because of this there can be no demographic payoff for alloparental care by nondiapause females early in the favorable season of a seasonal environment. With no quiescent phase of the life cycle, there is no structure for selection to favor placing a large number of gynes into a quiescence pool. Indeed, there are no gynes. Without the strength of selection that characterizes a seasonal life history, there has been only weak selection for alloparental care behaviors other than as a context for acquiring nourishment by the alloparental caregiver while she is developing her ovaries and waiting for an opportunity to reproduce.

Phylogenetic Implications

If the ancestor of *Polistes* was a multivoltine tropical wasp rather than a bivoltine wasp in a seasonal locale, then one argument for difference of origin of Stenogastrinae and Polistinae put forward in chapter 4—that Stenogastrinae originated from an aseasonal tropical ancestor, and Polistinae+Vespinae originated from a seasonal ancestor—becomes less compelling. Indeed, if the ancestor of *Polistes* was a multivoltine tropical wasp, then there are only three arguments against monophyly of (Stenogastrinae + (Polistinae+Vespinae)): (1)

the large number of autapomorphies that distinguish stenogastrines from polistines and vespines (table 4.2); (2) the extremely weak data and arguments currently put forward in support of monophyly of the social subfamilies; and (3) the implausibility of monophyly of Eumeninae (chapter 4). Those arguments retain their force whether the ancestor of *Polistes* was multivoltine or bivoltine, and I remain favorably disposed toward the propositions that sociality evolved twice in Vespidae and that Eumeninae is paraphyletic with regard to the social subfamilies.

The Selective Basis of Caste

The diapause ground plan underlies a life cycle that sets the demographic context in which selection can favor alloparental care by the females primed to reproduce now (the G1 females) if this leads to production of a large number of females prepared to diapause until the following season (the G2 females) (Hunt 1991; O'Donnell 1998). The Schaal-Leverich model for annual plants (chapter 7) is directly applicable. Another plant analogy clarifies and strengthens this perspective. Alloparental (worker) female offspring gather resources and thereby foster a larger "seed set" (= more gynes), and so they constitute vegetative growth rather than propagules. By virtue of their alloparental care activities, their mother will also be mother to a greater number of diapause females (" the propagules") and place these into quiescence later in the nesting cycle than would any wasp nesting and rearing offspring alone or with less abundant or less effective alloparental care. The demographic payoff of numerous diapause female offspring has selected for adaptations that shape the system. Where, then, is the focus of selection?

Much current thought would have it that the focus of selection is the workers themselves, with selection acting via relatedness to the reproductives that the workers care for. In cases of usurpation and inquilinism, however, worker females provide alloparental care to reproductive females to whom they are unrelated. Alloparental care by nonrelatives also occurs naturally in multifoundress colonies of *Polistes dominulus* (Queller et al. 2000). Phenomena such as these suggest that demographic drivers of the system can act independently of relatedness between alloparental caregivers and the brood that they tend. Selection will favor queens and their colonies that place large numbers of gyne females into quiescence, and the means to accomplish this are not directly linked to relatedness between the colony's workers and the queen's gyne offspring. The extent to which relatedness between worker and gyne offspring might enhance or accelerate selection that is driven primarily by demographic factors is unknown.

If one looks at elaboration of sociality beyond *Polistes*, however, relatedness between alloparental caregivers and the brood that they tend must play a role. Once the sociality threshold has been passed, morphological differences between nonreproductives and reproductives cannot evolve without kin selection. However, because "kin selection" is conflated with inclusive fitness maximization (chapter 9) in the minds of most students of sociality (Fletcher and Ross 1985), the term should be used with great care. Part III of this book deals with these questions.

Appendix: Simulation Studies

A simulation begins with a single foundress. She may lay zero to three eggs per day depending on the amount of food she can obtain (Miyano 1980). An egg hatches to become a small larva (corresponding to larval instars 1–3). Over 4 days the small larva (instars 1–3) must obtain enough food to molt into a large larva (instars 4–5) (Kudô 2003). If the larva fails, it is eliminated from the model (to simulate selective cannibalism of young brood, which is normal during unfavorable periods [Hunt 1991]). A surviving individual spends a minimum of 8 days as a large larva before it pupates, but the actual age at pupation is dependent on the individual's nutritional state, because a certain quantity of stored nutrients is needed to pupate, and individuals that do not obtain this quantity extend their large larval stage to get more food (Miyano 1990). Large larvae are never eliminated from the model, and as individuals pupate they are assigned an adult reproductive state based on the amount of nutrients they contain (O'Donnell 1998). Well-nourished individuals develop into gynes, whereas individuals that are more poorly nourished become workers (O'Donnell 1998; Karsai and Hunt 2002). The individuals emerge as adults 15 days later and are assigned a nest time-span. Workers are randomly given a nest time-span with a mean of 20 days (Strassmann 1985; Giannotti and Machado 1994). After the assigned number of days, the worker is eliminated from the model to simulate death. Gynes that are produced in the early phase of the colony cycle are given nest time-spans of 5 days on average to simulate early quiescence (Reeve et al. 1998), and gynes produced later in the season stay for an average of 20 days.

Each day, a random amount of food is made available for individuals that forage. The foundress forages only until she has produced two adult workers, and gynes do not forage. The inflow of nutrients is therefore linked to the number of workers on the nest. The total amount of food collected each day is divided between the adult wasps and the small and large larvae as follows. Each adult wasp has a nutritional need to be met, and for gynes this need is twice that of workers. If there is a surplus, it is asymmetrically divided between the small and large larvae: 20% and 80%, respectively. If the inflow of food is insufficient to satisfy the adults, nutrients are subtracted from the nutritional stores of the large larvae to simulate the food reserve role of larval trophallaxis (Maschwitz 1965). The small larvae are not fed in this case. (Simulation details from Hunt and Amdam [2005].)

PART III

Paradigm Lost—and Found?

For 30 years the field of study that encompasses social evolution in animals, including vespid wasps, has been dominated by a way of thinking that is not presented, except in a facetious footnote, anywhere in the first eight chapters of this book. The omission heretofore of this way of thinking has been intentional; however, it cannot be ignored. The time has come to bring this view forward for presentation and examination, to highlight differences between it and the view of vespid social evolution presented in this book, and to ask if difference can become synthesis or, at least, counterpoint in a broader view.

Perhaps the best way to introduce the dominant view of social evolution is by means of an example. The example that follows is in no way egregious; it is generic for the field. I selected it because it is recent, it was published in a prominent journal, and it touches directly on subjects and organisms addressed in this book. Many tens, if not hundreds, of similar examples on social insects could be brought forward, and if one included social vertebrates the number of examples would be hundreds, if not thousands.

Liebig et al. (2005) brought single-foundress colonies of the paper wasp Polistes dominulus *into the laboratory and established them in separate rearing cages. As an experimental treatment, eggs and larvae (but not pupae) were removed from nest cells of half of the colonies; no brood was removed from control colonies. The main result was that workers on the brood-removed colonies exhibited an increase in rate of egg laying, whereas workers on control colonies showed little or no egg laying. Increased egg laying by workers of brood-removed colonies occurred despite a simultaneous four-fold increase in egg laying by the queens of those colonies. Liebig et al. present this experiment as a test of two competing hypotheses for worker behavior, queen control versus queen signaling. According to the queen control hypothesis, the queen suppresses worker reproduction, against workers' interests (Michener and Brothers 1974; Breed and Gamboa 1977). Liebig et al. say (2005, p. 1342) that "this is clearly not the case in our experiment, where the foundress failed to control workers even though she was not manipulated." I agree that their data are evidence against a hypothesis that queens "control" their offspring workers through continuous behavioral domination. However, because Liebig et al. had framed their study as a contest between two hypotheses, they thereby set a logical pitfall in which they interpreted refutation of the queen control hypothesis as support for the queen signaling hypothesis, even though that hypothesis is not a null alternative. "This study shows that* Polistes *workers respond to brood depletion by reproducing and that brood abundance, possibly estimated through the frequency of empty cells in the nest, provides direct and highly reliable information on foundress fertility. Workers thus use the most reliable source of information [rather than indirect pheromonal cues] to decide whether to reproduce or not" (p. 1343). The key word in this diagnosis is* decide: *"our results are consistent with the general idea of the queen signalling hypothesis, that workers use the information about queen fertility to decide to refrain from reproducing and help the queen (Seeley 1985; Keller & Nonacs 1993; Monnin et al. 2002)" (p. 1342). Liebig et al. are unambiguous in their presentation of the decision-making as an exercise in freedom of choice among readily available reproductive options: "the lack of worker reproduction in natural colonies with a fully functioning queen is not due to queen control but to self-restraint by workers" (p. 1342). What basis is proposed for this self-restraint?*

> *Workers typically do not reproduce in the presence of the queen. Rather, they rear the queen's offspring when they gain sufficient indirect fitness benefits (Hamilton 1964[a, b]). Inclusive fitness is essentially a function of relatedness to the queen and of queen productivity. Typically, workers benefit from helping when the queen is fertile, but they favour producing their own sons when queen fertility decreases too much to be compensated by relatedness benefits (Bourke & Franks 1995). (Liebig et al. 2005, p. 1339)*

The only response variable that Liebig et al. quantified for the worker wasps was a fitness variable—oviposition rate. They did not quantify behaviors such as foraging. They did not quantify or control nourishment level; instead, the wasps were fed ad libitum. No consideration was given to the possibility that workers in the control colonies lay few or no eggs because physiological costs of brood care constrain their potential for ovarian development, nor was the possibility considered that workers in treatment colonies, freed from brood care and provided limitless nourishment, can show ovarian development in reflection of their idiobiont ancestral capacity to do so. In short, no mechanism was sought or, apparently, even envisioned. No reference is made to oviposition by worker wasps in circumstances such as queen replacement or satellite nest founding, even though both could occur (and the latter is likely to occur) when natal nests are filled with brood. At the same time, rejection of one hypothesis was interpreted as support for the non-null alternative that was put forward. In consequence of these methodological and conceptual lapses, Liebig et al. were unconstrained in applying the interpretation that they did—that offspring of a social wasp colony work voluntarily, because by doing so they secure inclusive fitness benefits for themselves.

This same interpretation, this same blindered view of social evolution focused exclusively on workers' inclusive fitness, is a key part of the ancestral form of behavioral ecology, which is sociobiology. In chapter 9 I look at the early history of sociobi-

ology and how it shaped perspectives on kin selection. I discuss some possible reasons that Hamilton's haplodiploidy hypothesis persisted for so long, and I shift attention to the role that haplodiploidy actually did play in Hymenopteran social evolution. After a brief look at sex ratio, I then look at Hamilton's rule and find it wanting as a tool for understanding social evolution. A future role for kin selection studies is proposed. In chapter 10 I look at the discipline of behavioral ecology and highlight characteristics of that discipline that are impediments to advancement and understanding. I do so in the spirit that the first step toward solving a problem is to recognize that a problem exists. In chapter 11 I look to the future. I propose both a general conceptual framework to study sociality and a specific research agenda for the future study of hymenopteran social evolution. I conclude with an encapsulation of social evolution in wasps as I currently understand it.

9 Kin Selection

W. D. Hamilton's formalization of inclusive fitness (Hamilton 1964a) and its application in understanding a range of issues in the biology of social taxa, including the evolution of social insects (Hamilton 1964b), was identified in the Preface to this book as the dividing line between initial and recent approaches to the study of social evolution. John Maynard Smith (1964) rearticulated Hamilton's concept of inclusive fitness, coined the term "kin selection," and contrasted kin and group selection. Edward O. Wilson was the first to strongly endorse Hamilton's ideas in their application to the evolution of insect sociality (Wilson 1966). Wilson's autobiographical recounting of reading Hamilton's papers and meeting Hamilton (Wilson 1994) gives good measure of the importance that he ascribed to Hamilton's work and the significance it had in shaping his thinking and writing. Wilson (1966) placed emphasis on Hamilton's proposition that haplodiploid sex determination would cause high relatedness among full sisters and asymmetries of relatedness among kin that could favor worker behavior by females that raise their reproductive sisters and thereby achieve high inclusive fitness. In a later, more tempered assessment in *The Insect Societies*, Wilson (1971, p. 328) characterized Hamilton's idea as "an audacious genetic theory of the origin of sociality which assigns the central role to haplodiploidy in a wholly different way [than Richards (1965)]." Wilson (1971) then evaluated evidence relevant to five predictions that could be drawn from Hamilton's hypothesis, and he granted the hypothesis provisional acceptance.

The major milestone that followed Wilson's endorsements was work by Trivers and Hare (1976), who were the first to pose a test of the hypothesis. Their conclusion that haplodiploidy played the role envisioned by Hamilton moved Wilson, in *Sociobiology: The New Synthesis* (Wilson 1975, p. 417), to say: "Trivers' remarkable result appears to confirm the operation of kin selection in ants as the controlling force, as opposed to individual selection leading to domination and exploitation by the queen."[1] The result put forward by Trivers and Hare was lauded in *Nature* by Krebs and May (1976, p. 10): "it puts the capstone to an edifice that draws together genetic principles and Darwin's theory of natural selection toward explaining the evolution of social insects." These assertions firmly established the prominence of Hamilton's ideas. The rest of the 1970s and early 1980s were marked by unprecedented levels of interest and activity in the field of social insect biology. Sociobiology was heralded as a new discipline, with inclusive fitness as its cornerstone; the journal *Behavioral Ecology and Sociobiology* was launched. The excitement was palpable.

Although endorsements by Wilson and others were central to the ascendancy of Hamilton's ideas, also at work may have been a tendency in "soft" sciences such as animal behavior that was captured in a humorous passage of an important review. In discussing Hamilton's haplodiploidy hypothesis, Andersson (1984, p. 182) wrote:

> A . . . reason for the popularity of this hypothesis could be the appeal of quantitative genetic arguments as compared to a morass of morphological, behavioral, and ecological preconditions, the relative importance of which are hard to quantify. (This may be a scientific version of "the street lamp temptation," which overwhelmed the couple who dropped their key one night when unlocking the door. After searching in vain in the darkness, they noticed the bright light beneath the street lamp and so searched there instead.)

Any momentum generated by the street lamp temptation, however, pales in comparison to the impact of Trivers and Hare.

Trivers and Hare (1976) asked whether social insect colonies exhibit "queen control" or "worker control." Equal production of males and reproductive females was proposed to evidence "queen control," because queens are equally related to their sons and daughters. Production of reproductive females in preference to males evidenced "worker control," because of the high mean relatedness between female workers and their reproductive sisters and low relatedness between sisters and brothers. The numbers of the sexes in reproductive broods don't favor either prediction, because males often outnumber reproductive females, especially in ants, whereas reproductive females often are larger, so each female represents more work. Trivers and Hare therefore examined sex and weight ratios (collectively, the "investment ratio") of reproductive brood for social, slave-making, and solitary species of Hymenoptera (mostly ants) and several termites. Using scatter diagrams where weight and sex ratios were the X and Y

1. In Wilson (1975), Trivers and Hare (1976) is cited as "Trivers, R. L., 1975, *Science*, in press."

axes (figure 9.1), they argued that data points for monogynous ants cluster around a 3:1 investment ratio line and therefore reflect worker control. Crozier and Pamilo (1996, p. 202) call this figure "the chief results of Trivers and Hare."

An important critique of Trivers and Hare appeared one year later. Among other criticisms, Alexander and Sherman (1977) pointed out a fault with the regression analysis implied by Trivers and Hare's figures. Male-biased sex ratios vary from one to infinity, but female-biased ratios are confined to zero to one. This generates unequal variance along the ratio lines and masks the true spread of the data. Trivers nonetheless reprinted the unmodified figure, without comment, in a textbook (Trivers 1985). Figure 9.2 shows the same data with axes intersecting at 1:1 ratios. In another criticism, Alexander and Sherman (1977) cited an ant species from Trivers and Hare's figure that had reproductives in only 12 of 20 colonies, 1 of which had males only, 1 of which had females only, 1 of which had a strongly female-biased sex ratio, and 9 of which had male-biased sex ratios ranging from 2:1 to 21:1. These were averaged to yield a sex ratio of 8:1 for the species. The effect of averaging is shown in figure 9.3: when plotted as individual colony data, there is no apparent central tendency toward either "investment ratio." The sex ratio literature remains active and is still focused on asymmetrical relatedness due to haplodiploidy.

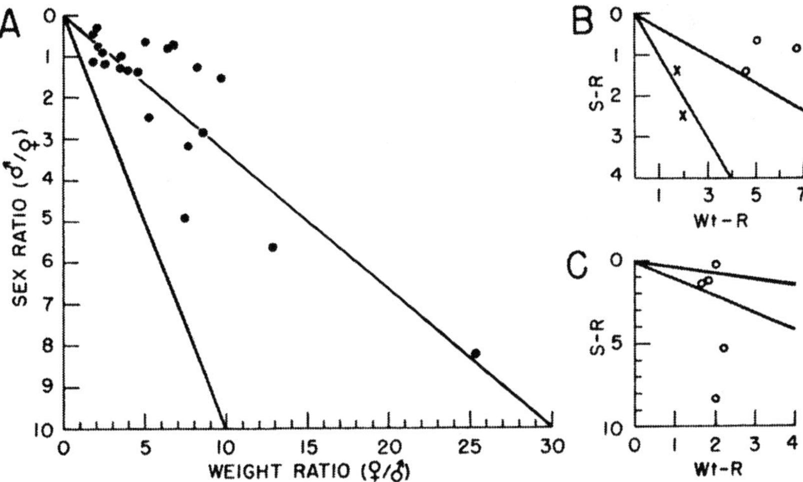

Figure 9.1
The sex ratio (male/female) and adult dry weight ratio (female/male) for (A) monogynous ants, (B) two slave-making (x) and three non–slave-making species of *Leptothorax* ants, and (C) two monogynous, two polygynous, and one mixed-gynous species of *Myrmica* ants. Lines indicate 1:1 and 1:3 "investment ratios." Reprinted with permission from Trivers, R. L., and H. Hare, 1976, "Haplodiploidy and the evolution of the social insects," *Science* 191:249–263. © 1976 AAAS.

The spotlight that beamed in 1975–1976 invited investigation of Hamilton's ideas. Hamilton (1964a) had defined inclusive fitness using an algebraic approach. However, population genetic allele frequency modeling (Craig 1979, 1980) showed that high relatedness among sisters need not favor worker evolution. Charlesworth (1978), Charnov (1978), and Craig (1982a) all asserted that haplodiploid male workers should evolve as easily as females, and, indeed, soldiers of gall-inhabiting social thrips (Order Thysanura, all of which are haplodiploid) are male as well as female (Crespi and Mound 1997). Cautionary assessments of the role of haplodiploidy as a factor in hymenopteran social evolution include those of Evans (1977b), Eickwort (1981), Crozier (1982), Andersson (1984), Stubblefield and Charnov (1985), Alexander et al. (1991), Hahn (1995), Jaffe (2001), and Gadagkar (2001). Both West-Eberhard (1975) and Brockmann (1984) outlined scenarios

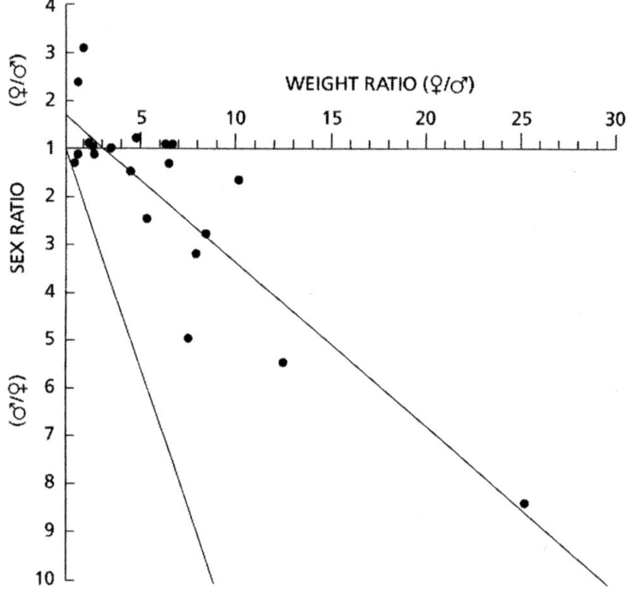

Figure 9.2
The same data as figure 9.1A replotted with axes intersecting at 1:1 ratios and equal variance along the ratio lines. For *Tetramorium caespitum*, sex ratio data come from Brian et al. (1967), who give a mean weight ratio for 17 colonies of 1.45. Trivers and Hare (1976) instead used a weight ratio of 4.0 from Peakin (1972), and the point for *T. caespitum* falls on Trivers and Hare's 3:1 investment line. However, Peakin (1972) reported a weight ratio of 1.49 for "young" reproductives, virtually identical to the mean given by Brian et al. Peakin's data also contained a 3.97 weight ratio for "old" reproductives, but this ratio reflects both weight gain by females and weight loss by males after emergence. Had a weight ratio of 1.45 been used in the original figure, the point would have fallen on the 1:1 investment line. A weight ratio of 1.45 is used here.

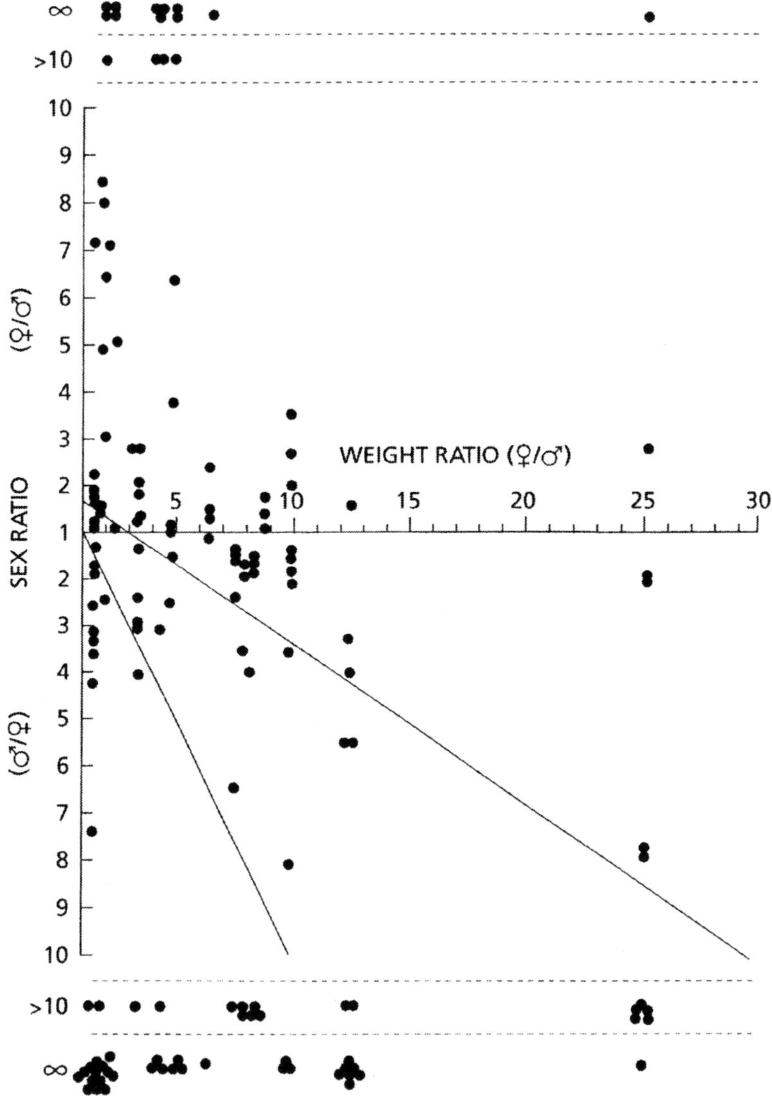

Figure 9.3

Sex and dry weight ratios for reproductives of 13 of the monogynous ant species used by Trivers and Hare (1976) to generate figure 9.1A. Each point shown here represents a single colony. Other species in their figure used unpublished or pooled data. Sex ratios > 10:1 are indicated as >10, and single-sex broods are indicated as ∞ at upper and lower margins. Species and sources: *Camponotus ferrugineus* and *C. pennsylvanicus* (Pricer 1908), *Prenolepis imparis* (Talbot 1943), *Aphaenogaster rudis* (Headley 1949; Talbot 1951), *Aphaenogaster treatae* (Talbot 1954), *Atta bisphaerica*, *A. laevigata* and *A. sexdens* (Autuori 1949–50), *Leptothorax curvispinosus* and *L. longspinosus* (Headley 1943), *Myrmica schencki* (Talbot 1945), *Myrmica sulcinodis* (Elmes 1974), *Solenopsis invicta* (Morrill 1974), *Tetramorium caespitum* (Brian et al. 1967). A weight ratio of 1.45 is used for *T. caespitum*.

in which relatedness asymmetries might ease evolution of helping behavior, but each gave greater weight to factors of phylogeny and ecology.

Despite the weight of evidence and opinion against it, Hamilton's haplodiploidy hypothesis was resilient and had great staying power. Haplodiploidy is still mentioned in occasional research papers, and it remains in some textbooks (e.g., Rose and Mueller 2006) as the explanatory factor of social evolution in Hymenoptera. A more tempered assessment is now current, as exemplified by Shreeves et al. (2003, p. 1617): "the celebrated haplodiploidy hypothesis of Hamilton (1964), that the decisive factor is the ¾-relatedness between female hymenopterans, has been largely superseded by models focusing on ecological costs and benefits." (The main such models are those of Queller [1989] and Gadagkar [1990].) A seemingly harsh, but arguably accurate evaluation is "the collapse of the 'haplodiploid hypothesis,' an early and once persuasive stanchion of the standard model, due to the discovery in recent years of enough phylogenetically separate lines (Choe and Crespi 1997) to render the association of haplodiploidy and eusociality originations statistically independent"[2] (Wilson and Hölldobler 2005, p. 13369). It is noteworthy, however, that there have been numerous and surprisingly diverse attempts to find a link between haplodiploidy and sociality, and only some of these place emphasis on relatedness. In fact, in the following section I argue that haplodiploidy does play a significant, central role in the evolution of sociality in Hymenoptera, but it is not the role proposed by Hamilton (1964b). A review of hypotheses about the possible role of haplodiploidy in social forms helps place its proper role into perspective.

Haplodiploidy

In a paper that accompanied his general model of inclusive fitness (Hamilton 1964a), Hamilton (1964b) proposed that social behavior in insects of the order Hymenoptera might be a phenomenon that exemplifies the principles of inclusive fitness in an unusual way. Hamilton (1964b) synthesized two observations. First, all then-known independently evolved lineages of social insects, except termites, were Hymenoptera. Second, workers in social Hymenoptera are exclusively female. Hamilton proposed that, due to haplodiploidy, female hymenopterans that rear reproductive full sisters, with whom they share an average three-quarters of their genome, can gain more fitness via inclusive fitness than by reproducing an equal number of reproductive daughters, to whom they are related by one-half. Termites had no known relatedness asymmetry that might favor worker behavior by one sex over another, and termite workers are both female and male. Thus haplodiploidy-engendered high relatedness between sisters and asymmetry of relatedness among relatives were thought to ease the evolution of worker behavior in Hymenoptera (of sisters rearing full sisters, that is),

2. The view of social evolution that I refer to as another view has been so pervasive and dominant in the field that these authors consider it to be the "standard model" of social evolution.

explaining the repeated evolution of sociality in the order and explaining why hymenopteran workers are exclusively female.

It has long been known that, in general, fertilized hymenopteran eggs produce females, and unfertilized eggs yield males (Dzierzon 1845). The underlying mechanism is not ploidy level itself; instead, it is zygosity at a sex determination locus (Cook 1993). Evidence from honey bees shows a single sex-determining locus with multiple alleles (Beye et al. 2003; Evans et al. 2004), although other mechanisms have been proposed (Beukeboom 1995; see Stahlhut and Cowan 2004). Heterozygosity at the sex-determining locus leads to development of a female. Homozygosity at the sex-determining locus produces diploid males. Diploid males, long known from honey bees (Mackensen 1951), may occur in most social Hymenoptera (Cook and Crozier 1995). However, diploid males, if not cannibalized as larvae (Woyke 1963), are usually sterile (cf. Krieger et al. 1999). Typical males of Hymenoptera are haploid and therefore hemizygous. Among the ramifications of haplodiploidy are that all sperm are identical, all loci act as if they are sex linked, all genes expressed in males act as dominants, males have a grandfather but no father, sex determination is under the control of an inseminated female at oviposition, and uninseminated females can lay only male eggs (although workers in some honey bees [Ruttner 1977] and some ants [Tschinkel and Howard 1978; Cagniant 1979; Schilder et al. 1999] can produce females parthenogenetically). A surprising ramification is that males of an ant in the *Myrmecia pilosula* complex with a diploid chromosome number of 2 have only one chromosome per cell (Crosland and Crozier 1986)!

The most famous ramification of haplodiploidy is, of course, the high relatedness it engenders among full sisters and the concomitant low relatedness between females and their brothers. In an article entitled, "The role of male parthenogenesis in the evolution of the social Hymenoptera," Snell (1932, p. 382) wrote: "there is a second interesting genetic consequence of male parthenogenesis which may have played some rôle in the evolution of the Hymenoptera. This is the uniformity which it causes in all the diploid individuals of a colony. Since the male, being haploid, produces only one kind of spermatozoon, his daughters are identical with respect to at least half of their germ-plasm. They are not only all sisters, they are half way to being identical twins." Snell's prescience passed unnoticed (the first citation of his article was in Eberhard [1974]) until a decade after Hamilton's independent and more compelling presentation of this idea and its possible genetic ramifications. In addition to this hypothesis of elevated full-sib sister relatedness, a number of other roles have been proposed for haplodiploidy in hymenopteran sociality.

Flanders (1946) proposed that male haploidy, with concomitant homozygosity at all loci, would constrain the evolution of male polymorphism. He implied that behavioral and morphological differences between queens and workers reflect genetic variability inherent in heterozygosity, which is found only in females. Homozygous males would lack allelic diversity and so could be constrained from expressing the diversity of morphologies and behaviors that differentiate castes among females. Lin and Michener (1972) and Michener (1974, pp. 251–252) proposed a similar idea.

Seger (1991) proposed that the ability of unfertilized haplodiploid females to lay viable male eggs might facilitate the origin of female workers. In taxa where an unmated working female retains some reproductive flexibility, working is not an absolute impediment to reproducing. Because males must mate in order to reproduce, however, this envisioned advantage (or, at least, lesser impediment) to working accrues only to females (Bourke and Franks 1995).

Reeve (1993) presented a population genetic hypothesis that if alloparental care genes appear in females of haplodiploid populations, they are less likely to be lost by genetic drift than would allopaternal care alleles in haploid males or alloparental care alleles in diploid taxa. Therefore, alleles that favor brood care of nondescendent kin by female workers are more likely to be found in living haplodiploid species than are any other kind of nondescendent brood-care alleles. Reeve's "protected invasion" hypothesis thus equals the three-quarters–relatedness hypothesis in addressing the concentration of social taxa and exclusivity of female workers in the haplodiploid Hymenoptera. Reeve's hypothesis additionally predicts female parental care in haplodiploid species, higher frequency of paternal care in diploid than haplodiploid species, and no sex bias in workers of social diploid species.

Saito (1994) emphasized the exposure of deleterious recessive alleles in haploid males, a topic introduced by Snell (1932) and also raised by Smith and Shaw (1980). Whereas Snell focused only on elimination of deleterious alleles per se, Saito reasoned that if such recessives are removed by selection, then a major constraint on inbreeding would be relaxed. Saito thus argued that inbreeding should be more common in haplodiploids than in diploids. If elevated relatedness due to inbreeding could facilitate kin selection, then higher rates of social evolution could result. Whether inbreeding can favor evolution of reproductive altruism was argued affirmatively by Wade and Breden (1981) and Wade (1980, 1982, 1985) and negatively by Michod (1980), Craig (1982b), Uyenoyama and Bengtsson (1982), Uyenoyama (1984), and Lacy (1984).

Maternal effect genes encode maternal messenger RNAs and maternally derived proteins in the egg and embryo that affect early embryonic development. Wade (2001) postulated the existence of antagonistic pleiotropy in maternal effect genes, whereby maternal transcripts have beneficial sib-social effects, but zygotic transcripts act negatively. Using population genetics models, Wade showed that antagonistic pleiotropy in maternal effect genes could evolve more easily in haplodiploid than in diploid populations. Key predictions of his models are female-biased sex ratios and relaxed constraints on evolution of maternal care.

Haplodiploidy could play a role in already social taxa by means of enhanced genetic diversity due to polyandry. Because each male can produce only one set of alleles in his sperm, the resulting highly similar genotypes in the offspring of one father (a "patriline") could enable selection favoring particular allele combinations that enhance colony-level efficiency. Workers of different patrilines in honey bee colonies do, in fact, perform particular tasks at different frequencies (Calderone and Page 1987; Robinson and Page 1988; Frumhoff and Baker 1988). Polyandry characterizes species with the largest colony sizes of their type: honey bees (Page 1980; Tarpy and Page 2001) and leaf cutter ants (Villesen et al. 1999). Task partitioning similar to that of honey bees has also been found for different matrilines in some polygynous ants (Snyder 1992, 1993; Carlin et al. 1993) and in artificially mixed colonies of a monogynous ant (Stuart and Page 1991). The monogynous western harvester ant, *Pogonomyrmex occidentalis*, shows a substantial fitness benefit to polyandry (Wiernasz et al. 2004). Tarpy and Page (2002) reviewed several ways in which intracolonial genetic diversity may have facilitated evolution of efficient colony organization.

In his advocacy of Hamilton's three-quarters–relatedness hypothesis, Wilson (1971) brushed aside a role for haplodiploidy in the origin of sociality that had been proposed by Richards (1953, 1965) and Brian (1965). This role was endorsed by Eickwort (1981) and Alexander (1974) and stated as simple fact by Michener (1974, p. 233) and Gauld and Bolton (1988, p. 30): "[haplodiploidy] bestows the ability to determine the sex of offspring and thus to adjust the sex ratio to suit the needs of the life history. Thus colonies exclusively of female workers can be produced until there is a need for sexuals." Protogyny characterizes all social Hymenoptera, including the sphecid wasp *Microstigmus* (Matthews 1991), all social bees (Michener 1974), and ants, in which all noninquiline taxa produce workers before they produce sexuals (Hölldobler and Wilson 1990).

Haplodiploidy is not necessary for sex-ratio bias, which can also be effected by mechanisms such as sex-dependent mortality of gametes, zygotes, embryos, or neonates. Haplodiploidy is distinctive, however, as a mechanism that can enable sex determination at the time of oviposition as a behavioral trait on which selection can act directly. Parasitoid Hymenoptera (chapter 1) lay fertilized or nonfertilized eggs as a function of host size (King 1987; Ueno 1999), host larval instar (Brunson 1937, 1938), host age (Ueno 1997), superparasitism (Holmes 1972; Werren 1980), or presence of conspecifics (King 1996). Solitary hunting wasps (chapter 2) lay fertilized or nonfertilized eggs as a function of host size (Dow 1942), nest cell size (Krombein 1967), or quantity of nest cell provision (Brockmann and Grafen 1992). Thus haplodiploidy has enabled gender determination at the time of oviposition as a life history adaptation in solitary forms throughout the order.

The exclusive production of females as first offspring in social Hymenoptera (chapter 3) is but one more manifestation of this widespread adaptive implementation of sex determination, but it is the critical and essential role played by haplodiploidy in the evolution of hymenopteran sociality. Social hymenopterans produce female offspring before producing males, reflecting selection that favors brood care behaviors by first-emerged offspring in an insect order in which maternal behavior occurs in all taxa that build nests and paternal behavior is unknown. Female hymenopterans will care for brood and males won't, with the consequence that colonies that include males among the initial brood will be at a selective disadvantage with regard to subsequent gyne production. Haplodiploidy, a sex-determination mechanism that is under control of inseminated females at the time of oviposition, enables protogyny as an adaptation that fosters colony growth that then can enable increased production of propagules late in the colony cycle.

Sex Ratio

Bourke and Franks (1995) and Crozier and Pamilo (1996) extensively reviewed sex ratio theory and data. They confirmed Fisher's (1930) sex ratio theorem (reviewed by Basolo 2001) that population sex investment will tend to be held equal by frequency-dependent selection. They also presented and reviewed models in which, for haplodiploid species in which workers control investment, population (not colony) ratios should be biased 3:1 in favor of females. The focus on population sex ratio rather than on colony sex ratio is necessitated because fitness advantage to a female rearing sisters rather than offspring disappears if males are rare in the breeding population (Maynard Smith and Szathmáry 1995). It is impossible, therefore, for sister-rearing altruists to represent a fixed end point of hymenopteran social evolution. Split sex ratio theory (Grafen 1986) now is invoked to mitigate this dilemma. Indeed, Maynard Smith and Szathmáry (1995, p. 267) assert that "it is hard to see how a causal connection between haplodiploidy and sociality could arise except through split sex ratios." Split sex ratio theory addresses sex ratios that differ by season within colony (Seger 1983) or among colonies within a population (Boomsma and Grafen 1990, 1991). Selection is supposed to favor females that work when sex ratio of the reproductive brood is female biased. However, in many ant species there are colonies that produce all male or all female reproductive broods (Forel 1874; Nonacs 1986), so current sex ratio studies focus at the population level rather than the colony level. Mehdiabadi et al. (2003) echo Queller and Strassmann (1998) when they assert that studies of sex ratio conflict have provided the most rigorous tests of kin selection theory. They also note (p. 91), however, that "the original paradigm of worker-controlled sex ratios clearly has been an oversimplification." Increasingly complex scenarios are required to keep recent empirical data within the theoretical construct of haplodiploidy-based maximization of inclusive fitness (Mehdiabadi et al. 2003). In a review of conceptual and methodological aspects of sex ratio studies, Alonso and Schuck-Paim (2002, p. 6843) concluded that "genes that act by biasing sex ratios to promote their own spread . . . depend on the social organization of the colonies where they are expressed, but . . . they are not, in any way, the precursors of these societies."

There are numerous possible reasons unrelated to kin selection whereby population sex ratios might not be one to one (Crozier and Pamilo 1993; Yanega

1996). Some possibilities include local mate competition (Hamilton 1967; Werren 1983), local resource enhancement (Schwarz 1988), inbreeding (Chapman and Stewart 1996), group selection (Frank 1986), genomic imprinting (Haig 1992), genomic intracellular sex ratio distorters (Werren et al. 1988), male mortality due to recessive lethal genes (Smith and Shaw 1980), sampling bias (Greene 1991), and infection by the parasitic microorganism *Wolbachia* (Strouthamer et al. 1990). However, although at least 50% of ant species are infected by *Wolbachia* (Wenseleers et al. 1998), this infection apparently does not cause female-biased sex ratios (Wenseleers and Billen 2000). In colonies of one ant species, sex ratio varies with queen number (Brown and Keller 2000). Demography and ecology may substantially affect individual colony sex ratios, especially of ants. Small, young, or poorly nourished ant colonies produce males more often than reproductive females, whereas larger, older, or well-nourished ant colonies more often produce reproductive females (Peakin 1972; Elmes and Wardlaw 1982; Brian et al. 1967; Herbers 1984; Nonacs 1986; Deslippe and Savolainen 1995; Ode and Rissing 2002). Food supplementation of the paper wasp *Polistes metricus* changed the males to gynes sex ratio from 3:1 to 1:1 in supplemented nests (Seal and Hunt 2004). Deslippe and Savolainen (1995, p. 375) assert that "food supply has an important proximate influence on sex investment, and may explain much of the natural variation in sex investment in populations of eusocial Hymenoptera."

Inclusive Fitness

If three-quarters relatedness and sex ratio fail to enlighten our quest to understand the evolutionary origin of sociality, what about the more general, encompassing hypothesis—what about kin selection? One early formulation of the concept of inclusive fitness, which is the basis for kin selection, is legend among students of social evolution. In the canonical version of the tale (but see Alexander et al. 1991), one evening in a pub, J. B. S. Haldane made a few calculations on a paper napkin and then boldly announced that he would gladly sacrifice his life to save the lives of more than two brothers, more than four half-brothers, or more than eight first cousins.[3] A decade or so later, Hamilton (1964a, p. 8) defined inclusive fitness as follows:

3. The paper napkin is central to the canonical version. When Robert E. Page, Jr., brought Gro V. Amdam and me together, we first spent an afternoon sightseeing in Berlin before settling down to discuss science over beer. Amdam produced a pen and notebook, whereupon Page threw out his hands and cried, "Stop! We have to do this on paper napkins!" Inquiry revealed that, lamentably, we had selected a cloth-napkin restaurant for our libations. With a collective sigh, we settled for the notebook paper, on which Amdam sketched out the dynamics of hormones, storage proteins, and longevity that characterize different sets of honey bee workers. The insight that now is the diapause ground plan hypothesis for polistine wasps emerged a few weeks following that conversation and in direct consequence of it.

> Inclusive fitness may be imagined as the personal fitness which an individual actually expresses in its production of adult offspring as it becomes after it has been first stripped and then augmented in a certain way. It is stripped of all components which can be considered as due to the individual's social environment, leaving the fitness which he would express if not exposed to any of the harms or benefits of that environment. This quantity is then augmented by certain fractions of the quantities of harm and benefit which the individual himself causes to the fitnesses of his neighbours. The fractions in question are simply the coefficients of relationship appropriate to the neighbours whom he affects: unity for clonal individuals, one-half for sibs, one-quarter for half-sibs, one-eighth for cousins, . . . and finally zero for all neighbours whose relationship can be considered negligibly small.

This definition sets up a contrast between fitness of the focal individual if it reproduced alone versus its fitness in a social setting. Because social insect workers reproduce little or not at all, the value of their direct reproduction if alone is diminished as a cost in the social setting, and the value of their indirect reproduction through kin that they help to rear is modulated by their relatedness to those kin. Therefore the envisioned selective framework for the evolution of "reproductive altruism" (Hamilton's term for worker/helping behavior with diminished reproduction) is when benefits to the beneficiary (b), modified by relatedness between beneficiary and actor (r), exceed costs to the actor (c). This relationship is now most commonly formulated as $rb - c > 0$ or more simply as $rb > c$. This inequality is "Hamilton's rule." It has been widely invoked as an explanation of social evolution in animals.

Noonan (1981) extensively documented the reproductive success of solitary foundresses and dominant and subordinate cofoundresses of *Polistes fuscatus*, and she found support for Hamilton's rule in the case of subordinate cofoundresses in small foundress associations—if cofoundresses are sisters (chapter 7). Noonan's work preceded the technological innovations that enable kinship identification, and she used kinship estimates. Metcalf and Whitt (1977) similarly used estimates of relatedness to reach the conclusion that inclusive fitness benefits accrue to subordinate cofoundresses of *Polistes metricus*. Gibo (1978) used estimates of relatedness in a study of *P. fuscatus* in which he concluded that subordinate cofoundresses benefit from inclusive fitness only under high levels of predation on nests. Queller and Strassmann (1988) used estimates of relatedness in a study of *P. annularis* in which they concluded that 86% of subordinate cofoundresses in one year and 100% of subordinate cofoundresses in another would have had higher fitness if they had nested alone. Thus, there have been few studies of social wasps that have directly quantified inclusive fitness (Queller and Strassmann 1989), and their support for Hamilton's rule has been mixed.

Subsequent to these studies, relatedness has become the most easily measured variable in Hamilton's rule. As is now generally recognized, however, the conclusion to be drawn from numerous studies is that there is no apparent link between degree of relatedness and insect sociality (Gadagkar 1991a). Attention thus has shifted to the benefit and cost terms of Hamilton's rule. This shift of attention is exemplified by the only study of social wasps that has quantified all three variables of Hamilton's rule. In an extensive and detailed study, Raghavendra Gadagkar and his associates have documented that Hamilton's rule is satisfied in the case of a female paper wasp *Ropalidia marginata* working in an existing colony versus initiating a nest on her own (Gadagkar 2001). No other study of social insects approaches the level of detail that Gadagkar and his group achieved. At the conclusion of reporting these investigations, Gadagkar (2001, p. 326) wrote: "Perhaps the single most important point I have made is that ecological, physiological, and demographic factors can be more important in promoting the evolution of eusociality than the genetic relatedness asymmetries potentially created by haplodiploidy. Put in another way, the benefit and cost terms in Hamilton's rule deserve more attention than the relatedness term."

Beyond this, however, attention also has recently been focused on Hamilton's rule itself. It can be argued that Edward O. Wilson is singularly responsible for elevating the inclusive fitness perspective on social evolution to prominence. Wilson, however, has made a 180-degree turn. He now believes that Hamilton's general theory of inclusive fitness is flawed by "three mistakes, which have led to the vitiation of his main thesis concerning altruism and the origin of sociality" (Wilson 2005, p. 161). The first mistake is the point that was the focus of Trivers and Hare (1976)—that in haplodiploid species a worker's mean relatedness to her reproductive sisters and brothers is the same as her relatedness to her offspring. Wilson states that Trivers' predictions on sex ratio control by workers have been proved correct,[4] but the consequence of this is that the force of kin selection is dissolutive rather than binding. That is, the evidence shows that kin selection in this case works in opposition to sociality rather than in its favor. The second mistake identified by Wilson (2005) is that the proposed strong support for the general theory of inclusive fitness that was provided by the special case of haplodiploid social species has now collapsed. The discovery since 1975 of "eusociality" in taxa such as an ambrosia beetle, snapping shrimp, and naked mole-rats (Choe and Crespi 1997) gives evidence that the association between sociality and haplodiploidy is no longer statistically viable.

The third mistake identified by Wilson (2005) is the most fundamental. Wilson (2005) proposes that Hamilton's rule is logically correct but incomplete. Wilson defines altruism as "behavior that benefits others at the cost of the lifetime production of offspring by the altruist" (Wilson and Hölldobler 2005, p. 13367). He then proposes that the correct formulation for Hamilton's rule should be $(rb_k + b_e) > c$, where b_k is the fitness benefit to an actor's collateral (nondescendent) kin, as proposed by Hamilton, and where b_e is the fitness benefit that the beneficiary would derive from selection acting at the colony level independently of kinship.

4. Given the diversity of mechanisms and selective forces that can affect sex ratio, I remain skeptical.

If the latter is overwhelmingly larger than the former, and Wilson (2005) and Wilson and Hölldobler (2005) argue that this is often the case, then the former will be too small to measure. Wilson (2005) proposes that this is the apparent condition in nature, in which case the inequality reduces to $b_e > c$. In other words, the selective effect of colony-level benefit outweighs costs of individual workers' fitness decrement. These assertions have not gone unchallenged (Foster et al. 2006), but the weight of evidence now is heavily on Wilson's side. The view of social evolution in Vespidae put forward in this book falls within the conceptual framework put forward by Wilson and Hölldobler (2005), whose paper should be read by everyone who reads this book.

A much earlier paper is also worth a look. Williams and Williams (1956) developed a population genetic model for the spread of an allele for donorism, including "total sacrifice of reproductive capacity by the social donors (workers and soldiers) to the benefit of their non-donor sibs (reproductives)" (p. 36). They continue, "it might seem unreasonable to expect that genes causing a large proportion of their carriers to be sterile would ever be favored by selection within a breeding population. Nevertheless, we feel that competition between sibships provides a way for such genes to be increased in frequency. They need only provide a large between sibship advantage in proportion to the within sibship disadvantage (A >> D)" (p. 36). Williams and Williams' formula is conceptually indistinguishable from Wilson's reformulation of Hamilton's rule. It predates Hamilton's rule by 8 years and Wilson's reformulation by half a century.

The Future of Kin Selection

One reason that fostered wide support for Hamilton's rule is that it was presented as a general rule with broad explanatory power. Invocation of the general rule then passed as an explanation in specific cases. Hamilton's rule, therefore, has blunted inquiry into mechanisms that foster and maintain sociality in the diverse lineages where sociality has evolved. The evolution of sociality in Vespidae embodies one or two of these lineages (depending on monophyly, or not, of Stenogastrinae and Polistinae+Vespinae), and knowing the general rule provides scant guidance to understanding the details of their social evolution. An analogy may perhaps clarify this position.

As a former ornithologist (Hunt 1971), for a good many years I taught a course in vertebrate evolution. One topic on which I lectured and in which I maintain an interest is the evolution of flight in birds. The Berlin specimen of *Archaeopteryx* is a few kilometers from where I now am writing, and I have made a pilgrimage to see it. I now teach entomology, and in that course I deal with the evolution of insect flight. There is a general theory of flight that covers both of these cases plus pterosaurs and bats as well, yet I have never heard it invoked in any discussion of the evolution of flight. In every case, thrust exceeds drag ($T > D$) and lift exceeds weight ($L > W$). If those two inequalities aren't satisfied, no animal gets off the ground and stays aloft very long. But what insight does that give into learning the evolution of flight in the lineages where it independently evolved? The general theory does, indeed, describe conditions that must be met for flight to occur, yet knowledge about the evolution of flight is sought not by quantifying the general rule but instead by dissecting the anatomy, physiology, life history, ecology, and phylogeny of the lineages where flight independently evolved. An important step in framing the future of kin selection, therefore, is to recognize that Hamilton's rule is a general rule and is incomplete as a conceptual framework within which to pursue an understanding of social evolution.

The next step in framing the future understanding of kin selection would be to identify when and where it can play a role. The inclusive fitness concept of Hamilton (1964a, b) inspired the commonly held view that kin selection plays a key role in the evolution of sterile workers. West-Eberhard (1987b, p. 370) holds a different view: "kin selection is not necessary to explain the origin of sterile workers." Instead, she suggests, kin selection can play a role in the maintenance and regulation of the expression of worker behavior—occurring after the origin of sterility: "kin selection would thus play a role, not in the spread of 'altruistic' alleles, but in the evolution of the *regulation* of the worker phenotype once it has originated as a side effect of selection in other contexts" (p. 370; emphasis in the original). A two-phase framework of this same form to explain the origin of vertebrate sociality has been put forward by Emlen (1994), and the same view of the role of kin selection in social evolution now is embraced by Wilson (2005; Wilson and Hölldobler 2005). This view of the role of kin selection also emerges from the framework for social evolution in vespid wasps that is put forward in this book.

To envision a role for kin selection at the origin of insect sterility fosters the widely held view that selfish altruists (an intentional oxymoron; recall that the worker wasps in the introduction to part III of this book "decide" to work) are bent on maximizing their inclusive fitness: "the kin selection theory of social evolution predicts that workers should try to increase their own evolutionary success by favouring closely related individuals" (Crozier and Pamilo 1996, p. 2). A changed view that places emphasis on elaboration rather than on origin of sociality seems more consonant with Darwin's (1859) family-level selection (chapter 10) and with the two-phase views cited in the preceding paragraph. In this changed view, kin selection encapsulates the genetic architecture by which a colony (family) with a potentially adaptive variation expressed in nonreproducing family members can achieve higher fitness than competing colonies (families) without that variation. In this view, the focus of analytical attention shifts from the inclusive fitness of a worker to a focus on colony and population genetic architectures that can mediate the spread of alleles for traits expressed in nonreproductives. This perspective is not equivalent to notions of "a gene for altruism" (West-Eberhard 1988, p. 123; see also Crozier 1992, West-Eberhard 1992b) being selected and spreading through a population. Instead, this perspective focuses attention on the spread of a worker phenotype (mandible shape, in Darwin's example; see chapter 10) through a population that confers competitive advantage to the initial colony that contains individuals bearing that phenotype.

Therefore, the third step toward the future of kin selection would be to investigate it appropriately. The kin-selection/inclusive-fitness literature has been narrowly focused on intracolonial reproductive contests and an envisioned role for these contests in the origin of nonreproductive worker behavior; there has been little consideration of the very real role that kin selection can play in the change in frequency of alleles in a population of colonies (families) that exhibit heritable variations among their nonreproductive worker members. This, then, is the frontier for kin selection.

Behavioral Ecology

10

In the late 1990s a seminar course at the University of California-Davis was entitled, "Is behavioral ecology dead?" (Dugatkin 2001; R. E. Page, Jr., personal communication). The question is a sign of the times. In the realm of social insect evolutionary studies, behavioral ecology approaches are characterized by the perspective that fitness variables are the only items of interest (Reeve and Sherman 1993). However, both phylogenetic (Hunt 1999; Danforth 2002) and ontogenetic (Hunt and Amdam 2005) approaches are shedding new light on the evolution of sociality, making it clear that a fitness-units approach, by itself, is inadequate at best. Thus, practitioners of behavioral ecology, narrowly applied, now run the risk of mostly talking to one another—of preaching to the choir. Behavioral ecology can be beguiling, however, and it has attracted many students to the study of social behavior. Because students are the future of any field, in this chapter I take a look at some practices in behavioral ecology that, in my view, have more often impeded than enhanced our quest for knowledge.

Altruism

Perhaps the greatest problem facing behavioral ecology as applied to social insects is its precept that altruism is a fundamental problem in evolution. For example, a leading textbook of evolution introduces one of the better treatments of social insect evolution among textbooks with the following assertion: "Darwin (1859) recognized that social insects represent the epitome of altruism" (Freeman and Herron 2003, p. 431). This textbook certitude is wrong. The statement is, however, reflective of writings on the evolution of insect sociality in general, which often begin with an invocation of Darwin's antecedence in such concerns. In a prominent example, Crozier and Pamilo (1996, p. 2) write: "Social insects have a special place in evolutionary biology, as already noted by Darwin (1859, p. 236) long before us. The problem from the evolutionary point of view is that only some of the individuals reproduce." Often the famous passage quoted in the preface to this book is also given: "He referred to 'the neuters or sterile castes in insect-communities . . . [which] from being sterile . . . cannot propagate their kind' as 'the one special difficulty, which at first appeared to me insuperable, and actually fatal to my whole theory'" (Alexander et al. 1991, p. 3). A corollary of these invocations is the assertion that the evolution of altruism has been a long-standing, very active area of concern that is central to the theory of evolution by natural selection (Barash 1982, p. 67). These assertions, in turn, have led to "textbook knowledge" that Darwin was concerned with the evolution of altruism as it relates to the origin of insect sociality.

Review of milestones in the study of social evolution (Espinas 1877; Morgan 1900; Roubaud 1916; Wheeler 1918, 1928; Fisher 1930; Evans 1958; Michener 1958; Hamilton 1964a, b) reveals that, with few exceptions (Wheeler 1923; Evans and West Eberhard 1970), before 1971 the passage from Darwin was almost never quoted or referenced, nor was his family-level selection hypothesis invoked in explanations of the evolution of social insects. The same is true for altruism. Although Haldane (1932) had characterized behaviors that place an individual at reproductive disadvantage as "altruistic," Barash (1992, p. 14) examined numerous pre-1964 textbooks and ascertained that "in fact—and contrary to my own above-cited assertion—before Hamilton's insight, evolutionary biologists were *not* very much troubled by the occurrence of apparently altruistic behavior" (emphasis in original). However, characterizations of altruism as a long-standing problem, accompanied by citation, and often quotation, of Darwin's famous passage, have been plentiful since the mid-1970s. The early history of these references is informative.

The first quotation of the passage was apparently by Edward O. Wilson in *The Insect Societies* (Wilson 1971, p 320). By means of mercurial phraseology, Wilson finessed a need to clarify the distinction between the origin of worker sterility and the origin of worker phenotypes (see below) in his reference to Darwin: "How, he asked, could the worker caste of insect societies have evolved if they are sterile and leave no offspring?" (p. 320). In *Sociobiology* (Wilson 1975) the famous lines from Darwin and much of the 1971 text are given again, but in this case the Darwin quotation immediately follows this assertion: "the concept of kin selection to explain such behavior [altruism] was originated by Charles Darwin in *The Origin of Species*" (p. 117). Emphasis was thus placed on the evolution of altruism. Copy-cat repetitions of Wilson's 1975 assertion and the ambiguous query ("how could the worker caste have evolved?") became widespread, and Darwin is now commonly referenced as having presaged kin selection in explanation of the evolution of altruism and, so, of insect sociality. To focus a rebuttal of this view, I quote from Darwin (1859, pp. 236–241) at greater length than is usual.

I will not here enter on these several cases, but will confine myself to one special difficulty, which at first appeared to me insuperable, and actually fatal to my whole theory. I allude to the neuters or sterile females in insect-communities; for these neuters often differ widely in instinct and in structure from both the males and fertile females, and yet, from being sterile, they cannot propagate their kind.

The subject well deserves to be discussed at great length, but I will here take only a single case, that of working or sterile ants. How the workers have been rendered sterile is a difficulty; but not much greater than that of any other striking modification of structure; for it can be shown that some insects and other articulate animals in a state of nature occasionally become sterile; and if such insects had been social, and it had been profitable to the community that a number should have been annually born capable of work, but incapable of procreation, I can see no very great difficulty in this being effected by natural selection. But I must pass over this preliminary difficulty. The great difficulty lies in the working ants differing widely from both the males and the fertile females in structure, as in the shape of the thorax, and in being destitute of wings and sometimes of eyes, and in instinct. . . . If a working ant or other neuter insect had been an animal in the ordinary state, I should have unhesitatingly assumed that all its characters had been slowly acquired through natural selection; namely by an individual having been born with some slight profitable modifications, this being inherited by its offspring; which again varied and were again selected, and so onwards. But with the working ant we have an insect differing greatly from its parents, yet absolutely sterile; so that it could never have transmitted successively acquired modification of structure or instinct to its progeny. It may well be asked how is it possible to reconcile this case with the theory of natural selection?

. . . I can see no great difficulty in any character becoming correlated with the sterile condition of certain members of insect-communities: the difficulty lies in understanding how such correlated modifications of structure could have been slowly accumulated by natural selection.

This difficulty, though appearing insuperable, is lessened, or, as I believe, disappears, when it is remembered that selection may be applied to the family, as well as to the individual, and may thus gain the desired end. . . .

But we have not as yet touched on the acme of the difficulty; namely, the fact that the neuters of several ants differ, not only from the fertile females and males, but from each other, sometimes to an almost incredible degree, and are thus divided into two or even three castes. . . .

It will indeed be thought that I have an overweening confidence in the principle of natural selection, when I do not admit that such

> wonderful and well-established facts at once annihilate the theory. In the simpler case of neuter insects all of one caste, which, as I believe, have been rendered different from the fertile males and females through natural selection we may conclude from the analogy of ordinary variations, that the successive, slight, profitable modifications did not first arise in all the neuters in the same nest, but in some few alone; and by the survival of the communities with females which produced most neuters having the advantageous modifications, all the neuters ultimately came to be thus characterized. . . .
>
> With these facts before me, I believe that natural selection, by acting on the fertile ants or parents, could form a species which should regularly produce neuters, all of large size with one form of jaw, or all of small size with widely different jaws; or lastly, and this is the greatest difficulty, one set of workers of one size and structure, and simultaneously another set of workers of a different size and structure.

From a reading of Darwin's *Origin*, therefore, it is clear that the *origin* of sterile worker insects is not Darwin's "one special difficulty" and that Darwin does not characterize worker social insects as altruistic. He proposes that the origin of sterility is "not much greater than that of any other striking modification of structure," and "if such insects had been social, and it had been profitable to the community that a number should have been annually born capable of work, but incapable of procreation," he could "see no very great difficulty in this being effected by natural selection." His one special difficulty is the "modifications of structure or instinct" in sterile workers versus their reproductive nestmates. That is, Darwin's one special difficulty is not the origin of sterility but instead the evolution of morphology and/or behavior in individuals that have already lost the ability to reproduce. Thus Darwin's one special difficulty concerns not the origin of workers but, instead, their adaptive elaboration.

The importance of understanding Darwin lies in how that understanding helps shape one's view of kin selection and its role in insect social evolution. Maynard Smith (1964, p. 1145) coined the term kin selection: "by kin selection I mean the evolution of characteristics which favour the survival of close relatives of the affected individual, by processes which do not require any discontinuities in population breeding structure" (i.e., kin selection does not require group selection). This definition encompassed not only "siblings of the affected individuals (for example, sterility in social insects . . .)" (p. 1145) but also phenomena as diverse as "the evolution of placentæ and of parental care" (p. 1145). Maynard Smith's original definition might be characterized as broad-sense kin selection. For the most part, however, a Hamiltonian, narrow-sense view has dominated discourse on social evolution: "kin selection . . . is the preferential favoring of collateral relatives (i.e., not including offspring) within groups according to their degree of relatedness" (Wilson and Hölldobler 2005, p. 13369). This narrow-sense view of kin selection is widely held but rarely defined, with the consequence that "kin selection," "inclusive fitness," and "altruism" have become confusingly commingled: "altruism . . . is an unnecessary concept; indeed, it is a fog that clouds other issues" (Fletcher and Ross 1985, p. 321).

The views of social evolution held by Williams and Williams (1956), West-Eberhard (1987b, c), and Wilson (2005; Wilson and Hölldobler 2005), described in the previous chapter, all are consonant with an interpretation that can be drawn from Darwin as quoted above: insect sociality is a two-phase evolutionary process. First, sterility occurs in some individuals, but their sterility occurs initially as a consequence of factors not involving family-level selection. Darwin did not dwell on this portion of the scenario, and he did not address whether the factors might be ecological, developmental, behavioral, or some combination. The second phase of the Darwinian view, then, is that selection can act on a consort of sterile and reproductive relatives (Darwin's "family") to yield adaptations expressed only in sterile members when those adaptations enhance the reproductive success of the sterile individuals' confamilial kin. Only in this second phase does family-level selection play a role. A shift in emphasis in the study and interpretation of kin selection might begin by recognizing that Darwin's family-level selection is central to the elaboration of worker phenotypes but not to the origin of sterility, and if such selection is labeled kin selection, it should be seen as kin selection in the original, broad sense.

A shift in emphasis in the study of kin selection might also begin with a rereading of Darwin, in accord with the perspective of Ghiselin (1969, p. 232): "to learn of the facts, one reads the latest journals. To understand biology, one reads Darwin." To this, however, I must add, in the complete, original version.

Teleology

One of the greatest impediments to progress and understanding in behavioral ecology is teleology. At every turn, organisms are imbued with motivation. In the preface to a major volume on *Polistes* (Turillazzi and West-Eberhard 1996), Hamilton (1996, p. v) described the experience of watching a *Polistes* colony for an hour: "It is a world human in its seeming motivations and activities far beyond all that seems reasonable to expect from an insect: constructive activity, duty, rebellion, mother care, violence, cheating, cowardice, unity in the face of threat—all these are there." Duty? Rebellion? Cowardice? If these are the interpretations of an hour's observation, then a perspective has been constructed into which subsequent data will be expected to fit, even before they have been gathered. Teleology shapes expectations and interpretations, and it constrains discovery that might emerge from investigation framed in more neutral terms. An investigator studying "investment ratios" is narrowly constrained by preconceptions, whereas an investigator studying patterns of reproduction is less constrained. An investigator studying "worker policing" will look for only a limited range of causes and explanations, whereas broader inquiry and interpretation are open to an investigator studying oophagy. An investigator studying "altruism" in social insects will be blind to the developmental and physiological bases for caste differentiation.

I did not engage in oophagy this morning, although I had a couple of eggs for breakfast. "Oophagy" is a term reserved to science, and it has no use or implied meaning in human culture. Thus it can be studied in objective neutralism. It is a word based on Greek roots and created to describe a specific biological phenomenon. It is an accepted and standard part of the entomological lexicon. William Morton Wheeler (1928) pilloried Deegener (1918) for neologisms such as "amphoterosynhesmium" and "heterosynepileium," used to describe categories of animal associations. Wilson (1971), in turn, took Wheeler to task for such terms as "phthisogyne" and "plerergate," used to describe variants among worker ants. It is easy to dismiss such contributions as excessive, but in doing so it would be wise not to throw out the baby with the bath water. For two and a half centuries systematists have known that the best name for an organism is a scientific binomial. It can be uniquely assigned to a species, it can be imbedded into any written or spoken language, and it is unencumbered by misconceptions that often accompany common names. Greek or Latin neologisms as names for processes or phenomena can serve similar ends. Coining neologisms is out of fashion, however. The best substitute, therefore, is to use terms that are as neutrally descriptive as possible. If a term implies motivation or cause, it should be replaced.

Jacques Monod made explicit efforts to eliminate teleology from the discipline of biochemistry (Keller 1983, p. 175). Some ecologists are attempting to clean up teleology in that discipline (Fauth et al. 1996). In animal behavior, discussion of teleology becomes entangled in one of the most famous assertions in comparative psychology, Lloyd Morgan's Canon: in no case is an animal activity to be interpreted in terms of higher psychological processes, if it can be fairly interpreted in terms of processes which stand lower in the scale of psychological evolution and development (Morgan 1894). Debate about what, precisely, Morgan intended by his statement (Thomas 2001) deflects attention from the most valuable extrapolation that can be drawn from it. Teleology hinders thinking and hampers investigation, and teleological thinking can lead to intellectual satisfaction based on partial knowledge.

Labeling

Medical patients undergoing diagnosis are generally comforted if the details of their clinical state can be encapsulated with a name. To be told that one has "gastritis" is somehow more comforting than being told that you have an ache in your mid-section (which you already knew) that might be caused by any of a number of things. The superficial diagnosis gives a sense of security despite its imprecision. In science, similar false security can come from interpreting a set of observations by giving them a name. Naming is necessary, but the utility of names for phenomena, like the utility of names for organisms, depends on the application. In social insect biology, a term that leads to a great deal of false comfort is "eusocial." As was noted in the introduction to this book, I choose not to use the term. There is a reason for this choice.

The terminology most widely used to categorize insect sociality (table 10.1) was set forth by Michener (1969) and broadly disseminated by Wilson (1971). At that time, "eusocial" was applicable to the then-known social taxa: termites, ants, social bees, and social wasps. Since then the term has been applied to some gall-inhabiting aphids (Aoki 1977; Itô 1989; Benton and Foster 1992), gall-inhabiting thrips (Crespi 1992b), a beetle (Kent and Simpson 1992), a spider (Vollrath 1986), at least two species of snapping shrimp (Duffy 1996, 1998), and to a mammal, the naked mole-rat (Jarvis 1981). Debate ensued over the appropriateness of applying the term to such diversity. Gadagkar (1994) and Sherman et al. (1995) argued for relaxation of the criteria defining eusociality, so that the term might be applied to a broader range of taxa, including cooperatively breeding birds. Crespi and Yanega (1995) argued for delimitation on the principle that vague terminology can hamper understanding. The "cooperative brood care" criterion illustrates this concern. In termites and in social hymenopterans, cooperative brood care is alloparental care. Solitary hymenopterans have maternal care, and pairs of proto-termites would have had biparental care (Nalepa 1994). Cooperative brood care in these taxa thus is modified parental care. Immature aphids and thrips are not helpless, so these insects don't have parental care; therefore, cooperative brood care in the gall-inhabiting social forms can't be alloparental care. Instead, morphologically specialized immature gall-inhabiting aphids and thrips act as defenders of the domicile, thus turning brood care on its head: the brood cares for the adults rather than the other way around. This is a transparently different form of sociality, and to give it the same label as the sociality of ants and termites is both facile and misleading.

Even among the alloparental social taxa, the term "eusocial" has encountered problems. Sufficient variation exists in eusocial hymenopterans alone to necessitate modification of the term, a problem anticipated in the very first use of the term eusocial by Batra (1966). From the outset, therefore, the term was given the modifiers "primitive" and "advanced." *Polistes* exemplifies primitive

Table 10.1
Degrees of sociality and their defining criteria as presented by Wilson (1971), after Michener (1969).

	Qualities of sociality		
Degrees of sociality	Cooperative brood care	Reproductive castes[a]	Overlap between generations
Solitary, subsocial, and communal	–	–	–
Quasi/social	+	–	–
Semisocial	+	+	–
Eusocial	+	+	+

[a]"Reproductive castes" is now more commonly stated as "division of reproductive labor." The criterion means that some members of the social group reproduce, whereas others do not.

eusociality; vespines exemplify advanced eusociality. In part to move away from connotations of advanced and primitive, and in part because even those two modifiers were inadequate to capture the variety of sociality, Kukuk (1994) proposed a revised set of modifiers, such as "morphologically, permanently eusocial." Myles (in Myles and Nutting 1988) proposed the term "ultrasocial" for taxa in which there are discrete morphological castes among workers. Neither Kukuk's nor Myles' contribution has been adopted. The term "facultatively eusocial" (Crespi and Yanega 1995) is now being used to describe Stenogastrinae (Field et al. 1998), and "facultatively social" has been applied to allodapine bees (Schwarz 1994) and carpenter bees (Dunn and Richards 2003). However, until such modifiers as "facultatively" come to be broadly accepted, and especially unless a new terminology is brought forward that differentiates alloparental eusociality from garrison eusociality, it is best to avoid using the term eusocial. It is preferable to go with simply "social" and then to describe the basic life history, development, and behavior of a taxon. Characterization of sociality then lies in the description and not in a term that has been predefined. Essentially this same conclusion was reached by Wcislo (1997, p. 11), who proposed that "authors [should] state in each paper precisely and explicitly how they operationally define the behavior in question for the specific hypothesis they wish to test. This option enables an author to bring to bear evidence from all relevant taxa, without psychological constraint from pre-existing (and possibly biased) definitions."

Adaptationism

James Watson placed the successful search for the structure of DNA in context by addressing the philosophy of his Ph.D. mentor, Salvador Luria: "While [Max] Delbruck kept hoping that purely genetic tricks could solve the problem [of how genes control heredity], Luria more often wondered whether the real answer would come only after the chemical structure of a virus (gene) had been cracked open. Deep down he knew that it is impossible to describe the behavior of something when you don't know what it is" (Watson 1968, p. 23). A lament that embodies this philosophy, and that is germane to the study of social evolution, was given form by Watson's collaborator, Francis Crick (quoted in Gould 1987, p. 21): "Why do you evolutionists always try to identify the value of something before you know how it is made?" The tendency to ascribe value (adaptation) without knowing underlying mechanisms is a recognized phenomenon: "most recent developments in behavioral ecology concern predictive models about current utility; they are silent on the issue of evolutionary origin" (Emlen et al. 1991, p. 259). The facility with which mechanismless adaptive scenarios are put forward (and published!) blunts efforts directed toward discovering underlying mechanisms. This has been and remains a persistent problem in behavioral ecology, even though many authors have emphasized the need to know mechanisms in order to understand the process of evolution (Williams 1966; Sober 1984; Jamieson 1986, 1989; Crawford 1993; James 1993; Tang-Martinez 2001; Autumn et al. 2002; Shuker and West 2004).

One component of facile adaptationist explanation in behavioral ecology is particularly symptomatic of the problem at large. Western culture has been shaped by the dualism of spiritual and material worlds. The spirit/matter dualism is embodied in concepts such as heaven/earth and mind/body, and in all such contrasts the spirit analog is viewed as higher in importance. In some areas of science, theoretical and empirical seem to be viewed as a higher/lower pair. Behavioral ecology, but almost no other branch of science, has its own variation on this higher/lower theme: ultimate and proximate. Christian creationists sometimes argue that evolution is a religion. It is not. However, whenever I encounter "ultimate factors" in a publication or presentation, I am reminded of that curious variant of Christian creationism known as "God in the gaps." God-in-the-gaps creationists concede the evidence of microevolution, but when a point arises on which data are scarce or nonexistent, the absence of empirical knowledge is taken as positive evidence of the existence of God. Invocation of ultimate factors serves a similar end. It gives intellectual comfort and the appearance of explanation in cases where material cause has yet to be found. The most common invocation of ultimate factors in the field of social evolution is to say that a trait, usually interpreted as altruistic, enhances the inclusive fitness of the trait bearer.

I was stunned by an assertion in a textbook of animal behavior that favors an ultimate factors approach: "proximate causes include all those factors that are not evolutionary in nature" (Dugatkin 2004, p. 105). One is left to wonder how hormones, neurons, sinew, genes, and developmental processes come to serve the roles in behavior that they do. My guess is that scientists who couch their research and thinking in terms of ultimate causes believe that they are working on questions of greater importance than scientists who frame their work in terms of proximate mechanisms. If so, they are wrong.

Asking Why

It is important to ask questions in a manner that can lead to a useful answer. Children regularly ask "why?" Parents sometimes give answers that provide more comfort than truth. It is more soothing for a six-year-old to hear, "because Grandma has a bigger yard than we do, and the dog will be happier there" than "the dog soils the carpet and chews the furniture, and I'm sick and tired of dealing with it." Adults often ask why. Comfort, if not an answer, is sometimes sought in religion. "She left this earth when called to a higher place" is much less jolting to bereaved family and friends than "she died of breast cancer at age forty-three, and we don't know what caused, could have prevented, or might have cured it." Asking why may, indeed, be one of the traits that distinguish *Homo* from all other taxa of life. Scientists, a set of professionally curious adults, regularly ask why. It is a beginning of the road to discovery. Problems ensue, however, when evolutionary biologists attempt a direct answer to the why interrogative. I have been challenged by a colleague that "in the end we must explain why lions hunt in groups and the solitary tiger stalks its prey in stealth." I believe that this challenge is framed in the wrong terms.

One pervasive problem is that an attempt at a direct answer to the why question opens the door to story telling and metaphysics. In the nonscientific world, such answers provide justification or consolation (Salmon 1998) as characterized by the simplistic examples in the preceding paragraph. In evolutionary biology, answers at this superficial level of analysis take the form of ultimate factor explanations as briefly described in the preceding section. Nonetheless, Salmon (1998) argues that why questions are appropriate in science because they call forth explanation of a phenomenon by virtue of (1) locating and identifying its cause or causes, and (2) subsumption under laws. A line of argumentation that Salmon calls Reichenbach's principle of the common cause (Reichenbach 1956) often frames thinking about broad questions. When eusociality is said to have evolved independently numerous times, this seems to be a set of "coincidences that require common causes for their explanation" (Salmon 1998, p. 132). In the study of social evolution, however, problems arise if the common cause that is invoked is that workers work in order to maximize their inclusive fitness according to Hamilton's rule. That is, in social evolution studies, it often has been the case that subsumption under a law (Hamilton's rule) has been mistaken as having identified a common cause. By that line of reasoning, there also should be a single common cause for the evolution of flight.

It must be made clear that there may, indeed, be common cause that underlies independent evolutions of insect sociality. Worker ontogeny in insects could be regulated by independent activation in each social lineage of common ancestral mechanisms of gene expression and hormone regulation. By this same line of reasoning it seems plausible that the development of wings from tetrapod forelimbs could have involved independent ontogenetic expression of genes held in common by all tetrapods. Heterochronic expression of genes that regulate bone elongation is a clear possibility. Wings of insects versus wings of tetrapods, however, constitute the classic example of analogous versus homologous structures. If insect wings evolved due to common cause with tetrapod wings, it was a reflection of truly ancient components of the shared genome of animals. In the case of social evolution, there may well be common cause components in the ontogenies that underlie the independently evolved worker phenotypes in insects. Worker phenotypes in wasps and naked mole-rats, however, seem as unlikely to be based on a common cause as the wings of insects and vertebrates.

The idea that conformity to Hamilton's rule constitutes a common cause has been a conceptual pitfall that has inhibited appropriate research to locate and identify the selective pressures, variation, and ontogeny—including possible common cause elements in the ground plan—that underlie multiple independent origins of sociality. To avoid that pitfall, evolutionary biologists (at least those in the field of social evolution) should never attempt to directly answer the question "why?" Instead, that question should be partitioned into an open-ended array of subquestions such as what, where, when, how, how often, and so on. Theoreticians can ask what if. Evolutionary explanations will arise from the convergence of answers to these questions.

Perhaps the most fundamental change in attitude toward the value of why questions will come about if a careful but useful distinction is drawn. Answers to why questions should be thought of as interpretation, and answers to how questions should be recognized as explanation. With this distinction drawn, it is easy to see that interpretation without explanation is vacuous. At the same time, it is clear to anyone who teaches that explanation without interpretation has little intellectual appeal and only modest pedagogical value, at least for introductory students. There is a place for both explanation and interpretation in understanding social evolution. There is no place, however, for interpretation without explanation.

Stationary populations of both lions and tigers will have net reproductive rates of 1.0, but they will have different variances around that mean. Environmental variables such as habitat and prey characteristics will have played major roles in favoring different foraging traits in the two species. Traits germane to the different fitness variances can be identified and mapped onto a phylogeny of Felidae. Those traits also can be dissected in terms of gene expression, neurobiology, endocrinology, ontogeny, and how these yield context-dependent differences in foraging behaviors and social interactions. Answers to investigations in these realms will explain how the different foraging strategies yield the same population net reproductive rate from the different foraging and social patterns in different ecological environments. These answers then may be interpreted in terms of why. However, attempts at direct interpretation of a why question that lack these kinds of basic data, as illustrated by the example that introduced part III of this book, should be seen for what they are—just-so stories.

My colleague Moushumi Sen Sarma insightfully observed that "the problem is not with the question; the problem is satisfaction with the answer!" Interpretation without explanation should satisfy no one, yet attempted direct answers to why questions plague the discipline of behavioral ecology in general and analyses of social evolution in particular. Why questions do, indeed, stimulate inquiry, but they should be used appropriately. They should be used to sketch the outlines of hypotheses. Those hypotheses then can be formalized, and rigorous tests can be designed to attempt falsification of those hypotheses. Those tests should be grounded in phylogeny and/or mechanisms. Once the data are in, they will address questions of how. Only then may interpretation (perhaps!) be couched in terms of why.

The Real Difficulty

During discussion at the conference at which E. O. Wilson first met W. D. Hamilton, Hamilton (1966, p. 108) said, "The real difficulty is explaining why the juveniles develop the altruistic trait." For the ensuing 30 years, maximization of inclusive fitness was put forward as the common cause direct answer to this why question of a teleological concept. This has been the real difficulty in understanding the evolution of sociality.

A Postmodern Synthesis

11

"How did [eusociality] originate in the Hymenoptera . . . is synonymous with the question, how did a sterile worker caste evolve?" (Fletcher and Ross 1985, p. 319). In this book I have focused on a single family of insects and tried to answer that question in two different ways. Part I outlined a historical scenario that extends from the earliest forms of Order Hymenoptera through the origin of thread-waisted Apocrita and nest-building Aculeata to the progressive provisioning, paper nest building, social Vespidae. Part II presented a dynamic scenario for the origin and maintenance of sterile workers in social Vespidae, constructed from individual, colony, and population perspectives, and concluded with a specific mechanistic hypothesis for the origin of the worker caste in *Polistes*. The two scenarios are complementary and mutually reinforcing, and together they paint a clear picture of how social wasps evolved. Part III then opened with a negative assessment of the paradigm that has dominated the study of social insect evolution for 30 years—a paradigm that, if followed exclusively, would never have led to the insights and hypotheses presented here. It therefore is incumbent on me to offer an alternative paradigm.

I believe one of the most fundamental reasons that the Hamiltonian paradigm was so strongly embraced is that the concept of inclusive fitness seemingly brought social insects into the Modern Synthesis of genetics and evolution. Perhaps not surprisingly, then, the failure of the Hamiltonian paradigm is due in part to shortcomings of the Modern Synthesis itself. Natural selection, unconstrained by a focus on the inclusive fitness of individual workers, is an appropriate and adequate paradigm with which to pursue knowledge about the evolution of insect sociality. Natural selection of the Modern Synthesis, however, requires incorporation of two additional components in order to fully encompass the origin and elaboration of sociality. One of these is phenotypic plasticity, and the other is natural selection acting at the level of the colony.

Phenotypic Plasticity

In the preface to this book I highlighted a rift among students of social insect evolution who had been dichotomized as "naturalists" and "geneticists." Their contrasting approaches to the study of sociality might more appropriately be called "phenotypic" and "genotypic." Plainly, the approach taken in this book has been phenotypic. Evolution of the Modern Synthesis, however, is largely genotypic, and Modern Synthesis thinking has long held hegemony over evolutionary thought. Phenotypes have been at the periphery of evolutionary research and theory for the past half century (West-Eberhard 2003).

Plastic phenotypes are organismal forms (e.g., morphology, development, physiology, behavior) that are altered by changes in environments. An argument is now being put forward that phenotypic plasticity has played a major evolutionary role that is not addressed by the Modern Synthesis (West-Eberhard 1987b, c, 1989, 1992a, 2003). Analytical methods to investigate phenotypic evolution are being advanced (Schlichting and Pigliucci 1998). Life history evolution has emerged as a focus of theory and research, and it incorporates similar analytical perspectives (Roff 1992; Stearns 1992). At the same time, "evo-devo" approaches are linking genotypes and phenotypes. Clearly, a Postmodern Synthesis is emerging. At its core lies the interplay of development, genetics, and evolution. Evolution without development is an incomplete story. Nowhere is this truer than in the evolution of insect sociality.

The diapause ground plan hypothesis rests on a foundation of phenotypic plasticity. If it is upheld, then the key feature in wasp social evolution is not the spread of genes for altruism but, instead, the context-dependent expression of pleiotropic gene networks that produce varied phenotypes in different environments. Figure 11.1 presents such a scenario. Although the figure is drawn in general terms, it is perhaps best described by using the organisms that are the focus of this book. At the center is the ground plan of a holometabolous insect, residing deep within the biology of idiobiont parasitoidism. The generic phenotype is that of a nest-provisioning hunting wasp. Environment A includes ambient and ecological variables (climate, predators, etc.) that act on the ground plan and generic phenotype to select traits that are expressed as maternal solitary life. Environment B has different environmental variables that select for varied forms of maternal traits such as nest architecture and provisioning behavior that engender nest-centered interindividual contacts other than courtship and mating. The initial expression of these interindividual contacts would have been nothing more than fleeting social behavior, but even so social life itself immediately becomes part of an individual's environment. Positive feedback via the social environment accelerates and accentuates selection for traits of social life. Thus the evolution of social life from solitary antecedents is initiated by changes in

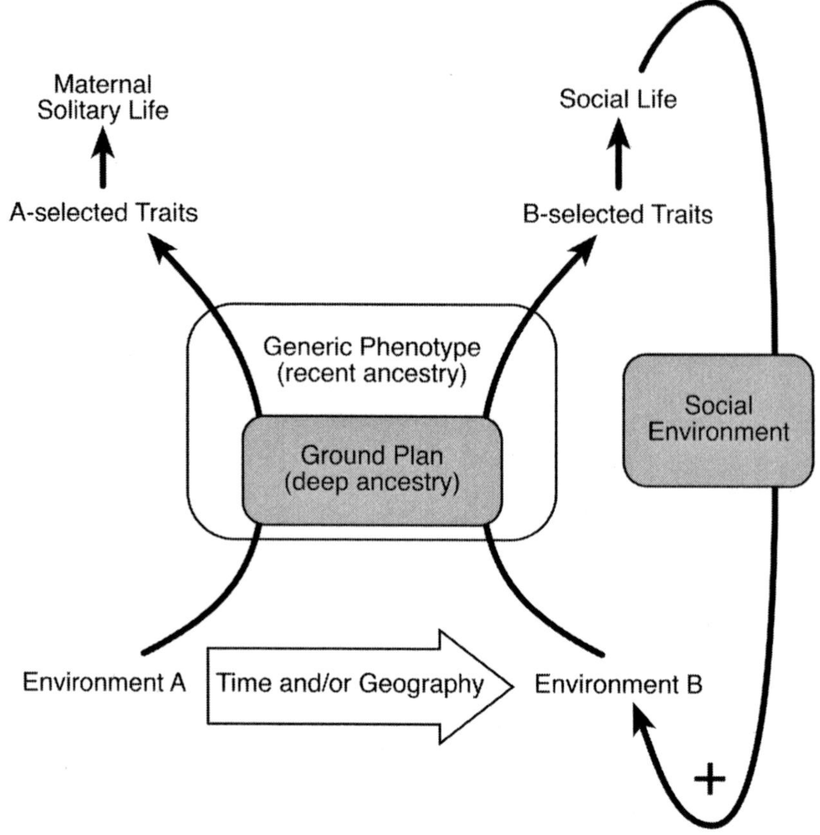

Figure 11.1
The evolution of social behavior reflects the selective agency of environmental changes affecting not merely the generic phenotype, which in this case is that of a progressively provisioning solitary wasp, but also physiology and gene expression contained within the ground plan inherited from ancestral forms. When an offspring of a species with maternal care remains, even briefly, at the maternal nest, it becomes a component of the environment for other individuals at the nest, including other larvae being reared. The social environment thereby augments, ameliorates, or in some other way modifies the ambient environment, leading to selected expression of traits reflecting components of both the ground plan and generic phenotype that differ from those of nonsocial maternal forms.

environment, either through time or as a function of geographic variables (or both), that act on the phenotype and underlying ground plan to select variables of behavior and development that have positive feedback in a social context. The model, then, is not a scenario for behavior changes per se; instead, it is a scenario of environmental changes acting as selective pressures on plastic phenotypes.

In the diapause ground plan hypothesis, genes held in common by solitary and social descendents of a shared ancestor are differentially expressed as a function of the environments in which each type of wasp lives. Genotype differences are certain to exist between and among the solitary and social forms, and these alleles would conform to rules of mutation, drift, and natural selection. However, differences in gene expression that respond to changed environments before adaptive changes in genotype are the foundation of phenotypic plasticity (West-Eberhard 2003). At the threshold of social evolution, gene expression is more important than genotype.

Colony-level Selection

When thinking beyond the threshold of social evolution, it is important to recognize that kin-selection, in the Darwin/Maynard Smith broad sense of the term, leads to expression of individual and collective attributes of social insect workers. The worker and colony-level properties constitute an extended phenotype of the genotypes of the queen and her mate(s). Selection acting via the workers of a colony on the genotypes of the reproductives is not group selection in one of the popular, pejorative conceptions of that term, nor is it narrow-sense kin selection in the popular conception of inclusive fitness maximization for the workers. Instead, it is natural selection acting, as Darwin (1859) proposed, on the family. We should acknowledge this as selection at the level of the colony, and we should think of it and study it as such.

The old concept of a superorganism (Wheeler 1928) has been recast in up-to-date evolutionary frameworks (Sober and Wilson 1998). Even so, the concept is anathema to some, perhaps in part because the superorganism label carries the stigma of the earlier simplistic characterizations and analogies. It is helpful, then, to take a broad view such as that of Buss (1987) and Maynard Smith and Szathmáry (1995, 1999). These authors highlight a major evolutionary milestone, the evolution of the individual as a biological entity. Earlier milestones, including chromosomal packaging of genomes and multicellularity, coalesced in such a way that a single genome is expressed as many cell types and tissues that collectively constitute an individual. Biologists are comfortable with the notion that individuals constitute a level of organization on which selection can operate or, more precisely, that an individual is a mosaic of phenotypic expressions of a single genotype on which selection operates collectively and simultaneously. We should strive to reach the same level of comfort with and understanding of the concept that selection in social insects acts on a mosaic of phenotypic expressions of the reproductives' genotypes. As West-Eberhard (1992a, p. 73) put it, "the castes of social insects—specialized workers and queens—are divergent but not dissociable: no one form . . . can reproduce on its own." In such a case, selection acts on the colony.

Rejection of the superorganism concept may be one impediment to thinking of colony-level selection. The contempt with which most biologists regard group selection is another. Perhaps the greatest impediment, however, is that population genetics models of the Modern Synthesis were not designed to address selection acting synchronously on multiple genotypes as a single selective unit. This must change. Wilson and Hölldobler (2005) discuss "the point of no return" in colony evolution. This is analogous to the threshold crossed in the evolution of individuality. To take an example from vespid wasps, no swarm-founding species can ever revert to solitary life. It would have the same chance of doing so as would a human leukocyte adopting independent amoeboid life in a pond. Just as population genetics can model the spread of a leukocyte-expressed major histocompatibility complex locus in a population of individual humans, so must it be able to model the spread of worker-expressed phenotypic traits in a population of swarm-founding wasp colonies—even though only some of the individuals in a given colony carry the genotype for a particular phenotypic trait. When appropriate colony-based population genetic models are constructed and data begin to accumulate, the concept of colony-level selection will become as central and integral to the study of social taxa as is the concept of individual-level selection to multicellular organisms of diverse tissue types. In the meantime, if existing population genetic models do not fully accommodate colony-level selection in social insects, we should not force the insects to conform to existing models; instead, we should craft new models to conform to the insects.

A colony of social insects is a remarkable level of evolutionary organization, the evolution of which is analogous to (and a step beyond) the evolution of multicellular individuality. We should celebrate it and study it as such.

An Action Plan for Research on Vespidae

The conceptual framework just presented is general, and so it should aid in the pursuit of understanding social evolution in any taxon, at least of insects. Details will differ in specific cases. What about the insects that are the focus of this book? In particular, how can we learn more about the evolution of sociality in Vespidae? Below I suggest three avenues.

Adopt a Population Perspective

Internecine contests on single nests have drawn attention away from where it needs to be placed—on the differential success of colonies in a population. The importance of variability in success among colonies is not a novel insight: "therefore, all indicators point to colony failure as a critical variable in the evolution of eusociality. . . . Future studies would do well to focus on colony failure in looking for advantages to eusociality" (Strassmann and Queller 1989, pp. 93–94). Few authors, including Strassmann and Queller, have done so, however. I proposed in chapter 7 that the Schaal-Leverich demographic model for annual plants is descriptive of *Polistes* populations in the temperate zone. Is it? What about other social wasp taxa, and what about any social wasp, including *Polistes*, in the tropics? Population demographic studies are an open and richly promising frontier.

In chapter 7 I proposed that the demographic pattern of placing large numbers of gynes into the quiescence pool would select for traits that foster delayed gyne production. But the summary statistics from the individual-based simulation model (Hunt and Amdam 2005; chapter 8) produced a pattern (figure 8.8, arrows) in which, if all of the colonies at that single food level are assumed to represent a model population, only a few colonies produced most of the gynes. Thus, the pattern that characterizes natural populations of *Polistes* emerged from the simulations, even though selection acting on varied genotypes was not a component of the model. This raises intriguing questions about the relationships among phenotypic plasticity, genotypically variable traits, stochastic factors that affect population demography, and the role played by each of these in determining which foundresses will be among the few that produce the largest numbers of late gynes.

Test the Diapause Ground Plan Hypothesis

Rapid progress toward understanding social evolution—in vespid wasps or in any taxon—can best be achieved if research is designed to attempt falsification of hypotheses (Platt 1964). The diapause ground plan for social wasps is currently only a hypothesis; therefore it should be the focus of attempts at falsification. Patterns of hormone activity, physiological markers such as storage protein, and differentially expressed genes can all be starting points for detailing the regulatory mechanisms in diapause and non-diapause solitary Eumeninae (or other wasps or, indeed, insects in other orders), and those same signatures should then be sought in gynes and workers of social Polistinae. Correspondence of signatures between non-diapause and diapause in solitary forms and the castes of social forms would be strong support for the hypothesis of common mechanisms underlying diapause in one case and social castes in the other. Noncorrespondence of diapause and caste signatures would be strong evidence against the hypothesis.

At the same time, predictions of the diapause ground plan hypothesis can be tested by natural history observations and experimental manipulations. Under appropriate conditions, social wasp workers of independent-founding species should show ovarian development and nest founding behavior, and gynes should not. When physiological and/or gene expression markers become available for G1/G2 signatures, those markers should coincide with observed patterns of worker/gyne emergence. At the same time, early dispersing *Polistes* should show a diapause signature, and late workers should show a nondiapause signature. The extended pupal period in putative diapause phenotype *Polistes* should be marked by increased synthesis of storage proteins. It should be possible to adjust rearing conditions and produce nondiapause versus diapause phenotypes under low versus high nourishment treatments. Such wasps should differ in response variables such as development time, storage protein, cold resistance, ovarian development, nest founding, and longevity, and the differences should coincide with nondiapause/diapause phenotype predictions. Existing natural history observations and experimental results foretell that the diapause ground plan hypothesis will be supported. Those observations and experiments now need to be repeated and expanded with nondiapause/diapause phenotypes as a framework explicitly being tested for refutation. Each of the following predictions, if unsupported by targeted investigations, could potentially falsify the diapause ground plan hypothesis:

- At the end of larval development and the onset of metamorphosis, gynes of *Polistes* will show higher production rates and higher levels of accumulation of hexameric storage proteins than workers.
- Over the course of development and pupation, gynes of *Polistes* will show hormonal signatures of diapause, and workers will show hormonal signatures of non-diapause.
- Gynes and workers of *Polistes* will differ in gene expression over the course of development. Differentially expressed genes will be linked to diapause regulation in other insects and/or to caste differentiation in advanced social hymenopterans such as Vespinae, *Apis*, or Formicidae.
- Developmental signatures of nondiapause and diapause pathways will be found in independent-founding polistine wasps in aseasonal tropics that have determinate colony cycles.
- Signatures of a diapause pathway will be absent from Stenogastrinae.

Test the Current Phylogeny

I have raised and reinforced a direct challenge to the prevailing phylogenetic hypothesis of relationships among the subfamilies of Vespidae (see chapters 4 and 8). I have hinted at intrigue involving the phylogeny of Vespinae. These should not be construed as challenges to phylogenetic systematics itself. Instead, they are calls in this specific case to expand the database and to directly test the current phylogenetic hypothesis for Vespidae. The six-subfamily hypothesis was initially based on only morphological data, and it has subsequently been supported by analyses that combine the morphology data set with molecular data. As for any hypothesis, the current phylogenetic hypothesis should be tested by attempting to falsify it. In particular, the hypothesis should be tested using new molecular data, especially from highly conserved genes that can reveal deep nodes in the clade. If the current hypothesis truly reflects evolutionary history, it will withstand such tests. If new data produce alternative phylogenetic hypotheses, then assessments of the competing hypotheses will have to be devised. Eumeninae is the largest and most diverse subfamily in the current classification of Vespidae. It holds the keys to the phylogeny of sociality in the family and, thereby, holds the keys to a full understanding of vespid social evolution.

In testing the current phylogeny, the hypotheses that (1) Eumeninae is paraphyletic with regard to social Vespidae and (2) Polistinae and Stenogastrinae do not share a recent common ancestor can be tested for rejection, using data from gene sequences and analyses conducted independently of currently existing phylogenetic data.

Conclusion

Ancestors of social Vespidae were hunting wasps with the anatomy and reproductive attributes of idiobiont parasitoid wasps. Such wasps feed as adults in order to mature a small number of large eggs laid serially over a long adult lifetime. The thread waist, which characterizes all adult Apocrita, restricts the wasp's nourishment to liquid forms. Nectar from flowers provides energy to sustain the wasp, and hemolymph of prey provides protein to synthesize her eggs.

Vespidae is one of several hymenopteran lineages in which nesting evolved as a mode of prey concealment. Larvae of the earliest vespids fed on prey in the relative safety of underground nest cells. Oviposition in a nest cell prior to provisioning enabled selection to act on sex sequence of serially laid eggs as a life history trait, and haplodiploidy as a sex determining mechanism enabled response to selection. Protogyny in all social hymenopterans is but one manifestation among many of adaptive sex determination in Hymenoptera.

Nesting behavior fostered central place foraging, and hunting for multiple small prey items to provision a single nest cell was ecologically advantageous. Nesting in natural chambers above ground led to adaptive radiation of nest architectures. Use of plastic materials, at first to modify existing chambers and then to wholly construct chambers, greatly advanced that radiation. The initial plastic material was mud, but many vespids incorporated plant products into the construction mix.

Progressive provisioning of a single larva reduced larval mortality caused by parasitoids. Wasps that mandibulated the prey items they collected thereby discovered and discarded prey containing parasitoids, further reducing losses. Mandibulating became malaxation, which destroyed parasitoids and made it unnecessary to discard prey items that contained them, and had two important consequences. Malaxation gives the adult wasp full access to the very nourishing prey hemolymph, and malaxated prey does not remain fresh and must be fed directly to larvae. Direct contact with larvae opens opportunities for reciprocal interactions between adults and larvae.

The primal polistine scraped wood fibers and mixed them with saliva to form paper nests of multiple open cells in which larvae were reared. Serially laid eggs became uneven-aged larvae, and multiple larvae were progressively provisioned simultaneously with thoroughly malaxated caterpillars. Young larvae were fed nectar and prey hemolymph; older larvae were fed caterpillar solids. Mouth-to-mouth contact between adults and larvae brought the adults into contact with larval saliva, perhaps secreted at first to moisten the caterpillar bolus from which hemolymph had been extracted. The saliva resembled nectar, and selection acted on both larvae that produce the saliva and adults that drink it to yield the copious and nutrient-rich saliva of living polistines. The saliva may appease adults that might cannibalize the larvae in times of low food availability. The saliva, even if offered in appeasement, can serve as a colony-level food reserve.

Larval saliva also is the stimulus that initiates mother–daughter colonial life, and it is the attractant that holds a colony together. The first female offspring of *Polistes* emerge disposed to enter into reproduction. However, newly emerged polistine adults remain on their natal nest for several days before flying, and during this time they encounter the saliva of the large larvae that are their nestmates. Amino acid-rich larval saliva would be attractive to a protein-hungry young idiobiont, and when the females begin to fly and forage, they return to their natal nest where the saliva is found. Behavioral interactions at the nest with adult and larval nestmates can affect these wasps' physiology, leading to expression of maternal care genes and behaviors. Foraging for prey, brood provisioning, and nest construction are behaviors that come naturally to a wasp disposed to reproduce, but the allure of larval saliva at the natal nest sets the framework in which these behaviors are transformed into alloparental care of larvae. The alloparental behaviors that these wasps perform exact a physiological cost in terms of the wasp's ability to reproduce. The wasp becomes a nonreproductive worker. Selection favored protogyny, enabled by haplodiploidy, because worker behavior is alloparental behavior, and no male hymenopteran builds nests or forages for prey with which it provisions larvae.

The primal polistine retained ground plan elements that in solitary forms can be expressed, sometimes facultatively, as two developmental pathways, one leading to emergence and reproduction in a single season and one leading to diapause and reproduction in the following season. In the primal polistine, all offspring completed development in a single season, but alloparental brood provisioning by early-emerging females leads to high levels of nourishment in later-developing female larvae. Elevated levels of larval nourishment activate the latent diapause machinery, although expression is as adult reproductive diapause rather than larval developmental diapause. Diapause-based patterning of reproductive physiology thus takes form as reproduce-now and reproduce-later phenotypes. Although some members of the reproduce-now cohort do, indeed, reproduce during that season, most become workers. None of the reproduce-later cohort wasps reproduce until the following season; they are gynes.

The ancestor of polistines may have lived in a benign tropical environment, as suggested by the paper nest. However, in an aseasonal tropical environment there would have been little selective advantage to producing a cohort of gynes. The primal polistine lived in a seasonal environment, which may have been tropical wet/dry or temperate warm/cold. In seasonal environments, high fitness accrues to queens that place large numbers of gynes into the pool of potential foundresses that can initiate nests the following season. Multiple foundresses, usurpation, and inquilinism are subsequently evolved adaptive strategies built upon a matrifilial foundation that can lead to increased probabilities of placing gynes into quiescence.

As workers bring provisions and construct the nest, their mother, now the queen, ceases almost all foraging. She feeds on larval saliva and on the hemolymph of prey foraged by her daughter workers. Her oviposition rate increases. The colony becomes an integrated unit of related individuals. Sociality in wasps reflects natural selection acting at the colony level. A colony of social wasps, as most clearly seen in (but not restricted to) swarm-founding wasps, is a polycorporal, polygenomic level of evolutionary organization from which regression to solitary life is impossible.

When the first working daughter fleetingly and imperfectly enhanced the fitness of her mother (and this was certainly only a slight enhancement), sociality had evolved.

* * *

Social life in the wasp family Vespidae evolved more than 63 million years ago. The route to sociality was not simple, and it cannot be captured in a mathematical formula or a sound-bite description. The essence of polistine sociality is the convergence of variables of nesting and individual development that resulted in the coincidental production of females disposed to reproduce yet that remained at their natal nest. The alloparental behaviors that these females performed had two consequences: the physiological costs of work diminished the workers' own capacity to reproduce, and their labor had the developmental consequence of triggering a latent diapause pathway in later-developing female larvae. This patterning of offspring into two phenotypic classes had the demographic consequence of increasing their mother's fitness as measured by foundresses that initiate the next generation. Thus a diapause ground plan of individual development in a seasonal environment was co-opted in a social context to foster a colony cycle that features vegetative growth (non-diapause workers) that can enable large numbers of propagules (diapause gynes) at the end of the favorable season. Intracolonial reproductive contests, which are the focus of inclusive fitness thinking, might have refined the system, but the principal selective regime leading to and maintaining sociality is differential reproductive success among colonies of a population. Colony-level selection acts on mother–daughter family groups to refine the components of sociality that now are expressed as an irreversible level of evolutionary organization characterized by polygenomic related colony members.

Although a great deal now is known about social evolution in insects, much remains to be learned. Chapter 11 and the "Questions Arising" sections of earlier chapters suggest some possible directions for future research. Much remains to be done in clarifying and solidifying the framework for social evolution put forward in this book—or, for that matter, any alternative that may displace it. The quest that led to my current understanding has at times been frustrating, but for the most part it has been exciting. Certainly, it has been intellectually both challenging and satisfying. Other investigators have shared this same quest, and surely they have enjoyed similar emotions. I hope that many more will join us. Curious naturalists will be richly rewarded.

References

Abe, Y., Y. Tanaka, H. Miyazaki, and Y. Y. Kawasaki. 1991. Comparative study of the composition of hornet larval saliva, its effect on behaviour and role of trophallaxis. *Comparative Biochemistry and Physiology C* 99:79–84.

Alam, S. M. 1958. Some interesting revelations about the nest of *Polistes hebroeus* Fabr. (Vespidae, Hymenoptera) – the common yellow wasp of India. *Proceedings of the Zoological Society of Calcutta* 11:113–122.

Alexander, R. D. 1974. The evolution of social behavior. *Annual Review of Ecology and Systematics* 5:325–383.

Alexander, R. D., K. M. Noonan, and B. J. Crespi. 1991. The evolution of eusociality. In *The biology of the naked mole-rat* (pp. 3–44), edited by P. W. Sherman, J. U. M. Jarvis and R. D. Alexander. Princeton, NJ: Princeton University Press.

Alexander, R. D., and P. W. Sherman. 1977. Local mate competition and parental investment in social insects. *Science* 196:494–500.

Alonso, W. J., and C. Schuck-Paim. 2002. Sex-ratio, kin selection, and the evolution of altruism. *Proceedings of the National Academy of Sciences U.S.A.* 99:6843–6847.

Amdam, G. V., A. Csondes, M. K. Fondrk, and R. E. Page, Jr. 2006. Complex social behaviour derived from maternal reproductive traits. *Nature* 439:76–78.

Amdam, G. V., K. Norberg, M. K. Fondrk, and R. E. Page, Jr. 2004. Reproductive ground plan may mediate colony-level selection effects on individual foraging behavior in honey bees. *Proceedings of the National Academy of Sciences U.S.A.* 101:11350–11355.

Amdam, G. V., K. Norberg, A. Hagen, and S. W. Omholt. 2003. Social exploitation of vitellogenin. *Proceedings of the National Academy of Sciences U.S.A.* 100:1799–1802.

Amdam, G. V., and S. W. Omholt. 2002. The regulatory anatomy of honeybee lifespan. *Journal of Theoretical Biology* 216:209–228.

Andersson, M. 1984. The evolution of eusociality. *Annual Review of Ecology and Systematics* 15:165–189.

Antonovics, J. 1987. The evolutionary dys-synthesis: which bottles for which wine? *The American Naturalist* 129:321–331.

Aoki, S. 1977. *Colophina clematis* (Homoptera, Pemphigidae), an aphid species with "soldiers". *Kontyû* 45:276–282.

Archer, M. E. 1972. The significance of worker size in the seasonal development of the wasps *Vespula vulgaris* (L.) and *Vespula germanica* (F.). *Journal of Entomology (A)* 46:175–183.

Arnold, S. J., and M. J. Wade. 1984. On the measurement of natural and sexual selection: theory. *Evolution* 38:709–719.

Askew, R. R., and M. R. Shaw. 1986. Parasitoid communities: their size, structure and development. In *Insect parasitoids* (pp. 225–264), edited by J. K. Waage and D. Greathead. London: Academic Press.

Autumn, K., M. J. Ryan, and D. B. Wake. 2002. Integrating historical and mechanistic biology enhances the study of adaptation. *The Quarterly Review of Biology* 77:383–408.

Autuori, M. 1949–50. Contribuição para o conhecimento da saúva (*Atta* spp.—Hymenoptera—Formicidae) V – Número de formas aladas e redução dos sauveiros iniciais. *Arquivos do Instituto Biologico* 19:325–331.

Baker, H. G., and I. Baker. 1968. Some anthecological aspects of the evolution of nectar-producing flowers, particularly amino acid production in nectar. In *Taxonomy and ecology* (pp. 243–264), edited by V. H. Heywood. New York: Academic Press.

Baker, H. G., and I. Baker. 1973. Amino acids in nectar and their evolutionary significance. *Nature* 241:543–545.

Baker, H. G., and P. Hurd. 1968. Intrafloral ecology. *Annual Review of Entomology* 13:385–414.

Barash, D. P. 1982. *Sociobiology and behavior*, 2nd ed. New York: Elsevier.

Barash, D. P. 1992. Anomalies in sociobiology. *Science* 256:14, 79.

Basolo, A. L. 2001. The effect of intrasexual fitness differences on genotype frequency stability at Fisherian sex ratio equilibrium. *Annales Zoologici Fennici* 38:297–304.

Batra, S. W. T. 1966. Nests and social behavior of halictine bees of India. *Indian Journal of Entomology* 28:375–393.

Beall, G. 1942. Mass movement of the wasp, *Polistes fuscatus* var. *pallipes* LeP. *The Canadian Field-Naturalist* 56:64–67.

Beani, L., M. F. Sledge, S. Maiani, F. Boscaro, M. Landi, A. Fortunato, and S. Turillazzi. 2002. Behavioral and chemical analyses of scent-marking in the lek system of a hover-wasp (Vespidae, Stenogastrinae). *Insectes Sociaux* 49:275–281.

Beani, L., and S. Turillazzi. 1999. Stripes display in hover-wasps (Vespidae: Stenogastrinae): a socially costly status badge. *Animal Behaviour* 57:1233–1239.

Belshaw, R., and D. L. J. Quicke. 2002. Robustness of ancestral state estimates: evolution of life history strategy in ichneumonoid parasitoids. *Systematic Biology* 51:450–477.

Belt, T. 1874. *The naturalist in Nicaragua*. London: John Murray.

Benton, T. G., and W. A. Foster. 1992. Altruistic housekeeping in a social aphid. *Proceedings of the Royal Society of London B* 247:199–202.

Berlese, A. 1900. Osservazioni su fenomeni che avvegnono durante la nifosi degli insetti metabolici. Parte I. Tessuto adiposo. *Revista di Patologia Vegetale* 9:177–344.

Beukeboom, L. W. 1995. Sex determination in Hymenoptera: a need for genetic and molecular studies. *BioEssays* 17:813–817.

Beye, M., M. Hasselmann, M. K. Fondrk, R. E. Page, Jr., and S. W. Omholt. 2003. The gene *csd* is the primary signal for sexual development in the honeybee and encodes an SR-type protein. *Cell* 114:419–429.

Boeve, J. L. 1991. Gregariousness, field distribution and defense in the sawfly larvae *Croesus varus* and *C. septentrionalis* (Hymenoptera, Tenthredinidae). *Oecologia* 85:440–446.

Bohart, R. M. 1989. A review of the genus *Euparagia* (Hymenoptera, Masaridae). *Journal of the Kansas Entomological Society* 62:462–467.

Bohart, R. M., F. D. Parker, and V. J. Tepedino. 1982. Notes on the biology of *Odynerus dilectus* [Hym.: Eumenidae], a predator of the alfalfa weevil, *Hypera postica* [Col.: Curculionidae]. *Entomophaga* 27:23–31.

Bohm, M. K. 1972a. Studies on reproduction in *Polistes*. Ph.D. dissertation, University of Kansas, Lawrence.

Bohm, M. K. 1972b. Effects of environment and juvenile hormone on ovaries of the wasp, *Polistes metricus*. *Journal of Insect Physiology* 18:1875–1883.

Boomsma, J. J., and A. Grafen. 1990. Intraspecific variation in ant sex ratios and the Trivers-Hare hypothesis. *Evolution* 44:1026–1034.

Boomsma, J. J., and A. Grafen. 1991. Colony-level sex-ratio selection in the eusocial Hymenoptera. *Journal of Evolutionary Biology* 4:383–407.

Bourke, A. F. G., and N. R. Franks. 1995. *Social evolution in ants*. Princeton, NJ: Princeton University Press.

Bouwma, P. E., A. M. Bouwma, and R. L. Jeanne. 2000. Social wasp swarm emigration: males stay behind. *Ethology, Ecology & Evolution* 12:35–42.

Brauns, H. 1910. Biologisches über südafrikanische Hymenopteren. *Zeitschrift für wissenschaftliche Insektenbiologie* 6:348–387, 445–447.

Bredekamp, H. 2002. Darwins Evolutionsdiagramm oder: Brauchen Bilder Gedanken? In *Grenzen und Grenzüberschreitungen: XIX. Deutscher Kongress fur Philosophie* (pp. 863–877), edited by W. Hogrebe and J. Bromand. Bonn: Sinclair Press.

Breed, M. D., and G. J. Gamboa. 1977. Behavioral control of workers by queens in primitively eusocial bees. *Science* 195:694–696.

Brian, M. V. 1965. Caste differentiation in social insects. In *Social Organization of Animal Communities* (pp. 13–38), edited by P. E. Ellis. *Symposia of the Zoological Society of London, no. 14.*

Brian, M. V., and A. D. Brian. 1952. The wasp *Vespula sylvestris* Scopoli, feeding, foraging and colony development. *Transactions of the Royal Entomological Society of London* 103:1–26.

Brian, M. V., G. Elmes, and A. F. Kelly. 1967. Populations of the ant *Tetramorium caespitum* Latreille. *Journal of Animal Ecology* 36:337–342.

Brimley, C. S. 1908. Male *Polistes annularis* survive the winter. *Entomological News* 8:107.

Brockmann, H. J. 1984. The evolution of social behaviour in insects. In *Behavioural ecology: An evolutionary approach*, 2nd ed. (pp. 340–361), edited by J. R. Krebs and N. B. Davies. Sunderland: Sinauer Associates.

Brockmann, H. J. 2004. Variable life-history and emergence patterns of the pipe-organ mud-daubing wasp, *Trypoxylon politum* (Hymenoptera: Sphecidae). *Journal of the Kansas Entomological Society* 77:503–527.

Brockmann, H. J., and A. Grafen. 1992. Sex ratios and life-history patterns of a solitary wasp, *Trypoxylon* (*Trypargilum*) *politum* (Hymenoptera: Sphecidae). *Behavioral Ecology and Sociobiology* 30:7–27.

Brooke, M. de L. 1981. The nesting biology and population dynamics of the Seychelles potter wasp *Eumenes alluaudi* Perez. *Ecological Entomology* 6:365–377.

Brooks, D. R., D. A. McLennan, J. M. Carpenter, S. G. Weller, and J. A. Coddington. 1995. Systematics, ecology, and behavior: integrating phylogenetic patterns and evolutionary mechanisms. *BioScience* 45:687–695.

Brothers, D. J. 1995. The vespoid families (except vespids and ants): Introduction. In *The Hymenoptera of Costa Rica* (pp. 504–512), edited by P. E. Hanson and I. D. Gauld. Oxford: Oxford University Press.

Brothers, D. J. 1999. Phylogeny and evolution of wasps, ants and bees (Hymenoptera, Chrysidoidea, Vespoidea and Apoidea). *Zoologica Scripta* 28:233–249.

Brothers, D. J., and J. M. Carpenter. 1993. Phylogeny of the Aculeata: Chrysidoidea and Vespoidea (Hymenoptera). *Journal of Hymenoptera Research* 2:227–304.

Brown, W. D., and L. Keller. 2000. Colony sex ratios vary with queen number but not relatedness asymmetry in the ant *Formica exsecta*. *Proceedings of the Royal Society of London B* 267:1751–1757.

Brunson, M. H. 1937. Influence of the instars of the host larvae on sex of *Tiphia popilliavora* Roh. *Science* 86:197.

Brunson, M. H. 1938. Influence of Japanese beetle instar on the sex and population of the parasite *Tiphia popilliavora*. *Journal of Agricultural Research* 57:379–386.

Bursell, E. 1963. Aspects of the metabolism of amino acids in the tsetse fly, *Glossinia* (Diptera). *Journal of Insect Physiology* 9:439–452.

Buss, L. W. 1987. *The evolution of individuality*. Princeton, NJ: Princeton University Press.

Buysson, R. du. 1903. Monographie des guêpes ou *Vespa*. *Annales de la Société Entomologique de France* 72:260–288.

Cagniant, H. 1979. La parthenogenese thelytoque et arrhenotoque chez la fourmi *Cataglyphis cursor* Fonsc. (Hym. Form.). Cycle biologique en elevage des colonies avec reine et des colonies sans reine. *Insectes Sociaux* 26:51–60.

Calderone, N. W., and R. E. Page. 1987. Evolutionary genetics of division of labor in colonies of the honey bee (*Apis mellifera*). *The American Naturalist* 138:69–92.

Cameron, S. A. 1985. Brood care by male bumble bees. *Proceedings of the National Academy of Sciences U.S.A.* 82:6371–6373.

Cameron, S. A. 1986. Brood care by males of *Polistes major* (Hymenoptera: Vespidae). *Journal of the Kansas Entomological Society* 59:183–185.

Camillo, E. 1999. A solitary mud-daubing wasp, *Brachymenes dyscherus* (Hymenoptera: Vespidae) fron [sic] Brazil with evidence of a life-cycle polyphenism. *Revista de Biología Tropical* 47:949–958.

Camillo, E. 2002. The natural history of the mud-dauber wasp *Sceliphron fistularium* (Hymenoptera: Sphecidae) in southeastern Brazil. *Revista de Biología Tropical* 50:127–134.

Carlin, N. F., H. K. Reeve, and S. P. Cover. 1993. Kin discrimination and division of labour among matrilines in the polygynous carpenter ant, *Camponotus planatus*. In *Queen number and sociality in insects* (pp. 362–401), edited by L. Keller. Oxford: Oxford University Press.

Carpenter, J. M. 1982. The phylogenetic relationships and natural classification of the Vespoidea (Hymenoptera). *Systematic Entomology* 7:11–38.

Carpenter, J. M. 1987. Phylogenetic relationships and classification of the Vespinae (Hymenoptera, Vespidae). *Systematic Entomology* 12:413–431.

Carpenter, J. M. 1988. The phylogenetic system of the Stenogastrinae (Hymenoptera, Vespidae). *Journal of the New York Entomological Society* 96:140–175.

Carpenter, J. M. 1991. Phylogenetic relationships and the origin of social behavior in the Vespidae. In *The social biology of wasps* (pp. 7–32), edited by K. G. Ross and R. W. Matthews. Ithaca, NY: Cornell University Press.

Carpenter, J. M. 1996a. Distributional checklist of species of the genus *Polistes* (Hymenoptera: Vespidae; Polistinae, Polistini). *American Museum Novitates* 3188:1–39.

Carpenter, J. M. 1996b. Phylogeny and biogeography of *Polistes*. In *Natural history and evolution of paper-wasps* (pp. 18–57), edited by S. Turillazzi and M. J. West-Eberhard. Oxford: Oxford University Press.

Carpenter, J. M. 2001. New generic synonymy in Stenogastrinae (Insecta: Hymenoptera; Vespidae). *Natural History Bulletin of Ibaraki University* 5:27–30.

Carpenter, J. M. 2003. On "Molecular phylogeny of Vespidae (Hymenoptera) and the evolution of sociality in wasps". *American Museum Novitates* 3389:1–20.

Carpenter, J. M. 2004. Synonymy of the genus *Marimbonda* Richards, 1978, with *Leipomeles* Möbius, 1856 (Hymenoptera: Vespidae; Polistinae), and a new key to the genera of paper wasps of the New World. *American Museum Novitates* 3465:1–16.

Carpenter, J. M., and J. M. Cumming. 1985. A character analysis of the North American potter wasps (Hymenoptera: Vespidae: Eumeninae). *Journal of Natural History* 19:877–916.

Carpenter, J. M., and B. R. Garcete-Barrett. 2002. A key to the neotropical genera of Eumeninae (Hymenoptera: Vespidae). *Boletin del Museo Nacional de Historia Natural de Paraguay* 14:52–73.

Carpenter, J. M., and J. I. Kojima. 1996. Checklist of the species in the subfamily Stenogastrinae (Hymenoptera: Vespidae). *Journal of the New York Entomological Society* 104:21–36.

Carpenter, J. M., and S. Mateus. 2004. Males of *Nectarinella* Bequaert (Hymenoptera, Vespidae, Polistinae). *Revista Brasiliera de Entomologia* 48:297–302.

Carpenter, J. M., and A. P. Rasnitsyn. 1990. Mesozoic Vespidae. *Psyche* 97:1–20.

Carpenter, J. M., and C. K. Starr. 2000. A new genus of hover wasps from Southeast Asia (Hymenoptera: Vespidae; Stenogastrinae). *American Museum Novitates* 3291:1–12.

Carpenter, J. M., J. W. Wenzel, and J. I. Kojima. 1996. Synonymy of the genera *Occipitalia* Richards, 1978, with *Clypearia* de Saussure, 1854, (Hymenoptera: Vespidae; Polistinae, Epiponini). *Journal of Hymenoptera Research* 5:157–165.

Cervo, R., and F. R. Dani. 1996. Social parasitism and its evolution in *Polistes*. In *Natural history and evolution of paper-wasps* (pp. 98–112), edited by S. Turillazzi and M. J. West-Eberhard. Oxford: Oxford University Press.

Cervo, R., V. Macinai, F. Dechigi, and S. Turillazzi. 2004. Fast growth of immature brood in a social parasite wasp: a convergent evolution between avian and insect cuckoos. *The American Naturalist* 164:814–820.

Cervo, R., and S. Turillazzi. 1985. Associative foundation and nesting sites in *Polistes nimpha*. *Naturwissenschaften* 72:48–49.

Chao, J. T., and H. R. Hermann. 1983. Spinning and external ontogenetic changes in the pupae of *Polistes annularis* (Hymenoptera, Vespidae, Polistinae). *Insectes Sociaux* 30:496–507.

Chapman, R. E., and A. F. G. Bourke. 2001. The influence of sociality on the conservation biology of social insects. *Ecology Letters* 4:650–662.

Chapman, R. F. 1998. *The insects: Structure and function*, 4th ed. Cambridge: Cambridge University Press.

Chapman, T. W., and S. C. Stewart. 1996. Extremely high levels of inbreeding in a natural population of the free-living wasp *Ancistrocerus antilope* (Hymenoptera; Vespidae: Eumeninae). *Heredity* 76:65–69.

Charlesworth, B. 1978. Some models of the evolution of altruistic behaviour between siblings. *Journal of Theoretical Biology* 72:297–319.

Charnov, E. L. 1978. Evolution of eusocial behavior: offspring choice or parental parasitism? *Journal of Theoretical Biology* 75:451–465.

Chen, B., T. Kayukawa, H. Jiang, A. Monteiro, S. Hoshizaki, and Y. Ishikawa. 2005. *DaTrypsin*, a novel clip-domain serine proteinase gene up-regulated during winter and summer diapauses of the onion maggot, *Delia antiqua*. *Gene* 347:115–123.

Choe, J. C., and B. J. Crespi, editors. 1997. *The evolution of social behavior in insects and arachnids*. Cambridge: Cambridge University Press.

Claude-Joseph, F. (H. Janvier). 1930. Recherches biologiques sur les prédateurs du Chile. *Annales des Sciences Naturelles, 10e série: Zoologie* 13:235–354.

Clement, S. L., and E. E. Grissell. 1968. Observations of the nesting habits of *Euparagia scutellaris* Cresson. *The Pan-Pacific Entomologist* 44:34–37.

Codella, S. G., and K. F. Raffa. 1995. Contributions of female oviposition patterns and larval behavior to group defense in conifer sawflies (Hymenoptera, Diprionidae). *Oecologia* 103:24–33.

Collins, A. M. 2004. Variation in time of egg hatch by the honey bee, *Apis mellifera* (Hymenoptera: Apidae). *Annals of the Entomological Society of America* 97:140–146.

Cook, J. M. 1993. Sex determination in the Hymenoptera: a review of models and evidence. *Heredity* 71:421–435.

Cook, J. M., and R. H. Crozier. 1995. Sex determination and population biology in the Hymenoptera. *Trends in Ecology and Evolution* 10:281–286.

Cooper, K. W. 1957. Biology of eumenine wasps. V. Digital communication in wasps. *Journal of Experimental Zoology* 134:469–514.

Cooper, K. W. 1966. Ruptor ovi, the number of moults in development, and method of exit from masoned nests. Biology of eumenine wasps, VII. *Psyche* 73:238–250.

Cooper, K. W. 1979. Plasticity in nesting behavior of a renting wasp, and its evolutionary implications. Studies on eumenine wasps, VIII (Hymenoptera, Aculeata). *Journal of the Washington Academy of Science* 69:151–158.

Corona, M., E. Estrada, and M. Zurita. 1999. Differential expression of mitochondrial genes between queens and workers during caste determination in the honeybee *Apis mellifera*. *The Journal of Experimental Biology* 202:929–938.

Costa, J. T., and R. W. Louque. 2001. Group foraging and trail following behavior of the red-headed pine sawfly *Neodiprion lecontei* (Fitch) (Hymenoptera : Symphyta : Diprionidae). *Annals of the Entomological Society of America* 94:480–489.

Coster-Longman, C., M. Landi, and S. Turillazzi. 2002. The role of passive defense (selfish herd and dilution effect) in the gregarious nesting of *Liostenogaster* wasps (Vespidae, Hymenoptera, Stenogastrinae). *Journal of Insect Behavior* 15:331–350.

Coster-Longman, C., and S. Turillazzi. 1995. Nest architecture in *Parischnogaster alternata* Sakagami (Vespidae, Stenogastrinae), intra-specific variability in building strategies. *Insectes Sociaux* 42:1–16.

Cowan, D. P. 1979. Sibling mating in a hunting wasp: adaptive inbreeding? *Science* 204:1403–1405.

Cowan, D. P. 1981. Parental investment in two solitary wasps *Ancistrocerus adiabatus* and *Euodynerus foraminatus* (Eumenidae: Hymenoptera). *Behavioral Ecology and Sociobiology* 9:95–102.

Cowan, D. P. 1986. Sexual behavior of eumenid wasps (Hymenoptera: Eumenidae). *Proceedings of the Entomological Society of Washington* 88:531–541.

Cowan, D. P. 1991. The solitary and presocial Vespidae. In *The social biology of wasps* (pp. 33–73), edited by K. G. Ross and R. W. Matthews. Ithaca, NY: Cornell University Press.

Cowan, D. P., and G. P. Waldbauer. 1984. Seasonal occurrence and mating at flowers by

Ancistrocerus antilope (Hymenoptera: Eumenidae). *Proceedings of the Entomological Society of Washington* 86:930–934.

Craig, R. 1979. Parental manipulation, kin selection, and the evolution of altruism. *Evolution* 33:319–334.

Craig, R. 1980. Sex ratio changes and the evolution of eusociality in the Hymenoptera: simulation and games theory studies. *Journal of Theoretical Biology* 87:55–70.

Craig, R. 1982a. Evolution of male workers in the Hymenoptera. *Journal of Theoretical Biology* 94:95–105.

Craig, R. 1982b. Evolution of eusociality by kin selection: the effect of inbreeding between siblings. *Journal of Theoretical Biology* 94:119–128.

Craig, S. F., L. B. Slobodkin, G. A. Wray, and C. H. Biermann. 1997. The 'paradox' of polyembryony: A review of the cases and a hypothesis for its evolution. *Evolutionary Ecology* 11:127–143.

Crawford, C. B. 1993. The future of sociobiology: counting babies or studying proximate mechanisms. *Trends in Ecology and Evolution* 8:183–186.

Cremonez, T. A., D. De Jong, and M. M. Bittondi. 1998. Quantification of hemolymph proteins as a fast method for testing protein diets for honeybees. *Journal of Economic Entomology* 91:1284–1289.

Crespi, B. J. 1992a. Cannibalism and trophic eggs in subsocial and eusocial insects. In *Cannibalism: Ecology and evolution among diverse taxa* (pp. 176–213), edited by M. A. Elgar and B. J. Crespi. Oxford: Oxford University Press.

Crespi, B. J. 1992b. Eusociality in Australian gall thrips. *Nature* 359:724–726.

Crespi, B. J., and L. A. Mound. 1997. Ecology and evolution of social behavior among Australian gall thrips and their allies. In: *Social behavior in insects and arachnids* (pp. 166–180), edited by J. C. Choe and B. J. Crespi. Cambridge: Cambridge University Press.

Crespi, B. J., and D. Yanega. 1995. The definition of eusociality. *Behavioral Ecology* 6:109–115.

Crosland, M. W. J., and R. H. Crozier. 1986. *Myrmecia pilosula*, an ant with only one pair of chromosomes. *Science* 231:1278.

Crozier, R. H. 1982. On insects and insects: twists and turns in our understanding of the evolution of eusociality. In *The biology of social insects* (pp. 4–9), edited by M. D. Breed, C. D. Michener and H. E. Evans. Boulder: Westview Press.

Crozier, R. H. 1992. The genetic evolution of flexible strategies. *The American Naturalist* 139:218–223.

Crozier, R. H., and P. Pamilo. 1993. Sex allocation in social insects: problems in prediction and estimation. In *Evolution and diversity of sex ratio in insects and mites* (pp. 369–383), edited by D. L. Wrensch and M. A. Ebbert. New York: Chapman and Hall.

Crozier, R. H., and P. Pamilo. 1996. *Evolution of social insect colonies: Sex allocation and kin selection*. Oxford: Oxford University Press.

Cruz, Y. P. 1981. A sterile defender morph in a polyembryonic hymenopterous parasite. *Nature* 294:446–447.

Danforth, B. N. 2002. Evolution of sociality in a primitively eusocial lineage of bees. *Proceedings of the National Academy of Sciences U.S.A.* 99:286–290.

Danforth, B. N., L. Conway, and S. Q. Ji. 2003. Phylogeny of eusocial *Lasioglossum* reveals multiple losses of eusociality within a primitively eusocial clade of bees (Hymenoptera : Halictidae). *Systematic Biology* 52:23–36.

Dani, F. R., and R. Cervo. 1992. Reproductive strategies following nest loss in *Polistes gallicus* (L.) (Hymenoptera Vespidae). *Ethology, Ecology & Evolution* Special Issue 2:49–53.

Dapporto, L., E. Palagi, and S. Turillazzi. 2005. Sociality outside the nest: helpers in pre-hibernating clusters of *Polistes dominulus*. *Annales Zoologici Fennici* 42:135–139.

Darwin, C. R. 1859. *On the origin of species by means of natural selection, or the preservation of favoured races in the struggle for life*. London: John Murray.

Deegener, P. 1918. *Die Formen der Vergesellschaftung im Tierreiche*. Leipzig: Veit & Co.

Dejean, A., E. Francescato, and S. Turillazzi. 1994. Food sources and alimentary behaviour of *Polybioides tabidus* (Vespidae, Polistinae). *Journal of African Zoology* 108:251–260.

Deleurance, É.-P. 1949. Sur le déterminisme de l'apparition des ouvrières et des fondatrices-filles chez les *Polistes* (Hyménoptères: Vespides). *Comptes rendus hebdomadaires des Séances de l'Académie des Sciences* Paris 229:303–304.

Deleurance, É.-P. 1952 Le polymorphisme sociale et son déterminisme chez les guêpes. *Colloques internationaux du Centre National de la Recherche scientifique* 34:141–155.

Deleurance, É.-P. 1955. Contribution à l'étude biologique des *Polistes* (Hyménoptères – Vespides). II. Le cycle évolutif du couvain. *Insectes Sociaux* 2:285–302.

Deleurance, É.-P. 1956. Analyse du comportement bâtisseur chez "Polistes" (Hyménoptères vespides). L'activité bâtisseuse d'origine "interne". In *L'instinct dans le comportement des animaux et de l'homme*. Paris: Masson.

DeMarco, B. B. 1982. Population studies of a paper wasp, *Polistes metricus*. M.S. thesis, University of Missouri-St. Louis, St. Louis.

Denlinger, D. L. 2002. Regulation of diapause. *Annual Review of Entomology* 47:93–122.

Deslippe, R. J., and R. Savolainen. 1995. Sex investment in a social insect: the proximate role of food. *Ecology* 76:375–382.

Dew, H. E., and C. D. Michener. 1981. Division of labor among workers of *Polistes metricus* (Hymenoptera: Vespidae): laboratory foraging activities. *Insectes Sociaux* 28:87–101.

Dow, R. 1942. The relation of the prey of *Sphecius speciosus* to the size and sex of the adult wasp (Hym.: Sphecidae). *Annals of the Entomological Society of America* 35:310–317.

Downing, H. A. 1991. The function and evolution of exocrine glands. In *The social biology of wasps* (pp. 540–569), edited by K. G. Ross and R. W. Matthews. Ithaca, NY: Cornell University Press.

Downing, H. A. 2004. Effect of mated condition on dominance interactions and nesting behavior in the social wasp, *Polistes fuscatus* (Hymenoptera: Vespidae). *Journal of the Kansas Entomological Society* 77:288–291.

Downing, H. A., and R. L. Jeanne. 1983. Correlation of season and dominance status with activity of exocrine glands in *Polistes fuscatus* (Hymenoptera: Vespidae). *Journal of the Kansas Entomological Society* 56:387–397.

Downing, H. A., and R. L. Jeanne. 1987. A comparison of nest construction behavior in two species of *Polistes* paper wasps (Insecta, Hymenoptera: Vespidae). *Journal of Ethology* 5:53–66.

Downing, H. A., and R. L. Jeanne. 1990. The regulation of complex building behaviour in the paper wasp, *Polistes fuscatus* (Insecta, Hymenoptera, Vespidae). *Animal Behaviour* 39:105–124.

Dowton, M., and A. D. Austin. 1994. Molecular phylogeny of the insect order Hymenoptera: apocritan relationships. *Proceedings of the National Academy of Sciences U.S.A.* 91:9911–9915.

Dowton, M., and A. D. Austin. 2001. Simultaneous analysis of 16S, 28S, COI and morphology in the Hymenoptera: Apocrita—evolutionary transitions among parasitic wasps. *Biological Journal of the Linnean Society* 74:87–111.

Duffy, J. E. 1996. Eusociality in a coral-reef shrimp. *Nature* 381:512–514.

Duffy, J. E. 1998. On the frequency of eusociality in snapping shrimps (Decapoda: Alpheidae), with description of a second eusocial species. *Bulletin of Marine Science* 63:387–400.

Dugatkin, L. A., editor. 2001. *Model systems in behavioral ecology: Integrating conceptual, theoretical, and empirical approaches*. Princeton, NJ: Princeton University Press.

Dugatkin, L. A. 2004. *Principles of animal behavior*. New York: W. W. Norton & Co.

Duncan, C. D. 1928. Plant hairs as building material for *Polistes* (Hymenoptera, Vespidæ). *The Pan-Pacific Entomologist* 5:90.

Duncan, C. D. 1939. A contribution to the biology of North American vespine wasps. *Stanford University Publications (University Series, Biological Sciences)* 8:1–272.

Dunn, T., and M. H. Richards. 2003. When to bee social: interactions among environmental constraints, incentives, guarding, and relatedness in a facultatively social carpenter bee. *Behavioral Ecology* 14:417–424.

Dzierzon, J. 1845. Gutachten über die von Herrn Direktor Stöhr im ersten und zweiten Kapitel des General-Gutachtens aufgestellten Fragen. *Eichstädter Bienenzeitung* 1:109–113, 119–121.

Eberhard, W. G. 1974. The natural history and behaviour of the wasp *Trigonopsis cameronii* Kohl (Sphecidae). *Transactions of the Royal Entomological Society, London* 125:295–328.

Edwards, R. 1980. *Social wasps: Their biology and control*. East Grinstead: Rentokill Limited.

Eggleton, P., and R. Belshaw. 1992. Insect parasitoids: an evolutionary overview. *Philosophical Transactions of the Royal Society of London B* 337:1–20.

Eickwort, G. C. 1981. Presocial insects. In *Social insects* (pp. 199–280), edited by H. R. Hermann. New York: Academic Press.

Eickwort, K. R. 1969a. Differential variation of males and females in *Polistes exclamans*. *Evolution* 23:391–405.

Eickwort, K. 1969b. Separation of the castes of *Polistes exclamans* and notes on its biology (Hym.: Vespidae). *Insectes Sociaux* 16:67–72.

Elmes, G. W. 1974. Colony populations of *Myrmica sulcinodis* Nyl. (Hym. Formicidae). *Oecologia* 15:337–343.

Elmes, G. W., and J. C. Wardlaw. 1982. A population study of the ants *Myrmica sabuleti* and *Myrmica scabrinodis*, living at two sites in the south of England. I. A comparison of colony populations. *Journal of Animal Ecology* 51:651–664.

Emlen, S. T. 1994. Benefits, constraints and the evolution of the family. *Trends in Ecology and Evolution* 9:282–285.

Emlen, S. T., H. K. Reeve, P. W. Sherman, P. H. Wrege, F. L. W. Ratnieks, and J. Shellman-Reeve. 1991. Adaptive versus nonadaptive explanations of behavior: the case of alloparental helping. *The American Naturalist* 138:259–270.

Engels, W., and H. Fahrenhorst. 1974. Alters- und kastenspezifische Veeränderungen ser Haemolymph-Protein-Spectren bei *Apis mellifera*. *Wilhelm Roux' Archiv* 174:285–296.

Espelie, K. E., and D. S. Himmelsbach. 1990. Characterization of pedicel, paper, and larval silk from nest of *Polistes annularis*. *Journal of Chemical Ecology* 16:3467–3477.

Espinas, A. 1877. *Les sociétés animales: Etude de psychologie comparée*. New York: G. E. Stechert & Co. [reprint 1924].

Evans, H. E. 1956. Notes on the biology of four species of ground-nesting Vespidae (Hymenoptera). *Proceedings of the Entomological Society of Washington* 58:265–270.

Evans, H. E. 1958. The evolution of social life in wasps. *Proceedings of the 10th International Congress of Entomology (Montreal, 1956)* 2:449–457.

Evans, H. E. 1970. Ecological-behavioral studies of the wasps of Jackson Hole, Wyoming. *Bulletin of the Museum of Comparative Zoology, Harvard* 140:451–511.

Evans, H. E. 1973. Notes on the nests of *Montezumia* (Hymenoptera, Eumenidae). *Entomological News* 84:285–290.

Evans, H. E.1977a. Observations on the nests and prey of eumenid wasps (Hymenoptera, Eumenidae). *Psyche* 83:255–259.

Evans, H. E.1977b. Extrinsic versus intrinsic factors in the evolution of insect sociality. *BioScience* 27:613–617.

Evans, H. E., and R. W. Matthews. 1974. Notes on nests and prey of two species of ground-nesting Eumenidae from So. America (Hymenoptera). *Entomological News* 85:149–153.

Evans, H. E., and M. J. West Eberhard. 1970. *The wasps*. Ann Arbor: University of Michigan Press.

Evans, J. D., D. C. A. Shearman, and B. P. Oldroyd. 2004. Molecular basis of sex determination in haplodiploids. *Trends in Ecology and Evolution* 19:1–3.

Evans, J. D., and D. E. Wheeler. 1999. Differential gene expression between developing queens and workers in the honey bee, *Apis mellifera. Proceedings of the National Academy of Sciences U.S.A.* 96:5575–5580.

Fauth, J. E., J. Bernardo, M. Camara, W. J. Resetarits, Jr., J. Van Buskirk, and S. A. McCollum. 1996. Simplifying the jargon of community ecology: a conceptual approach. *The American Naturalist* 147:282–286.

Ferguson, C. S., and J. H. Hunt. 1989. Near-nest behavior of a solitary mud-daubing wasp, *Sceliphron caementarium* (Hymenoptera: Sphecidae). *Journal of Insect Behavior* 2:315–323.

Field, J., and W. Foster. 1999. Helping behaviour in facultatively eusocial hover wasps: an experimental test of the subfertility hypothesis. *Animal Behaviour* 57:633–636.

Field, J., W. Foster, G. Shreeves, and S. Sumner. 1998. Ecological constraints on independent nesting in facultatively eusocial hover wasps. *Proceedings of the Royal Society of London B* 265:973–977.

Field, J., G. Shreeves, and S. Sumner. 1999. Group size, queuing and helping decisions in facultatively eusocial hover wasps. *Behavioral Ecology and Sociobiology* 45:378–385.

Fisher, R. A. 1930. *The genetical theory of natural selection*. New York: Dover [reprint, 1958].

Flanders, S. E. 1946. Haploidy as a factor in the polymorphic differentiation of the Hymenoptera. *Science* 103:555–556.

Flanders, S. E. 1950. Regulation of ovulation and egg disposal in the parasitic Hymenoptera. *Canadian Entomologist* 82:134–140.

Flannagan, R. D., S. P. Tammariello, K. H. Joplin, R. A. Cikra-Ireland, G. D. Yocum, and D. L. Denlinger. 1998. Diapause-specific gene expression in pupae of the flesh fly *Sarcophaga crassipalpis*. *Proceedings of the National Academy of Sciences U.S.A.* 95:5616–5620.

Fletcher, D. J. C., and K. G. Ross. 1985. Regulation of reproduction in eusocial Hymenoptera. *Annual Review of Entomology* 30:319–343.

Flowers, R. W., and J. T. Costa. 2003. Larval communication and group foraging dynamics in the red-headed pine sawfly, *Neodiprion lecontei* (Fitch) (Hymenoptera: Symphyta: Diprionidae). *Annals of the Entomological Society of America* 96:336–343.

Fong, D. W., T. C. Kane, and J. C. Culver. 1995. Vestigialization and loss of nonfunctional characters. *Annual Review of Ecology and Systematics* 26:249–268.

Forel, A. 1874. *Les fourmis de la Suisse*. Zurich: Imprimerie Zurcher & Furrer.

Foster, K. R., T. Wenseleers, and F. L. W. Ratnieks. 2006. Kin selection is the key to altruism. *Trends in Ecology and Evolution* 21:57–60.

Francke-Grossman, H. 1967. Ectosymbiosis in wood-inhabiting insects. In *Symbiosis* (pp. 142–205), edited by S. M. Henry. New York: Academic Press.

Frank, S. A. 1986. Hierarchical selection theory and sex ratios. I. General solutions for structured populations. *Theoretical Population Biology* 29:312–342.

Free, J. B., and Y. Spencer-Booth. 1959. The longevity of worker honey bees (*Apis mellifera*). *Proceedings of the Royal Entomological Society London* 34A:141–150.

Freeman, B. E. 1973. Preliminary studies on the population dynamics of *Sceliphron assimile* Dahlbom (Hymenoptera: Sphecidae) in Jamaica. *Journal of Animal Ecology* 42:173–182.

Freeman, B. E. 1977. Aspects of the regulation of size of the Jamaican population of *Sceliphron assimile* (Dahlbom) (Hymenoptera: Sphecidae). *Journal of Animal Ecology* 46:231–247.

Freeman, B. E. 1980. A population study in Jamaica on adult *Sceliphron assimile* Dahlbom (Hymenoptera: Sphecidae). *Ecological Entomology* 5:19–30.

Freeman, B. E., and B. Y. Johnston. 1978. The biology in Jamaica of the adults of the sphecid wasp *Sceliphron assimile* Dahlbom. *Ecological Entomology* 2:39–52.

Freeman, S., and J. C. Herron. 2003. *Evolutionary analysis, third edition*. Upper Saddle River, NJ: Prentice-Hall, Inc.

Frumhoff, P. C., and J. Baker. 1988. A genetic component to division of labour within honey bee colonies. *Nature* 333:358–361.

Fukuda, H., J. Kojima, and R. L. Jeanne. 2003. Colony specific morphological caste differences in an Old World, swarm-founding polistine, *Ropalidia romandi* (Hymenoptera: Vespidae). *Entomological Science* 6:37–47.

Gadagkar, R. 1990. Evolution of eusociality: the advantage of assured fitness returns. *Philosophical Transactions of the Royal Society of London B* 329:17–25.

Gadagkar, R. 1991a. On testing the role of genetic asymmetries created by haplodiploidy in the evolution of eusociality in the Hymenoptera. *Journal of Genetics* 70:1–31.

Gadagkar, R. 1991b. *Belonogaster, Mischocyttarus, Parapolybia,* and independent-founding *Ropalidia*. In *The social biology of wasps* (pp. 149–190), edited by K. G. Ross and R. W. Matthews. Ithaca, NY: Cornell University Press.

Gadagkar, R. 1994. Why the definition of eusociality is not helpful to understand its evolution and what should we do about it. *Oikos* 70:485–488.

Gadagkar, R. 1996. The evolution of eusociality, including a review of the social status of *Ropalidia marginata*. In *Natural history and evolution of paper-wasps* (pp. 248–271), edited by S. Turillazzi and M. J. West-Eberhard. Oxford: Oxford University Press.

Gadagkar, R. 2001. *The social biology of Ropalidia marginata: Toward understanding the evolution of eusociality*. Cambridge, MA: Harvard University Press.

Gadagkar, R., S. Bhagavan, K. Chandrashekara, and C. Vinutha. 1991. The role of larval nutrition in pre-imaginal biasing of caste in the primitively eusocial wasp *Ropalidia marginata* (Hymenoptera: Vespidae). *Ecological Entomology* 16:435–440.

Gadagkar, R., S. Bhagavan, R. Malpe, and C. Vinutha. 1990. On reconfirming the evidence for pre-imaginal caste bias in a primitively eusocial wasp. *Proceedings of the Indian Academy of Sciences (Animal Science)* 99:141–150.

Gadagkar, R., C. Vinutha, A. Shanubhogue, and A. P. Gore. 1988. Pre-imaginal biasing of caste in a primitively eusocial insect. *Proceedings of the Royal Society of London B* 233:175–189.

Gamboa, G. J. 1978. Intraspecific defense: advantage of social cooperation among paper wasp foundresses. *Science* 199:1463–1465.

Gamboa, G. J. 1980. Comparative timing of brood development between multiple- and single-foundress colonies of the paper wasp, *Polistes metricus*. *Ecological Entomology* 5:221–225.

Gamboa, G. J., and J. A. Dropkin. 1979. Comparisons of behaviors in early vs. late foundress associations of the paper wasp, *Polistes metricus* (Hymenoptera: Vespidae). *The Canadian Entomologist* 111:919–926.

García A., R. 1974. Observaciones sobre "*Polistes peruvianus*" Bequaert (Hym., Vespidae) en los alrededores de Lima. *Biota* 10:11–27.

Gauld, I. D. 1986. Taxonomy, its limitations and its role in understanding parasitoid biology. In *Insect parasitoids* (pp. 1–22), edited by J. K. Waage and D. Greathead. London: Academic Press.

Gauld, I., and B. Bolton, editors. 1988. *The Hymenoptera*. Oxford: British Museum (Natural History)/Oxford University Press.

Gervet, J. 1964. Le comportement d'oophagie differentielle chez *Polistes gallicus* L. (Hyménoptère, Vespide). *Insectes Sociaux* 11:343–382.

Gess, F. W., and S. K. Gess. 1976. An ethological study of *Parachilus insignis* (Saussure) (Hymenoptera: Eumenidae) in the Eastern Cape Province of South Africa. *Annals of the Cape Town Provincial Museums (Natural History)* 11:83–102.

Gess, F. W., and S. K. Gess. 1992. Ethology of three Southern African ground nesting Masarinae, two *Celonites* species and a silk spinning *Quartinia* species, with a discussion of nesting by the subfamily as a whole (Hymenoptera: Vespidae). *Journal of Hymenoptera Research* 1:145–155.

Gess, S. K. 1996. *The pollen wasps: Ecology and natural history of the Masarinae*. Cambridge, MA: Harvard University Press.

Ghiselin, M. T. 1969. *The triumph of the Darwinian method*. Berkeley: University of California Press.

Giannotti, E., and V. L. L. Machado. 1994. Longevity, life table and age polyethism in *Polistes lanio lanio* (Hymenoptera, Vespidae), a primitive eusocial wasp. *Journal of Advanced Zoology* 15:95–101.

Gibo, D. L. 1972. Hibernation sites and temperature tolerance of two species of *Vespula* and one species of *Polistes* (Hymenoptera: Vespidae). *Journal of the New York Entomological Society* 80:105–108.

Gibo, D. L. 1974. A laboratory study on the selective advantage of foundress associations in *Polistes fuscatus* (Hymenoptera: Vespidae). *The Canadian Entomologist* 106:101–106.

Gibo, D. L. 1976. Cold-hardiness in fall and winter of adults of the social wasp *Polistes fuscatus* (Hymenoptera: Vespidae) in southern Ontario. *The Canadian Entomologist* 108:801–806.

Gibo, D. L. 1978. The selective advantage of foundress associations in *Polistes fuscatus* (Hymenoptera: Vespidae): a field study of the effects of predation on productivity. *The Canadian Entomologist* 110:519–540.

Giray, T., M. Giovanetti, and M. J. West-Eberhard. 2005. Juvenile hormone, reproduction, and worker behavior in the neotropical social wasp *Polistes canadensis*. *Proceedings of the National Academy of Sciences U.S.A.* 102:3330–3335.

Gobbi, N., H. G. Fowler, J. C. Netto, and S. L. Nazareth. 1993. Comparative colony productivity of *Polistes simillimus* and *Polistes versicolor* (Hymenoptera: Vespidae) and the evolution of paragyny in the Polistinae. *Zoologische Jahrbücher. Abteilung für allgemeine Zoologie und Physiologie der Tiere* 97:239–243.

Gobbi, N., and V. L. L. Machado. 1985. Material capturado e utilizado na alimentação de *Polybia (Myrapetra) paulista* Ihering, 1896 (Hymenoptera – Vespidae). *Anais da Sociedade Entomológica do Brasil* 14:189–195.

Godfray, H. C. J. 1987. The evolution of clutch size in parasitic wasps. *The American Naturalist* 129:221–233.

González, J. M., J. Piñango, E. Blanco D., and R. W. Matthews. 2005. On the mass aggregations of *Polistes versicolor* (Oliver) (Hymenoptera: Vespidae) along the northern cordillera of Venezuela, South America. *Journal of Hymenoptera Research* 14:15–21.

Gould, S. J. 1987. Freudian slip. *Natural History* 96 (2):14–21.

Gould, S. J., and R. C. Lewontin. 1979. The spandrels of San Marco and the Panglossian paradigm: a critique of the adaptationist programme. *Proceedings of the Royal Society of London B* 205:581–598.

Grafen, A. 1986. Split sex ratios and the evolutionary origins of eusociality. *Journal of Theoretical Biology* 122:95–121.

Grandi, G. 1961. Studi di un entomologo sugli Imenotteri superiori. *Bollettino dell'Istituto di Entomologia dell'Università di Bologna* 25:141–144.

Grbic, M., P. J. Ode, and M. R. Strand. 1992. Sibling rivalry and brood sex-ratios in polyembryonic wasps. *Nature* 360:254–256.

Greene, A. 1991. *Dolichovespula* and *Vespula*. In *The social biology of wasps* (pp. 263–305), edited by K. G. Ross and R. W. Matthews. Ithaca, NY: Cornell University Press.

Greenstone, M. H., and J. H. Hunt. 1993. Determination of prey antigen half-life in *Polistes metricus* using a monoclonal antibody-based immunodot assay. *Entomologia Experimentalis et Applicata* 68:1–7.

Grinfel'd, E. K. 1977. The feeding of the social wasp *Polistes gallicus* (Hymenoptera, Vespidae). *Entomological Review* 56:24–29.

Grogan, D. E., and J. H. Hunt. 1977. Digestive proteases of two species of wasps of the genus *Vespula*. *Insect Biochemistry* 7:191–196.

Grogan, D. E., and J. H. Hunt. 1979. Pollen proteases: their potential role in insect digestion. *Insect Biochemistry* 9:309–313.

Grogan, D. E., and J. H. Hunt. 1980. Age correlated changes in midgut protease activity of the honeybee, *Apis mellifera* (Hymenoptera: Apidae). *Experientia* 36:1347–1348.

Grogan, D. E., and J. H. Hunt. 1984. Chymotrypsin-like activity in the honeybee midgut: patterns in a three year study. *Journal of Apicultural Research* 23:61–63.

Grogan, D. E., and J. H. Hunt. 1986. Midgut endopeptidase activities of the hornet *Vespa crabro germana* Christ (Hymenoptera: Vespidae). *Insectes Sociaux* 33:486–489.

Haeselbarth, E. 1979. Zur Parasitierung der Puppen von Forleule (*Panolis flammea* [Schiff.]), Kiefernspanner (*Bupalus piniarius* [L.]) und Heidelbeerspanner (*Boarmia bistortana* [Goezel]) in bayerischen Kiefernwaeldern. *Zeitschrift für Angewandte Entomologie* 87:186–202, 311–322.

Haeseler, V. 1980. Zum Necktarraub solitärer Faltenwespen (Hymenoptera: Vespoidea: Eumenidae). *Entomologia Generalis* 6:49–55.

Haggard, C. M., and G. J. Gamboa. 1980. Seasonal variation in body size and reproductive condition of a paper wasp, *Polistes metricus* (Hymenoptera: Vespidae). *The Canadian Entomologist* 112:239–248.

Hahn, P. D. 1995. Relatedness asymmetry and the evolution of eusociality. *Sociobiology* 26:1–32.

Haig, T. 1992. Intragenomic conflict and the evolution of eusociality. *Journal of Theoretical Biology* 156:401–403.

Haldane, J. B. S. 1932. *The causes of evolution*. New York: Harper Bros.

Hamilton, W. D. 1964a. The genetical evolution of social behaviour. I. *Journal of Theoretical Biology* 7:1–16.

Hamilton, W. D. 1964b. The genetical evolution of social behaviour. II. *Journal of Theoretical Biology* 7:17–52.

Hamilton, W. D. 1966. [Audience discussion following Kennedy, J. S., Some outstanding questions in insect behaviour]. In *Insect behaviour*, vol. 3 (p. 108), edited by P. T. Haskell. London: Symposium of the Royal Entomological Society of London.

Hamilton, W. D. 1967. Extraordinary sex ratios. *Science* 156:477–488.

Hamilton, W. D. 1996. Foreword. In *Natural history and evolution of paper-wasps*, edited by S. Turillazzi and M. J. West-Eberhard. Oxford: Oxford University Press.

Handlirsch, A. 1907. *Die Fossilen Insekten und die Phylogenie der rezenten Formen. Ein Handbuch für Pälaontologen und Zoologen*. Leipzig: Engelmann.

Hansell, M. H. 1981. Nest construction in the subsocial wasp *Parischnogaster mellyi* (Saussure) Stenogastrinae (Hymenoptera). *Insectes Sociaux* 28:208–216.

Hansell, M. H. 1982. Brood development in the subsocial wasp *Parischnogaster mellyi* (Saussure) (Stenogastrinae, Hymenoptera). *Insectes Sociaux* 29:3–14.

Hansell, M. 1985. The nest material of Stenogastrinae (Hymenoptera Vespidae) and its effect on the evolution of social behaviour and nest design. *Actes Colloques des Insectes Sociaux* 2:57–63.

Hansell, M. H. 1986. Colony biology of the stenogastrine wasp *Holischnogaster gracilipes* (van der Vecht) (Hum.) on Mount Kinabalu (Borneo). *Entomologists Monthly Magazine* 122:31–36.

Hansell, M. 1987a. Nest building as a facilitating and limiting factor in the evolution of eusociality in the Hymenoptera. *Oxford Surveys in Evolutionary Biology* 4:155–181.

Hansell, M. H. 1987b. Elements of eusociality in colonies of *Eustenogaster calyptodoma* (Sakagami & Yoshikawa) (Stenogastrinae, Vespidae). *Animal Behaviour* 35:131–141.

Hansell, M. H. 1989. Wasp papier-mâché. *Natural History* 8/89:52–61.

Hansell, M. H., C. Samuel, and J. I. Furtado. 1982. *Liostenogaster flavolineata*: social life in the small colonies of an Asian tropical wasp. In *The biology of social insects* (pp. 192–195), edited by M. D. Breed, C. D. Michener and H. E. Evans. Boulder, CO: Westview Press.

Hanson, P. E., and I. D. Gauld, editors. 1995. *The Hymenoptera of Costa Rica*. Oxford: Oxford University Press.

Hardy, I. C. W., P. J. Ode, and M. R. Strand. 1993. Factors influencing brood sex-ratios in polyembryonic Hymenoptera. *Oecologia* 93:343–348.

Harrison, J. N., and S. P. Roberts. 2000. Flight respiration and energetics. *Annual Review of Physiology* 62:179–205.

Hartman, C. 1905. Observations on the habits of some solitary wasps of Texas. *Bulletin of the University of Texas* 65:9–10.

Headley, A. E. 1943. Population studies of two species of ants, *Leptothorax longispinosus* Roger and *Leptothorax curvispinosus* Mayr. *Annals of the Entomological Society of America* 36:743–753.

Headley, A. E. 1949. A population study of the ant *Aphaenogaster fulva* ssp. *aquia* Buckler (Hymenoptera, Formicidae). *Annals of the Entomological Society of America* 42:265–272.

Heithaus, E. R. 1979. Community structure of neotropical flower visiting bees and wasps: diversity and phenology. *Ecology* 60:190–202.

Heldmann, G. 1936. Über das Leben auf Waben mit meheren überwinterten Weibchen von *Polistes gallica* (L.). *Biologisches Zentralblatt* 56:389–400.

Henshaw, M. T., J. E. Strassmann, S. Q. Quach, and D. C. Queller. 2000. Male production in *Parachartergus colobopterus*, a neotropical, swarm-founding wasp. *Ethology, Ecology & Evolution* 12:161–174.

Hepperle, C., and K. Hartfelder. 2001. Differentially expressed regulatory genes in honey bee caste development. *Naturwissenschaften* 88:113–116.

Herbers, J. M. 1984. Queen-worker conflict and eusocial evolution in a polygynous ant species. *Evolution* 38:631–643.

Hermann, H. R., D. Gerling, and T. F. Dirks. 1974. The cohibernation and mating activity of five polistine wasp species (Hymenoptera: Vespidae: Polistinae). *Journal of the Georgia Entomological Society* 9:203–204.

Hingston, R. W. G. 1926. The mason wasp (*Eumenes conica*). Part 1. Architecture. *Journal of the Bombay Natural History Society* 31:241–247.

Hirose, Y., and M. Yamasaki. 1984. Dispersal of females for colony founding in *Polistes jadwigae* Dalla Torre (Hymenoptera, Vespidae). *Kontyû* 52:65–71.

Hölldobler, B., and E. O. Wilson. 1990. *The ants*. Cambridge, MA: Belknap Press of Harvard University Press.

Holmes, H. B. 1972. Genetic evidence for fewer progeny and a higher percent males when *Nasonia vitripennis* oviposits in previously parasitized hosts. *Entomophaga* 17:79–88.

Hook, A. 1982. Observations on a declining nest of *Polistes tepidus* (F.) (Hymenoptera: Vespidae). *Journal of the Australian Entomological Society* 21:277–278.

Hoshikawa, T. 1979. Observations on the polygynous nests of *Polistes chinensis antennalis* Pérez (Hymenoptera: Vespidae) in Japan. *Kontyû* 47:239–243.

Howard, K. J., A. R. Smith, S. O'Donnell, and R. L. Jeanne. 2002. Novel method of swarm emigration by the epiponine wasp, *Apoica pallens* (Hymenoptera Vespidae). *Ethology, Ecology & Evolution* 14:365–371.

Hughes, C. R. 1987. Group nesting and reproductive conflict in primitively eusocial wasps. Ph.D. dissertation, Rice University, Houston, TX.

Hughes, D. P., L. Beani, S. Turillazzi, and J. Kathirithamby. 2003. Prevalence of the parasite Strepsiptera in *Polistes* as detected by dissection of immatures. *Insectes Sociaux* 50:62–68.

Hughes, D. P., J. Kathirithamby, S. Turillazzi, and L. Beani. 2004a. Social wasps desert the colony and aggregate outside if parasitized: parasite manipulation? *Behavioral Ecology* 15:1037–1043.

Hughes, D. P., J. Kathirithamby, and L. Beani. 2004b. Prevalence of the parasite strepsiptera in adult *Polistes* wasps: field collections and literature overview. *Ethology, Ecology & Evolution* 16:363–75.

Hunt, J. H. 1971. A field study of the Wrenthrush, *Zeledonia coronata*. *The Auk* 88:1–20.

Hunt, J. H. 1975. On the evolution of eusocial wasps and ants. *Bulletin of the Ecological Society of America* 56 (2):54.

Hunt, J. H. 1982. Trophallaxis and the evolution of eusocial Hymenoptera. In *The biology of social insects* (pp. 201–205), edited by M. D. Breed, C. D. Michener and H. E. Evans. Boulder, CO: Westview.

Hunt, J. H. 1984. Adult nourishment during larval provisioning in a primitively eusocial wasp, *Polistes metricus* Say. *Insectes Sociaux* 31:452–460.

Hunt, J. H. 1988. Lobe erection behavior and its possible social role in larvae of *Mischocyttarus* paper wasps. *Journal of Insect Behavior* 1:379–386.

Hunt, J. H. 1991. Nourishment and the evolution of the social Vespidae. In *The social biology of wasps* (pp. 426–450), edited by K. G. Ross and R. W. Matthews. Ithaca, NY: Cornell University Press.

Hunt, J. H. 1993. Survivorship, fecundity, and recruitment in a mud dauber wasp, *Sceliphron assimile* (Hymenoptera: Sphecidae). *Annals of the Entomological Society of America* 86:51–59.

Hunt, J. H. 1994. Nourishment and evolution in wasps *sensu lato*. In *Nourishment and evolution in insect societies* (pp. 221–254), edited by J. H. Hunt and C. A. Nalepa. Boulder: Westview Press.

Hunt, J. H. 1999. Trait mapping and salience in the evolution of eusocial vespid wasps. *Evolution* 53:225–237.

Hunt, J. H. 2000. European hornet. In *Handbook of household and structural insect pests* (pp. 26–29), edited by R. E. Gold and S. C. Jones. Lanham, MD: Entomological Society of America.

Hunt, J. H. 2003. Cryptic herbivores of the rainforest canopy. *Science* 300:916–917.

Hunt, J. H., and G. V. Amdam. 2005. Bivoltinism as an antecedent to eusociality in the paper wasp genus *Polistes*. *Science* 308: 264–267.

Hunt, J. H., I. Baker, and H. G. Baker. 1982. Similarity of amino acids in nectar and larval saliva: the nutritional basis for trophallaxis in social wasps. *Evolution* 36:1318–1322.

Hunt, J. H., R. J. Brodie, T. P. Carithers, P. Z. Goldstein, and D. H. Janzen. 1999. Dry season migration by Costa Rican lowland paper wasps to high elevation cold dormancy sites. *Biotropica* 31:192–196.

Hunt, J. H., P. A. Brown, K. M. Sago, and J. A. Kerker. 1991. Vespid wasps eat pollen (Hymenoptera: Vespidae). *Journal of the Kansas Entomological Society* 64:127–130.

Hunt, J. H., N. A. Buck, and D. E. Wheeler. 2003. Storage proteins in vespid wasps: characterization, developmental pattern, and occurrence in adults. *Journal of Insect Physiology* 49:785–794.

Hunt, J. H., and J. M. Carpenter. 2004. Intra-specific nest form variation in some neotropical swarm-founding wasps of the genus *Parachartergus* (Hymenoptera: Vespidae, Epiponini). *Journal of the Kansas Entomological Society* 77:448–456.

Hunt, J. H., R. D. Cave, and G. R. Borjas. 2001a. First records from Honduras of a yellowjacket wasp, *Vespula squamosa* (Drury) (Hymenoptera: Vespidae, Vespinae). *Journal of the Kansas Entomological Society* 74:118–119.

Hunt, J. H., and M. A. Dove. 2002. Nourishment affects colony demographics in the paper wasp *Polistes metricus*. *Ecological Entomology* 27:467–474.

Hunt, J. H., and G. J. Gamboa. 1978. Joint nest use by two paper wasp species. *Insectes Sociaux* 25:373–374.

Hunt, J. H., R. L. Jeanne, I. Baker, and D. E. Grogan. 1987. Nutrient dynamics of a swarm-founding social wasp species, *Polybia occidentalis* (Hymenoptera: Vespidae). *Ethology* 75:291–305.

Hunt, J. H., R. L. Jeanne, and M. G. Keeping. 1995. Observations on *Apoica pallens*, a nocturnal neotropical social wasp (Hymenoptera: Vespidae, Polistinae, Epiponini). *Insectes Sociaux* 42:223–236.

Hunt, J. H., and C. A. Nalepa. 1994. Nourishment, evolution and insect sociality. In *Nourishment and evolution in insect societies* (pp. 1–19), edited by J. H. Hunt and C. A. Nalepa. Boulder, CO: Westview Press.

Hunt, J. H., and K. C. Noonan. 1979. Larval feeding by male *Polistes fuscatus* and *Polistes metricus* (Hymenoptera: Vespidae). *Insectes Sociaux* 26:247–251.

Hunt, J. H., S. O'Donnell, N. Chernoff, and C. Brownie. 2001b. Observations on two neotropical swarm-founding wasps, *Agelaia yepocapa* and *A. panamaensis* (Hymenoptera: Vespidae). *Annals of the Entomological Society of America* 94:555–562.

Hunt, J. H., A. M. Rossi, N. J. Holmberg, S. R. Smith, and W. R. Sherman. 1998. Nutrients in social wasp (Hymenoptera: Vespidae, Polistinae) honey. *Annals of the Entomological Society of America* 91:466–472.

Hunt, J. H., D. K. Schmidt, S. S. Mulkey, and M. A. Williams. 1996. Caste dimorphism in the wasp *Epipona guerini* (Hymenoptera: Vespidae; Polistinae, Epiponini): further evidence for larval determination. *Journal of the Kansas Entomological Society* 69:362–369.

Hunt, S. 1970. Amino acid composition of silk from the pseudoscorpion *Neobisium maritimum* (Leach): a possible link between the silk fibroins and the keratins. *Comparative Biochemistry and Physiology* 34:773–776.

Huxley, J. 1964. New introduction by the author. In *Evolution: The modern synthesis*. New York: John Wiley & Sons, Inc.

Ikan, R., E. D. Bergmann, J. Ishay, and S. Gitter. 1968. Proteolytic enzyme activity in the various colony members of the oriental hornet, *Vespa orientalis* F. *Life Sciences* 7:929–934.

Ikan, R., and J. Ishay. 1966. Larval wasp secretions and honeydew of the aphids, *Chaitophorus populi*, feeding on *Populus euphratica* as sources of sugars in the diet of the oriental hornet, *Vespa orientalis*, F. *Israel Journal of Zoology* 15:64–68.

Inagawa, K., J. Kojima, K. Sayama, and K. Tsuchida. 2001. Colony productivity of the paper wasp *Polistes snelleni*: comparison between cool-temperate and warm-temperate populations. *Insectes Sociaux* 48:259–265.

Isely, D. 1913. The biology of some Kansas Eumenidae. *The University of Kansas Science Bulletin* 8:235–309.

Ishay, J. 1975a. Orientation by pupating larvæ of *Vespa orientalis* (Hymenoptera: Vespidæ). *Insectes Sociaux* 1:67–73.

Ishay, J. 1975b. Caste determination by social wasps: cell size and building behaviour. *Animal Behaviour* 23:425–431.

Ishay, J., and R. Ikan. 1968a. Food exchange between adults and larvae in *Vespa orientalis* F. *Animal Behaviour* 16:289–303.

Ishay, J., and R. Ikan. 1968b. Gluconeogenesis in the oriental hornet, *Vespa orientalis* F. *Ecology* 49:169–171.

Itino, T. 1986. Comparison of life tables between the solitary eumenid wasp *Anterhynchium flavomarginatum* and the subsocial eumenid wasp *Orancistrocerus drewseni* to evaluate the adaptive significance of maternal care. *Researches on Population Ecology* 28:185–199.

Itô, Y. 1989. The evolutionary biology of sterile soldiers in aphids. *Trends in Ecology and Evolution* 4:69–73.

Iwata, K. 1938a. Habits of four species of *Odynerus* (*Ancistrocerus*) in Japan. *Tenthredo* 2:19–32.

Iwata, K. 1938b. Habits of eight species of Eumeninae (*Rhygchium*, *Lionotus* and *Symmorphus*) in Japan. *Mushi* 11:110–132.

Iwata, K. 1939. Habits of a paper-making potter wasp (*Eumenes architectus* Smith) in Japan. *Mushi* 12:83–85.

Iwata, K. 1953. Biology of *Eumenes* in Japan (Hymenoptera: Vespidae). *Mushi* 25:25–45.

Iwata, K. 1967. Report of the fundamental research on the biological control of insect pests in Thailand. II. The report on the bionomics of aculeate wasps—Bionomics of subsocial wasps of Stenogastrinae (Hymenoptera, Vespidae). *Nature and Life in Southeast Asia* 5:259–293.

Iwata, K. 1976. *Evolution of instinct: Comparative ethology of Hymenoptera (translated from the Japanese)*. New Delhi: Amerind Publishing Co., Pvt. Ltd.

Jacobson, E. 1935. Aanteekeningen over Stenogastrinae (Hym., Vespidae). *Entomologische Mededeelingen van Nederlandsch-Indie* 1:15–19.

Jaffe, K. 2001. On the relative importance of haplo-diploidy, assortative mating and social synergy on the evolutionary emergence of social behavior. *Acta Biotheoretica* 49:29–42.

James, W. H. 1993. Continuing confusion. *Nature* 365:8.

Jamieson, I. B. 1986. The functional approach to behavior: is it useful? *The American Naturalist* 127:195–208.

Jamieson, I. B. 1989. Levels of analysis or analyses at the same level. *Animal Behaviour* 37:696–697.

Janet, C. 1903. *Observations sur les Guêpes*. Paris: Carré et Naud.

Jarvis, J. U. M. 1981. Eusociality in a mammal: cooperative breeding in naked mole-rat colonies. *Science* 212:571–573.

Jayakar, S. D. 1963. 'Proterandry' in solitary wasps. *Nature* 198:208–209.

Jayakar, S. D., and H. Spurway. 1965. Normal and abnormal nests of *Eumenes emarginatus conoideus* (Gmelin) including notes on crépissage in this and other members of the genus (Vespoidea, Hymenoptera). *Journal of the Bombay Natural History Society* 62:193–200.

Jayakar, S. D., and H. Spurway. 1966. Re-use of cells and brother-sister mating in the Indian species *Stenodynerus miniatus* (Sauss.) (Vespidae: Eumeninae). *Journal of the Bombay Natural History Society* 63:378–398.

Jayasingh, D. B. 1980. A new hypothesis on cell provisioning in solitary wasps. *Biological Journal of the Linnean Society* 13:167–170.

Jeanne, R. L. 1970. Chemical defense of brood by a social wasp. *Science* 168:1465–1466.

Jeanne, R. L. 1972. Social biology of the neotropical wasp *Mischocyttarus drewseni*. *Bulletin of the Museum of Comparative Zoology, Harvard University* 144:63–150.

Jeanne, R. L. 1980. Evolution of social behavior in the Vespidae. *Annual Review of Entomology* 25:371–396.

Jeanne, R. L. 1981. Chemical communication during swarm emigration in the social wasp *Polybia sericea* (Olivier). *Animal Behaviour* 29:102–113.

Jeanne, R. L. 1986. The organization of work in *Polybia occidentalis*: the costs and benefits of specialization in a social wasp. *Behavioral Ecology and Sociobiology* 19:333–341.

Jeanne, R. L. 1991. The swarm-founding Polistinae. In *The social biology of wasps* (pp. 191–231), edited by K. G. Ross and R. W. Matthews. Ithaca, NY: Cornell University Press.

Jeanne, R. L., and R. Fagen. 1974. Polymorphism in *Stelopolybia areata* (Hymenoptera, Vespidae). *Psyche* 81:155–166.

Jeanne, R. L., C. A. Graf, and B. S. Yandell. 1995a. Non-size-based morphological castes in a social insect. *Naturwissenschaften* 82:296–298.

Jeanne, R. L., and J. H. Hunt. 1992. Observations on the social wasp *Ropalidia montana* from peninsular India. *Journal of Biosciences* 17:1–14.

Jeanne, R. L., J. H. Hunt, and M. G. Keeping. 1995b. Foraging in social wasps: *Agelaia* lacks recruitment to food. *Journal of the Kansas Entomological Society* 68:279–289.

Jeanne, R. L., and M. G. Keeping. 1995. Venom spraying in *Parachartergus colobopterus*: a novel defensive behavior in a social wasp (Hymenoptera: Vespidae). *Journal of Insect Behavior* 8:433–442.

Jenner, R. A. 2004. The scientific status of metazoan cladistics: why current research practice must change. *Zoologica Scripta* 33:293–310.

Jervis, M. A., G. E. Heimpel, P. N. Ferns, J. A. Harvey, and N. A. C. Kidd. 2001. Life-history strategies in parasitoid wasps: a comparative analysis of 'ovigeny'. *Journal of Animal Ecology* 70:442–458.

Jervis, M. A., and A. C. Kidd. 1986. Host-feeding strategies in hymenopteran parasitoids. *Biological Reviews of the Cambridge Philosophical Society* 61:395–434.

Jervis, M., and L. Vilhelmsen. 2000. Mouthpart evolution in adults of the basal, 'symphytan', hymenopteran lineages. *Biological Journal of the Linnean Society* 70:121–146.

Jukes, T. H. 1990. Females at work. *Science* 249:1359.

Karsai, I. 1999. Decentralized control of construction behavior in paper wasps: an overview of the stigmergy approach. *Artificial Life* 5:117–136.

Karsai, I., and J. H. Hunt. 2002. Food quantity affects traits of offspring in the paper wasp *Polistes metricus* (Hymenoptera: Vespidae). *Environmental Entomology* 31:99–106.

Karsai, I., and Z. Penzes. 1998. Nest shapes in paper wasps: can the variability of forms be deduced from the same construction algorithm? *Proceedings of the Royal Society of London B* 265:1261–1268.

Karsai, I., Z. Penzes, and J. W. Wenzel. 1996. Dynamics of colony development in *Polistes dominulus*: a modeling approach. *Behavioral Ecology and Sociobiology* 39:97–105.

Kasuya, E. 1981a. Male mating territory in a Japanese paper wasp, *Polistes jadwigae* Dalla Torre (Hymenoptera, Vespidae). *Kontyû* 49:607–614.

Kasuya, E. 1981b. Polygyny in the Japanese paper wasp, *Polistes jadwigae* Dalla Torre (Hymenoptera, Vespidae). *Kontyû* 49:306–313.

Kathirithamby, J. 1991. Strepsiptera. In *The Insects of Australia: A textbook for students and research workers*, 2nd ed. (pp. 684–695), edited by I. D. Naumann, P. B. Carne, J. F. Lawrence, E. S. Nielsen, J. P. Spradberry, R. W. Taylor, M. J. Whitten and M. J. Littlejohn. Melbourne: CSIRO, Melbourne University Press.

Kaufman, S., M. Peña Claros, S. A. Morehead, J. H. Hunt, D. Craig, M. Singer, and J. Apple. 1995. A survivorship schedule for *Polybia occidentalis* (Vespidae) in Palo Verde Marsh. In *Tropical biology, an ecological approach: Coursebook 95–1* (pp. 91–94), edited by E. Olson. San Jose, Costa Rica: Organization for Tropical Studies.

Kayes, B. M. 1978. Digestive proteases in four species of *Polistes* wasps. *Canadian Journal of Zoology* 56:1454–1459.

Keegans, S. J., E. D. Morgan, S. Turillazzi, B. D. Jackson, and J. Billen. 1993. The Dufour gland and the secretion placed on eggs of 2 species of social wasps, *Liostenogaster flavolineata* and *Parischnogaster jacobsoni* (Vespidae, Stenogastrinae). *Journal of Chemical Ecology* 19:279–290.

Keeping, M. G. 1991. Nest construction by the social wasp, *Belonogaster petiolata* (DeGeer) (Hymenoptera: Vespidae). *Journal of the Entomological Society of South Africa* 54:17–28.

Keeping, M. G. 2002. Reproductive and worker castes in the primitively eusocial wasp *Belonogaster petiolata* (DeGeer) (Hymenoptera: Vespidae): evidence for pre-imaginal differentiation. *Journal of Insect Physiology* 48:867–879.

Keller, E. F. 1983. *A feeling for the organism: The life and work of Barbara McClintock*. New York: Freeman.

Keller, L., and P. Nonacs. 1993. The role of queen pheromones in social insects: queen control or queen signal. *Animal Behaviour* 64:477–485.

Kemp, W. P., J. Bosch, and B. Dennis. 2004. Oxygen consumption during the life cycles of the prepupa-wintering bee *Megachile rotundata* and the adult-wintering bee *Osmia lignaria* (Hymenoptera: Megachilidae). *Annals of the Entomological Society of America* 97:161–170.

Kent, D. S., and J. A. Simpson. 1992. Eusociality in the beetle *Austroplatypus incompertus* (Coleoptera: Curculionidae). *Naturwissenschaften* 79:86–87.

King, B. H. 1987. Offspring sex ratios in parasitoid wasps. *The Quarterly Review of Biology* 62:367–396.

King, B. H. 1996. Sex ratio responses to other parasitoid wasps: multiple adaptive explanations. *Behavioral Ecology and Sociobiology* 39:367–374.

King, L. L., and T. H. Jukes. 1969. Non-Darwinian evolution. *Science* 164:788–798.

Klahn, J. E. 1979. Philopatric and nonphilopatric foundress associations in the social wasp *Polistes fuscatus*. *Behavioral Ecology and Sociobiology* 5:417–424.

Klahn, J. E. 1981. Alternative reproductive tactics of single foundresses of a social wasp, *Polistes fuscatus*. Ph.D. dissertation, University of Iowa, Iowa City.

Klahn, J. 1988. Intraspecific comb usurpation in the social wasp *Polistes fuscatus*. *Behavioral Ecology and Sociobiology* 23:1–8.

Klahn, J. E., and G. J. Gamboa. 1983. Social wasps: discrimination between kin and nonkin brood. *Science* 221:482–484.

Kłudkiewicz, B., J. Godlewski, K. Grzelak, B. Cymborowski, and Z. Lasota. 1996. Influence of low temperature on the synthesis of some *Galleria mellonella* larvae proteins. *Acta Biochimica Polonica* 43:639–644.

Knisley, C. B. 1985. Utilization of tiger beetle larval burrows by a nest-provisioning wasp, *Leucodynerus russatus* (Bohart) (Hymenoptera: Eumenidae). *Proceedings of the Entomological Society of Washington* 87:481.

Kojima, J. 1983. Peritrophic sac extraction in *Ropalidia fasciata* (Hymenoptera, Vespidae). *Kontyû* 51:502–508.

Kojima, J. 1990. Immatures of hover wasps (Hymenoptera, Vespidae, Stenogastrinae). *Japanese Journal of Entomology* 58:506–522.

Kojima, J. 1993. Feeding of larvae by males of an Australian paper wasp, *Ropalidia plebiana* Richards (Hymenoptera, Vespidae). *Japanese Journal of Entomology* 1:213–215.

Kojima, J. 1996. Colony cycle of an Australian swarm-founding paper wasp, *Ropalidia romandi* (Hymenoptera: Vespidae). *Insectes Sociaux* 43:411–420.

Kojima, J. 1998. Larvae of social wasps (Insects: Hymenoptera; Vespidae). *Natural History Bulletin of Ibaraki University* 2:7–227.

Kojima, J., S. Hartini, W. A. Noerdjito, S. Kahono, N. Fujiyama, and H. Katakura. 2001. Descriptions of pre-emergence nests and mature larvae of *Vespa fervida*, with a note on a multiple-foundress colony of *V. affinis* in Sulawesi (Hymenoptera: Vespidae; Vespinae). *Entomological Science* 4:355–360.

Kort, C. A. D. de, and A. B. Koopmanschap. 1994. Nucleotide and deduced amino acid sequence of a cDNA clone encoding diapause protein 1, an arylphorin-type storage heaxamer of the Colorado potato beetle. *Journal of Insect Physiology* 40:527–535.

Krebs, H. 1964. Gluconeogenesis. *Proceedings of the Royal Society of London B* 159:545–564.

Krebs, J., and R. M. May. 1976. Social insects and the evolution of altruism. *Nature* 260:9–10.

Krieger, M. J. B., K. G. Ross, C. W. Y. Chang, and L. Keller. 1999. Frequency and origin of triploidy in the fire ant *Solenopsis invicta*. *Heredity* 82:142–150.

Kristensen, N. P. 1999. Phylogeny of endopterygote insects, the most successful lineage of living organisms. *European Journal of Entomology* 96:237–253.

Krombein, K. V. 1967. *Trap-nesting wasps and bees: Life histories, nests, and associates*. Washington, DC: Smithsonian Press.

Krombein, K. V. 1978. Biosystematic studies of Ceylonese wasps III. Life history, nest and associates of *Paraleptomenes mephitis* (Cameron) (Hymenoptera: Eumenidae). *Journal of the Kansas Entomological Society* 51:721–734.

Kudô, K. 2000. Variable investments in nests and worker production by the foundresses of *Polistes chinensis* (Hymenoptera: Vespidae). *Journal of Ethology* 18:37–41.

Kudô, K. 2002. Daily foraging activities and changes of allocation pattern of proteinaceous resources in pre-emergence laboratory foundresses of *Polistes chinensis* (Hymenoptera; Vespidae). *Sociobiology* 39:243–257.

Kudô, K. 2003. Growth rate and body weight of foundress-reared offspring in a paper wasp, *Polistes chinensis* (Hymenoptera, Vespidae): no influence of food quantity on the first offspring. *Insectes Sociaux* 50:77–81.

Kudô, K., Sô. Yamane, and S. Miyano. 1996. Occurrence of a binding matrix in a nest of a primitively eusocial wasp, *Eustenogaster calyptodoma* (Hymenoptera, Vespidae). *Japanese Journal of Entomology* 64:891–895.

Kudô, K., Sô. Yamane, and H. Yamamoto. 1998. Physiological ecology of nest construction and protein flow in pre-emergence colonies of *Polistes chinensis* (Hymenoptera Vespidae): effects of rainfall and microclimates. *Ethology, Ecology & Evolution* 10:171–183.

Kukuk, P. F. 1994. Replacing the terms 'primitive' and 'advanced': new modifiers for the term 'eusocial'. *Animal Behaviour* 47:1475–1478.

Kumano, N., and E. Kasuya. 2001. Why do workers of the primitively eusocial wasp *Polistes chinensis antennalis* remain at their natal nest? *Animal Behaviour* 61:655–660.

Kundu, H. L. 1967. Observations on *Polistes hebraeus* (Hymenoptera). *Birla Institute of Technological Science, Journal (Pilani)* 1:152–161.

Lacy, R. C. 1984. The evolution of termite eusociality: reply to Leinaas. *The American Naturalist* 123:876–878.

Lai-Fook, J., and M. J. Wiley. 1976. The amino acid composition of the silk of *Calpodes ethuliuis* (Hesperiidae, Lepidoptera). *Comparative Biochemistry and Physiology B* 53:545–547.

Landes, D. A., and J. H. Hunt. 1988. Occurrence of *Chalybion zimmermanni zimmermanni* Dahlbom in a mixed sleeping aggregation with *Chalybion californicum* (Saussure) in Missouri (Hymenoptera: Sphecidae). *Journal of the Kansas Entomological Society* 61:230–231.

Landes, D. A., M. S. Obin, A. B. Cady, and J. H. Hunt. 1987. Seasonal and latitudinal variation in spider prey of the mud dauber *Chalybion californicum* (Hymenoptera: Sphecidae). *The Journal of Arachnology* 15:249–256.

Latreille, P. 1802. *Histoire naturelle, générale et particulière des crustacés et insectes. Tome Troisiè. Familles naturelles des genres*. Paris: F. Dufart.

Lewis, D. K., D. Spurgeon, T. W. Sappington, and L. L. Keeley. 2002. A hexamerin protein, AgSP-1, is associated with diapause in the boll weevil. *Journal of Insect Physiology* 48:887–901.

Liddell, H. G., and R. Scott [revised and augmented by H. Stuart Jones and R. McKenzie]. 1978. *A Greek-English lexicon*, new (ninth) edition. Oxford: Clarendon Press.

Liebig, J., T. Monnin, and S. Turillazzi. 2005. Direct assessment of queen quality and lack of worker suppression in a paper wasp. *Proceedings of the Royal Society of London B* 272:1339–1344.

Liebert, A. E., R. N. Johnson, G. T. Switz, and P. T. Starks. 2004. Triploid females and diploid males: underreported phenomena in *Polistes* wasps? *Insectes Sociaux* 51:205–211.

Lin, N., and C. D. Michener. 1972. Evolution of sociality in insects. *The Quarterly Review of Biology* 47:131–159.

Linksvayer, T. A., and M. J. Wade. 2006. The evolutionary origin and elaboration of sociality in the aculeate Hymenoptera: maternal effects, sib-social effects, and heterochrony. *The Quarterly Review of Biology* 80:317–336.

Litte, M. 1977. Behavioral ecology of the social wasp, *Mischocyttarus mexicanus*. *Behavioral Ecology and Sociobiology* 2:229–246.

Litte, M. 1979. *Mischocyttarus flavitarsis* in Arizona: social and nesting biology of a polistine wasp. *Zeitschrift für Tierpsychologie* 50:282–312.

Litte, M. 1981. Social biology of the polistine wasp *Mischocyttarus labiatus*: survival in a Colombian rain forest. *Smithsonian Contributions to Zoology* 327:1–27.

Littledyke, M., and J. M. Cherrett. 1976. Direct ingestion of plant sap from cut leaves by the leaf-cutting ants *Atta cephalotes* (L.) and *Acromyrmex octospinosus* (Reich) (Formicidae, Attini). *Bulletin of Entomological Research* 66:205–217.

Lombardi, S. J., and D. L. Kaplan. 1990. The amino-acid composition of major ampullate gland silk (dragline) of *Nephila clavipes* (Araneae, Tetragnathidae). *The Journal of Arachnology* 18:297–306.

London, K. B., and R. L. Jeanne. 1998. Envelopes protect social wasps' nests from phorid infestation (Hymenoptera: Vespidae, Diptera: Phoridae). *Journal of the Kansas Entomological Society* 71:175–182.

Longair, R. W. 1985. Male behavior in *Euparagia richardsi* Bohart (Hymenoptera: Vespidae). *Pan-Pacific Entomologist* 61:318–320.

Lorenzi, M. C., and S. Turillazzi. 1986. Behavioral and ecological adaptations to the high mountain environment of *Polistes biglumis bimaculatus*. *Ecological Entomology* 11:199–204.

Lucas, F., J. T. B. Shaw, and S. G. Smith. 1957. Amino-acid composition of the silk of *Chrysopa* egg-stalks. *Nature* 179:906–907.

Machado, V. L. L., N. Gobbi, and D. Simões. 1987. Material capturado e utilizado na alimentação de *Stelopolybia pallipes* (Olivier, 1791) (Hymenoptera—Vespidae). *Anais da Sociedade Entomológica do Brasil* 16:73–79.

Mackensen, O. 1951. Viability and sex determination in the honey bee (*Apis mellifera* L.). *Genetics* 36:500–509.

Makino, S. 1983. Larval feeding by *Polistes biglumis* males (Hymenoptera, Vespidae). *Kontyû* 51:487.

Makino, S. 1989. Usurpation and nest rebuilding in *Polistes riparius*: two ways to reproduce after the loss of the original nest (Hymenoptera: Vespidae). *Insectes Sociaux* 36:116–128.

Makino, S. 1993. Sexual differences in larval feeding behavior in a paper wasp, *Polistes jadwigae* (Hymenoptera, Vespidae). *Journal of Ethology* 11:73–75.

Makino, S., and K. Sayama. 1991. Comparison of intraspecific nest usurpation between two haplometrotic paper wasp species (Hymenoptera: Vespidae: *Polistes*). *Journal of Ethology* 9:121–128.

Maneval, H. 1939. Notes sur les Hyménoptères. *Annales de la Société Entomologique de France* 108:49–108.

Marchal, P. 1896. La reproduction et l'évolution des guêpes sociales. *Archives de zoologie expérimentale et générale* 4 (Série 3):1–100.

Marchal, P. 1897. La castration nutriciale chez les Hyménoptères sociaux. *Comptes rendus des séances de la Société de Biologie (Paris)* 1897:556–557.

Marino Piccioli, M. T. 1968. The extraction of the larval peritrophic sac by the adults in *Belonogaster*. *Monitore zoologico italiano (Nuova serie)* 2:203–206.

Marino Piccioli, M. T., and L. Pardi. 1978. Studies on the biology of *Belonogaster* (Hymenoptera Vespidae). 3. The nest of *Belonogaster griseus* (Fab.). *Monitore zoologico italiano (Nuova serie)* 10:179–228.

Markin, G. P., and A. R. Gittins. 1967. Biology of *Stenodynerus claremontensis* (Cameron) (Hymenoptera: Vespidae). *University of Idaho College of Agriculture Research Bulletin* 74:1–25.

Martin, M. M. 1987. *Invertebrate-microbial interactions: Ingested fungal enzymes in arthropod biology*. Ithaca, NY: Cornell University Press.

Maschwitz, U. 1965. Larven als Nahrungsspeicher im Wespenvolk: Ein Beitrag zum Trophallaxisproblem. *Verhandlungen der Deutschen Zoologischer Gesellschaft (Zoologischer Anzeiger 29 Supplementband)* 59:530–534.

Maschwitz, U. 1966. Das Speichelsekret der Wespenlarven und seine biologische Bedeutung. *Zeitschrift für vergleischende Physiologie* 53:228–252.

Matsuura, M. 1977. The life of the paper wasps (in Japanese). *Shizen* 32:26–36.

Matsuura, M. 1984. Comparative biology of the five Japanese species of the genus *Vespa* (Hymenoptera, Vespidae). *Bulletin of the Faculty of Agriculture, Mie University* 69:1–131.

Matsuura, M. 1991. *Vespa* and *Provespa*. In *The social biology of wasps* (pp. 232–262), edited by K. G. Ross and R. W. Matthews. Ithaca, NY: Cornell University Press.

Matsuura, M., and Sk. Yamane. 1990. *Biology of the vespine wasps*. Berlin: Springer-Verlag.

Matthews, R. W. 1968. *Microstigmus comes*: sociality in a sphecid wasp. *Science* 160:787–788.

Matthews, R. W. 1991. Evolution of social behavior in sphecid wasps. In *The social biology of wasps* (pp. 570–602), edited by K. G. Ross and R. W. Matthews. Ithaca, NY: Cornell University Press.

Matthews, R. W., and I. D. Naumann. 1988. Nesting biology and taxonomy of *Arpactophilus mimi*, a new species of social sphecid (Hymenoptera: Sphecidae). *Australian Journal of Zoology* 36:585–597.

Maurizio, A. 1950. The influence of pollen feeding and brood rearing on the length of life and physiological conditions of the honeybee. *Bee World* 31:9–12.

Maurizio, A. 1975. How bees make honey. In *Honey: A comprehensive survey* (pp. 71–105), edited by E. Crane. London: Heinemann.

Mayhew, P. J. 1998. The evolution of gregariousness in parasitoid wasps. *Proceedings of the Royal Society of London B* 265:383–389.

Mayhew, P. J., and T. M. Blackburn. 1999. Does development mode organize life-history traits in the parasitoid Hymenoptera? *Journal of Animal Ecology* 68:906–916.

Mayhew, P. J., and J. J. M. Van Alphen. 1999. Gregarious development in alysiine parasitoids evolved through a reduction in larval aggression. *Animal Behaviour* 58:131–141.

Maynard Smith, J. 1964. Group selection and kin selection. *Nature* 201:1145–1147.

Maynard Smith, J., and E. Szathmáry. 1995. *The major transitions in evolution*. Oxford: Oxford University Press.

Maynard Smith, J., and E. Szathmáry. 1999. *The origins of life: From the birth of life to the origin of language*. Oxford: Oxford University Press.

McCorquodale, D. B., and I. D. Naumann. 1988. A new Australian species of communal ground nesting wasp, in the genus *Spilomena* Shuckard (Hymenoptera: Sphecidae: Pemphredoninae). *Journal of the Australian Entomological Society* 27:221–231.

Mead, F., D. Gabouriaut, and C. Habersetzer. 1995. Nest-founding behavior induced in the first descendants of *Polistes dominulus* Christ (Hymenoptera: Vespidae) colonies. *Insectes Sociaux* 42:385–396.

Mead, F., C. Habersetzer, D. Gabouriaut, and J. Gervet. 1994. Dynamics of colony development in the paper wasp *Polistes dominulus* Christ (Hymenoptera, Vespidae): the influence of prey availability. *Journal of Ethology* 12:43–51.

Mead, F., and M. Pratte. 2002. Prey supplementation increases productivity in the social wasp *Polistes dominulus* Christ (Hymenoptera Vespidae). *Ethology, Ecology & Evolution* 14:111–128.

Medler, J. T. 1964. Biology of *Rygchium foraminatum* in trap-nests in Wisconsin (Hymenoptera: Vespidae). *Annals of the Entomological Society of America* 57:56–60.

Medler, J. T., and R. E. Fye. 1956. Biology of *Ancistrocerus antilope* (Panzer) (Hymenoptera, Vespidae) in trap nests in Wisconsin. *Annals of the Entomological Society of America* 49:97–102.

Mehdiabadi, N. J., H. K. Reeve, and U. G. Mueller. 2003. Queens versus workers: sex-ratio conflict in eusocial Hymenoptera. *Trends in Ecology and Evolution* 18:88–93.

Melo, G. A. R. de, and L. A. O. Campos. 1993. Trophallaxis in a primitively social sphecid wasp. *Insectes Sociaux* 40:107–110.

Menke, A. S., and F. D. Parker. 1996. Phenology of ammophiline wasps in a premontane wet forest in Costa Rica (Hymenoptera, Sphecidae, Ammophilini). *Journal of Hymenoptera Research* 5:184–189.

Menke, A. S., and L. A. Stange. 1986. *Delta campaniforme rendalli* (Birmingham) and *Zeta argillaceum argillaceum* (Linnaeus) established in southern Florida, and comments on generic discretion in *Eumenes s.l.* (Hymenoptera: Vespidae; Eumeninae). *Florida Entomologist* 69:697–702.

Metcalf, R. A., and G. S. Whitt. 1977. Relative inclusive fitness in the social wasp *Polistes metricus*. *Behavioral Ecology and Sociobiology* 2:353–360.

Michener, C. D. 1953. Comparative morphological and systematic studies of bee larvae with a key to the families of hymenopterous larvae. *The University of Kansas Science Bulletin* 35:987–1102.

Michener, C. D.1958. The evolution of social behavior in bees. *Proceedings Tenth International Congress of Entomology (Montreal, 1956)*, 2:441–447.

Michener, C. D. 1969. Comparative social behavior of bees. *Annual Review of Entomology* 14:299–342.

Michener, C. D. 1974. *The social behavior of the bees: A comparative study*. Cambridge, MA: Belknap Press of Harvard University Press.

Michener, C. D., and D. J. Brothers. 1974. Were workers of eusocial Hymenoptera initially altruistic or oppressed? *Proceedings of the National Academy of Sciences U.S.A.* 71:671–674.

Micheu, S., K. Crailsheim, and B. Leonhard. 2000. Importance of proline and other amino acids during honeybee flight (*Apis mellifera carnica* Pollmann). *Amino Acids* 18:157–175.

Michod, R. E. 1980. Evolution of interactions in family-structured populations: mixed mating models. *Genetics* 96:275–296.

Miller, P. L. 1966. The regulation of breathing in insects. *Advances in Insect Physiology* 3:279–354.

Miotk, P. 1979. Zur Biologie und Ökologie von *Odynerus spinipes* (L.) und *O. reniformis* (Gmel.) and den Lößwänden des Kaiserstuhls (Hymenoptera: Eumenidae). *Zoologische Jahrbücher. Abteilung für Systematik* 106:374–405.

Mitchell, P. S., and J. H. Hunt. 1984. Nutrient and energy assays of larval provisions and feces in the black and yellow mud dauber, *Sceliphron caementarium* (Drury) (Hymenoptera: Sphecidae). *Journal of the Kansas Entomological Society* 57:700–704.

Miyano, S. 1980. Life tables of colonies and workers in a paper wasp, *Polistes chinensis antennalis*, in central Japan (Hymenoptera: Vespidae). *Researches on Population Ecology* 22:69–88.

Miyano, S. 1983. Number of offspring and seasonal changes of their body weight in a paperwasp, *Polistes chinensis antennalis* Pérez (Hymenoptera: Vespidae), with reference to male production by workers. *Researches on Population Ecology* 25:198–209.

Miyano, S. 1986. Colony development, worker behavior and male production in orphan colonies of a Japanese paper wasp, *Polistes chinensis antennalis* Perez (Hymenoptera, Vespidae). *Researches on Population Ecology* 28:347–361.

Miyano, S. 1990. Number, larval durations and body weights of queen-reared workers of a Japanese paper wasp, *Polistes chinensis antennalis* (Hymenoptera, Vespidae). *Natural History Research* 1:93–97.

Miyano, S. 1994. Some ecological observations on social wasps (Insecta: Hymenoptera: Vespidae) in the Northern Mariana Islands, Micronesia. In *Biological expedition to the Northern Mariana Islands, Micronesia* (Natural History Research, special issue no. 1, pp. 237–245), edited by A. Asakura and T. Furuki. Chiba: Natural History Museum and Institute.

Monnin, T., F. L. W. Ratnieks, G. R. Jones, and R. Beard. 2002. Pretender punishment induced by chemical signalling in a queenless ant. *Nature* 419:61–65.

Moore, W. S. 1975. Observations on the egg laying and sleeping habits of *Euparagia scutellaris* Cresson (Hymenoptera: Masaridae). *The Pan-Pacific Entomologist* 51:286.

Morgan, C. L. 1894. *An introduction to comparative psychology*. London: W. Scott.

Morgan, C. L. 1900. *Animal behaviour*. London: Edward Arnold.

Morimoto, R. 1954. On the development of *Polistes chinensis antennalis* Perez. Studies on the social Hymenoptera of Japan, III. *Science Bulletin of the Faculty of Agriculture, Kyushu University* 14:337–353.

Morrill, W. L. 1974. Production and flight of alate red imported fire ants. *Environmental Entomology* 3:265–271.

Morris, D. C., M. P. Schwarz, S. J. B. Cooper, and L. A. Mound. 2002. Phylogenetics of Australian *Acacia* thrips: the evolution of behaviour and ecology. *Molecular Phylogenetics and Evolution* 25:278–292.

Myles, T. G., and W. L. Nutting. 1988. Termite eusocial evolution—a re-examination of Bartz hypothesis and assumptions. *The Quarterly Review of Biology* 63:1–23.

Nalepa, C. A. 1994. Nourishment and the origin of termite eusociality. In *Nourishment and evolution in insect societies* (pp. 57–104), edited by J. H. Hunt and C. A. Nalepa. Boulder, CO: Westview Press.

Naumann, M. G. 1975a. Swarming behavior: evidence for communication in social wasps. *Science* 189:642–644.

Naumann, M. G.1975b. The nesting behavior of *Protopolybia pumila* in Panama (Hymenoptera: Vespidae). Ph.D. dissertation, University of Kansas, Lawrence.

Nelson, J. M. 1968. Parasites and symbionts of nests of *Polistes* nests. *Annals of the Entomological Society of America* 61:1528–1539.

Nijhout, H. F. 1999. When developmental pathways diverge. *Proceedings of the National Academy of Sciences U.S.A.* 96:5348–5350.

Noll, F. B., J. W. Wenzel, and R. Zucchi. 2004. Contrasting pre-imaginal and post-imaginal caste determination in neotropical swarm-founding wasps (Hymenoptera: Vespidae; Epiponini). *American Museum Novitates* 3467:1–24.

Noll, F. B., and R. Zucchi. 2000. Increasing caste differences related to life cycle progression in some neotropical swarm-founding polygynic polistine wasps (Hymenoptera Vespidae Epiponini). *Ethology, Ecology & Evolution* 12:43–65.

Nonacs, P. 1986. Ant reproductive strategies and sex allocation theory. *The Quarterly Review of Biology* 61:1–21.

Nonacs, P., and H. K. Reeve. 1993. Opportunistic adoption of orphaned nests in papers wasps as an alternative reproductive strategy. *Behavioural Processes* 30:47–59.

Nonacs, P., and H. K. Reeve. 1995. The ecology of cooperation in wasps: causes and consequences of alternative reproductive decisions. *Ecology* 76:953–967.

Noonan, K. M. 1979. Individual strategies of inclusive fitness maximizing in the social wasp, *Polistes fuscatus* (Hymenoptera: Vespidae). Ph.D. dissertation, University of Michigan, Ann Arbor.

Noonan, K. M. 1981. Individual strategies of inclusive-fitness-maximizing in *Polistes fuscatus* foundresses. In *Natural selection and social behavior: Recent research and new theory* (pp. 18–44), edited by R. D. Alexander and D. W. Tinkle. New York: Chiron Press.

O'Brien, M. F., and F. E. Kurczewski. 1982a. Ethology and overwintering of *Podalonia luctuosa* (Hymenoptera: Sphecidae). *Great Lakes Entomologist* 15:261–275.

O'Brien, M. F., and F. E. Kurczewski. 1982b. Nesting and overwintering behavior of *Liris argentata* (Hymenoptera: Larridae). *Journal of the Georgia Entomological Society* 17:60–68.

Ochiai, S. 1960. Comparative studies on embryology of the bees – *Apis, Polistes, Vespula* and *Vespa*, with special reference to the development of the silk gland. *Bulletin of the Faculty of Agriculture, Tamagawa University* 1:13–45.

Ode, P. J., and J. A. Rosenheim. 1998. Sex allocation and the evolutionary transition between solitary and gregarious parasitoid development. *The American Naturalist* 152:757–761.

Ode, P. J., and S. W. Rissing. 2002. Resource abundance and sex allocation by queen and workers in the harvester ant, *Messor pergandei*. *Behavioral Ecology and Sociobiology* 51:548–556.

O'Donnell, S. 1994. Nestmate copulation in the Neotropical eusocial wasp *Polistes instabilis* De Saussure (Hymenoptera: Vespidae). *Psyche* 101:33–36.

O'Donnell, S. 1995a. Necrophagy by neotropical swarm-founding wasps (Hymenoptera: Vespidae, Epiponini). *Biotropica* 27:133–136.

O'Donnell, S. 1995b. Division of labor in post-emergence colonies of the primitively eusocial wasp *Polistes instabilis* de Saussure (Hymenoptera: Vespidae). *Insectes Sociaux* 42:17–29.

O'Donnell, S. 1996. Reproductive potential and division of labor in wasps: are queen and worker behavior alternative strategies? *Ethology, Ecology & Evolution* 8:305–308.

O'Donnell, S. 1998. Reproductive caste determination in eusocial wasps (Hymenoptera: Vespidae). *Annual Review of Entomology* 43:323–346.

O'Donnell, S., and R. L. Jeanne. 1990. Notes on an army ant (*Eciton burchelli*) raid on a social wasp colony (*Agelaia yepocapa*) in Costa Rica. *Journal of Tropical Ecology* 6:507–509.

O'Donnell, S., and R. L. Jeanne. 1991. Interspecific occupation of a tropical social wasp colony (Hymenoptera: Vespidae: *Polistes*). *Journal of Insect Behavior* 4:397–400.

O'Donnell, S., and F. J. Joyce. 2001. Seasonality and colony composition in a montane tropical eusocial wasp. *Biotropica* 33:727–732.

Ohgushi, R., S. F. Sakagami, Sô. Yamane, and N. D. Abbas. 1983. Nest architecture and related notes of stenogastrine wasps in the province of Sumatera Barat, Indonesia (Hymenoptera, Vespidae). *Science Reports of Kanazawa University* 28:27–58.

Ohgushi, R.-I., Sô. Yamane, and N. D. Abbas. 1986. Additional descriptions and records of stenogastrine nests collected in Sumatera Barat, Indonesia, with some biological notes (Hymenoptera, Vespidae). *Kontyû* 54:1–11.

Omholt, S. W. 1988. Relationships between worker longevity and the intracolonial population dynamics of the honeybee. *Journal of Theoretical Biology* 130:275–284.

O'Neill, K. M. 2001. *Solitary wasps: Behavior and natural history*. Ithaca, NY: Cornell University Press.

Oster, G. F., and E. O. Wilson. 1978. *Caste and ecology in the social insects*. Princeton, NJ: Princeton University Press.

Pagden, H. T. 1958. Some Malayan social wasps. *Malayan Nature Journal* 12:131–148.

Pagden, H. T. 1962. More about *Stenogaster*. *Malayan Nature Journal* 16:95–102.

Page, R. E. 1980. The evolution of multiple mating by honey bee queens (*Apis mellifera* L.). *Genetics* 96:263–273.

Page, R. E., and M. K. Fondrk. 1995. The effects of colony level selection on the social organization of honey bee (*Apis mellifera* L.) colonies—colony level components of pollen hoarding. *Behavioral Ecology and Sociobiology* 36:135–144.

Page, R. E., Jr., D. C. Post, and R. A. Metcalf. 1989. Satellite nests, early males, and plasticity of reproductive behavior in a paper wasp. *The American Naturalist* 134:731–748.

Pant, R., and B. Unni. 1978. Amino-acid composition of fibres of some silk worm species (*Philosamia ricini*, *Antheraea mylitta* and *Attacus atalus*). *Current Science* 47:681–682.

Pardi, L. 1939. I corpi grassi degli insetti. *Redia* 25:87–288.

Pardi, L. 1942. Ricerche sui Polistini V. la poliginia iniziale in *Polistes gallicus* (L.). *Bollettino dell'Istituto di Entomologia dell'Università di Bologna* 14:1–106.

Pardi, L. 1946. Ricerche sui Polistini. VII. La "dominazione" e il ciclo ovarico annuale in *Polistes gallicus* (L.). *Bollettino dell'Istituto di Entomologia dell'Università di Bologna* 15:25–84.

Pardi, L. 1948. Dominance order in *Polistes* wasps. *Physiological Zoology* 21:1–13.

Pardi, L. 1951. Studio dell'attività e della divisione di lavoro in una società di *Polistes gallicus* (L.) dopo la comparsa delle operaie (Ricerche sui Polistini 12). *Archivo Zoologico Italiano* 36:363–431.

Pardi, L., and M. Cavalcanti. 1951. Esperienze sul meccanismo della monoginia funzionale in *Polistes gallicus* (L.). *Bollettino di zoologia / Unione Zoologica Italiana* 18:247–252.

Pardi, L., and M. T. Marino Piccioli. 1981. Studies on the biology of *Belonogaster* (Hymenoptera Vespidae). 4. On caste differences in *Belonogaster griseus* (Fab.) and the position of this genus among social wasps. *Monitore zoologico italiano (Nuova serie)* 14:131–146.

Pardi, L., and S. Turillazzi. 1982. Biologia delle Stenogastrinae (Hymenoptera, Vespoidea). *Atti dell'Accademia Nazionale Italiana de Entomologia, Rendiconti* 30:3–21.

Parker, F. D. 1966. A revision of the North American species in the genus *Leptochilus* (Hymenoptera: Eumenidae). *Miscellaneous Publications of the Entomological Society of America* 5:153–229.

Peakin, G. J. 1972. Aspects of productivity in *Tetramorium caespitum* L. *Ecologia Polska* 20:55–63.

Pexton, J. J., D. J. Rankin, C. Dytham, and P. J. Mayhew. 2003. Asymmetric larval mobility and the evolutionary transition from siblicide to nonsiblicidal behavior in parasitoid wasps. *Behavioral Ecology* 14:182–193.

Pianka, E. R. 1970. On r- and K-selection. *The American Naturalist* 104:592–597.

Pickering, J. 1980. Sex ratio, social behavior and ecology in *Polistes* (Hymenoptera, Vespidae), *Pachysomoides* (Hymenoptera, Ichneumonidae) and *Plasmodium* (Protozoa, Haemosporida). Ph.D. dissertation, Harvard University, Cambridge.

Pickett, K. M., D. M. Osborne, D. Wahl, and J. W. Wenzel. 2001. An enormous nest of *Vespula squamosa* from Florida, the largest social wasp nest reported from North America, with notes on colony cycle and reproduction. *Journal of the New York Entomological Society* 109:408–415.

Pierce, W. D. 1909. A monographic revision of the twisted winged insects comprising the order Strepsiptera Kirby. *Bulletin of the United States National Museum* 66:1–232.

Platt, J. R. 1964. Strong inference. *Science* 146:347–53.

Post, D. C., R. L. Jeanne, and E. H. Erickson, Jr. 1988. Variation in behavior among workers of the primitively social wasp *Polistes fuscatus variatus*. In *Interindividual behavioral variability in social insects* (pp. 283–321), edited by R. L. Jeanne. Boulder, CO: Westview Press.

Powell, J. A., and W. J. Turner. 1975. Observations on oviposition behavior and host selection in *Orussus occidentalis* (Hymenoptera: Siricoidea). *Journal of the Kansas Entomological Society* 48:299–307.

Prashad, B., A. K. Saund, and N. K. Mathur. 1972. Amino-acid composition of spider silk. *Indian Journal of Biochemistry and Biophysics* 9:351–352.

Prezoto, F., and N. Gobbi. 2003. Patterns of honey storage in nests of the neotropical paper wasp *Polistes simillimus* Zikán, 1951 (Hymenoptera, Vespidae). *Sociobiology* 41:437–442.

Pricer, J. L. 1908. The life history of the carpenter ant. *Biological Bulletin (Woods Hole)* 14:177–218.

Queller, D. C. 1989. The evolution of eusociality: reproductive head starts of workers. *Proceedings of the National Academy of Sciences U.S.A.* 86:3224–3226.

Queller, D. C. 1996. The origin and maintenance of eusociality: the advantage of extended parental care. In *Natural history and evolution of paper-wasps* (pp. 218–234), edited by S. Turillazzi and M. J. West-Eberhard. Oxford: Oxford University Press.

Queller, D. C., J. A. Negrón-Sotomayor, J. E. Strassmann, and C. R. Hughes. 1993. Queen number and genetic relatedness in a neotropical wasp, *Polybia occidentalis*. *Behavioral Ecology* 4:7–13.

Queller, D. C., and J. E. Strassmann. 1988. Reproductive success and group nesting in the paper wasp, *Polistes annularis*. In *Reproductive success: Studies of individual variation in contrasting breeding systems* (pp. 76–96), edited by T. H. Clutton-Brock. Chicago: University of Chicago Press.

Queller, D. C., and J. E. Strassmann. 1989. Measuring inclusive fitness in social wasps. In *The genetics of social evolution* (pp. 103–122), edited by M. D. Breed and R. E. Page, Jr. Boulder: Westview Press.

Queller, D. C., and J. E. Strassmann. 1998. Kin selection and social insects. *BioScience* 48:165–175.

Queller, D. C., F. Zacchi, R. Cervo, S. Turillazzi, M. T. Henshaw, L. A. Santorelli, and J. E. Strassmann. 2000. Unrelated helpers in a social insect. *Nature* 405:784–787.

Quicke, D. L. J. 1997. *Parasitic wasps*. London: Chapman & Hall.

Quicke, D. L. J., and M. R. Shaw. 2004. Cocoon silk chemistry in parasitic wasps (Hymenoptera, Ichneumonoidea) and their hosts. *Biological Journal of the Linnean Society* 81:161–170.

Rabb, R. L. 1960. Biological studies of *Polistes* in North Carolina (Hymenoptera: Vespidae). *Annals of the Entomological Society of America* 53:111–121.

Rau, P. 1928. The honey-gathering habits of *Polistes* wasps. *The Biological Bulletin* 54:503–519.

Rau, P. 1929. Feeding experiments on *Polistes* wasps. *The Canadian Entomologist* 61:25–30.

Rau, P. 1930a. Ecological and behavior notes on the wasp, *Polistes pallipes*. *The Canadian Entomologist* 62:143–147.

Rau, P. 1930b. The behavior of hibernating *Polistes* wasps. *Annals of the Entomological Society of America* 23:461–466.

Rau, P. 1930c. Mortality of *Polistes annularis* wasps during hibernation. *The Canadian Entomologist* 62:81–83.

Rau, P. 1931. *Polistes* wasps and their use of water. *Ecology* 12:690–693.

Rau, P. 1935. The courtship and mating of the wasp, *Monobia quadridens* (Hymen.: Vespidae). *Entomological News* 46:57–58.

Rau, P. 1941. The swarming of *Polistes* wasps in temperate regions. *Annals of the Entomological Society of America* 34:580–584.

Rau, P. 1945a. The size of the cell and the sex of the wasp in *Ancistrocerus catskillensis* De Sauss. (Hymenoptera). *Annals of the Entomological Society of America* 38:88.

Rau, P. 1945b. The carnivorous habits of the adult wasp, *Odynerus dorsalis* Fab. *Bulletin of the Brooklyn Entomological Society* 40:29–30.

Rau, P., and N. Rau. 1918. *Wasp studies afield*. Princeton, NJ: Princeton University Press.

Rawlings, G. B. 1957. *Guiglia schauinslandi* (Ashmead) (Hym. Orussidae), a parasite of *Sirex noctilio* (Fabricius) in New Zealand. *The Entomologist* 90:35–26.

Réaumur, R. A. F. de. 1742a. Sixieme memoire. Histoire des guespes in general, et en particulier de celles qui vivent sous terre en sociéte. In *Memoires pour servir a l'histoire des insectes,* vol. 6. Paris: Imprimerie Royale (reprint 1748 by Pierre Mortier, Amsterdam).

Réaumur, R. A. F. de. 1742b. Septieme memoire. Des frelons, des guespes cartonnieres, et de quelques autres guespes qui vivent en sociéte. In *Memoires pour servir a l'histoire des insectes,* vol. 6. Paris: Imprimerie Royale (reprint 1748 by Pierre Mortier, Amsterdam).

Reed, H. C., J. Gallego, and J. Nelson. 1988. Morphological evidence for polygyny in post-emergence colonies of the red paper wasp, *Polistes perplexus* Cresson (Hymenoptera, Vespidae). *Journal of the Kansas Entomological Society* 61:453–463.

Reed, H. C., and P. J. Landolt. 1991. Swarming of paper wasp (Hymenoptera, Vespidae) sexuals at towers in Florida. *Annals of the Entomological Society of America* 84:628–35.

Reeve, H. K. 1991. *Polistes*. In *The social biology of wasps* (pp. 99–148), edited by K. G. Ross and R. W. Matthews. Ithaca, NY: Cornell University Press.

Reeve, H. K. 1993. Haplodiploidy, eusociality and absence of male parental and alloparental care in Hymenoptera: a unifying genetic hypothesis distinct from kin selection theory. *Philosophical Transactions of the Royal Society of London B* 342:335–352.

Reeve, H. K., J. M. Peters, P. Nonacs, and P. T. Starks. 1998. Dispersal of first "workers" in social wasps: causes and implications of an alternative reproductive strategy. *Proceedings of the National Academy of Sciences U.S.A.* 95:13737–13742.

Reeve, H. K., and P. W. Sherman. 1993. Adaptation and the goals of evolutionary research. *The Quarterly Review of Biology* 68:1–32.

Reichenbach, H. 1956. *The direction of time*. Berkeley: University of California Press.

Reid, J. A. 1942. On the classification of the larvae of the Vespidae (Hymenoptera). *Transactions of the Royal Entomological Society, London* 92:285–331.

Richards, O. W. 1953. *The social insects*. London: MacDonald & Co.

Richards, O. W. 1965. Concluding remarks on the social organization of insect communities. In *Social organization of animal communities* (pp. 169–172), edited by P. E. Ellis. Symposia of the Zoological Society of London. Symposium No. 14.

Richards, O. W. 1971. The biology of the social wasps (Hymenoptera, Vespidae). *Biological Reviews of the Cambridge Philosophical Society* 46:483–528.

Richards, O. W. 1978. *The social wasps of the Americas, excluding the Vespinae*. London: British Museum (Natural History).

Richards, O. W., and A. H. Hamm. 1939. The biology of the British Pompilidae. *Transactions of the Society for British Entomology* 6:51–114.

Richards, O. W., and M. J. Richards. 1951. Observations on the social wasps of South America (Hymenoptera Vespidae). *Transactions of the Royal Entomological Society of London* 102:1–169.

Robinson, G. E., and R. E. Page. 1988. Genetic determination of guarding and undertaking in honey-bee colonies. *Nature* 333:356–358.

Roff, D. A. 1992. *The evolution of life histories: Theory and analysis*. New York: Chapman & Hall.

Ronquist, F. 1999. Phylogeny of the Hymenoptera (Insecta): the state of the art. *Zoologica Scripta* 28:3–11.

Ronquist, F., A. P. Rasnitsyn, A. Roy, K. Eriksson, and M. Lindgren. 1999. Phylogeny of the Hymenoptera: a cladistic reanalysis of Rasnitsyn's (1988) data. *Zoologica Scripta* 28:13–50.

Rose, M. R., and L. D. Mueller. 2006. *Evolution and ecology of the organism*. Upper Saddle River, NJ: Pearson Prentice Hall.

Röseler, P. F., and I. Röseler. 1989. Dominance of ovarectomized foundresses of the paper wasp, *Polistes gallicus*. *Insectes Sociaux* 36:219–234.

Rosenheim, J. A. 1993. Single-sex broods and the evolution of nonsiblicidal parasitoid wasps. *The American Naturalist* 141:90–104.

Ross, K. G., and R. W. Matthews. 1989a. Population genetic structure and social evolution in the sphecid wasp *Microstigmus comes*. *The American Naturalist* 134:574–598.

Ross, K. G., and R. W. Matthews. 1989b. New evidence for eusociality in the sphecid wasp *Microstigmus comes*. *Animal Behaviour* 38:613–619.

Ross, K. G., and R. W. Matthews, editors. 1991. *The social biology of wasps*. Ithaca, NY: Cornell University Press.

Rossi, A. M., and J. H. Hunt. 1988. Honey supplementation and its developmental consequences: evidence for food limitation in a paper wasp, *Polistes metricus*. *Ecological Entomology* 13:437–442.

Roubaud, E. 1908. Gradation et perfectionnement de l'instinct chez les Guêpes solitaires d'Afrique, du genre *Synagris*. *Comptes rendus des séances de l'Académie des Sciences, Paris* 147:695–697.

Roubaud, E. 1911. The natural history of the solitary wasps of the genus *Synagris*. *Smithsonian Institution Annual Report 1910*:507–525.

Roubaud, E. 1916. Recherches biologiques sur les guêpes solitaires et sociales d'Afrique.

La genèse de la vie sociale et l'évolution de l'instinct maternel chez les vespides. *Annales des Sciences Naturelles, 10ᵉ série: Zoologie* 1:1–160.

Ruttner, F. 1977. The problem of the Cape bee (*Apis mellifera capensis* Escholtz): parthenogenesis, size of population, evolution. *Apidologie* 8:281–294.

Sacktor, B., and C. C. Childress. 1967. Metabolism of proline in insect flight muscle and its significance in stimulating the oxidation of pyruvate. *Archives of Biochemistry and Biophysics* 120:583–588.

Saito, Y. 1994. Is sterility by deleterious recessives an origin of inequalities in the evolution of eusociality? *Journal of Theoretical Biology* 166:113–115.

Sakagami, S. F., and Y. Maeta. 1987a. Multifemale nests and rudimentary castes of an "almost" solitary bee *Ceratina flavipes*, with additional observation on multifemale nests of *Ceratina japonica* (Hymenoptera, Apoidea). *Kontyû* 55:391–409.

Sakagami, S. F., and Y. Maeta. 1987b. Sociality, induced and/or natural, in the basically solitary small carpenter bees (*Ceratina*). In *Animal societies: Theories and facts* (pp. 1–16), edited by Y. Itô, J. L. Brown and J. Kikkawa. Tokyo: Scientific Societies Press, Ltd.

Salmon, W. C. 1998. Why ask, "Why?"? An inquiry concerning scientific explanation. In *Causality and explanation* (pp. 125–141), edited by W. C. Salmon. New York: Oxford University Press. [Original edition, 1978. *Proceedings and Addresses of the American Philosophical Association* 51: 683–705.]

Samuel, C. T. 1987. Factors affecting colony size in the stenogastrine wasp *Liostenogaster flavolineata*. Ph.D. dissertation, University of Malaya, Kuala Lumpur.

Schaal, B. A., and W. J. Leverich. 1981. The demographic consequences of two-stage life cycles: survivorship and the time of reproduction. *The American Naturalist* 118:135–138.

Schatton-Gadelmayer, K., and W. Engels. 1988. Hemolymph proteins and body weight in newly emerged worker honeybees according to the different rates of parasitation by brood mites. *Entomologia Generalis* 14:93–101.

Schiff, N. 2004. *Pseudoperga guerini* (Hymenoptera: Pergidae). A female protecting her young on a eucalyptus leaf. *American Entomologist* 50:129.

Schilder, K., J. Heinze, and B. Hölldobler. 1999. Colony structure and reproduction in the thelytokous parthenogenetic ant *Platythyrea punctata* (F. Smith) (Hymenoptera, Formicidae). *Insectes Sociaux* 46:150–158.

Schlichting, C. D., and M. Pigliucci. 1998. *Phenotypic evolution: A reaction norm perspective*. Sunderland, MA: Sinauer Associates.

Schmitz, J., and R. F. A. Moritz. 1998. Molecular phylogeny of Vespidae (Hymenoptera) and the evolution of sociality in wasps. *Molecular Phylogenetics and Evolution* 9:183–191.

Schulmeister, S. 2003. Simultaneous analysis of basal Hymenoptera (Insecta): introducing robust-choice sensitivity analysis. *Biological Journal of the Linnean Society* 79:245–275.

Schwarz, M. P. 1988. Local resource enhancement and sex ratios in a primitively social bee. *Nature* 331:346–348.

Schwarz, M. P. 1994. Female-biased sex-ratios in a facultatively social bee and their implications for social evolution. *Evolution* 48:1684–1697.

Schwarz, M. P., N. J. Bull, and S. J. B. Cooper. 2003. Molecular phylogenetics of allodapine bees, with implications for the evolution of sociality and progressive rearing. *Systematic Biology* 52:1–14.

Seal, J. N., and J. H. Hunt. 2004. Food supplementation affects colony-level life history traits in the annual social wasp *Polistes metricus*. *Insectes Sociaux* 51:239–242.

Seeley, T. D. 1985. *Honeybee ecology: A study of adaptation in social life*. Princeton, NJ: Princeton University Press.

Seger, J. 1983. Partial bivoltinism may cause alternating sex-ratio biases that favour eusociality. *Nature* 301:59–62.

Seger, J. 1991. Cooperation and conflict in social insects. In *Behavioural ecology: An evolutionary approach*, 3rd ed. (pp. 338–373), edited by J. R. Krebs and N. B. Davies. Oxford: Blackwell.

Sherman, P. W., E. A. Lacey, H. K. Reeve, and L. Keller. 1995. The eusociality continuum. *Behavioral Ecology* 6:102–108.

Shima, S. N., Sô. Yamane, and R. Zucchi. 1994. Morphological caste differences in some neotropical swarm-founding polistine wasps I. *Apoica flavissima* (Hymenoptera, Vespidae). *Japanese Journal of Entomology* 62:811–822.

Shreeves, G., M. A. Cant, A. Bolton, and J. Field. 2003. Insurance-based advantages for subordinate co-foundresses in a temperate paper wasp. *Proceedings of the Royal Society of London B* 270:1617–1622.

Shreeves, G., and J. Field. 2002. Group size and direct fitness in social queues. *The American Naturalist* 159:81–95.

Shuker, D. M., and S. A. West. 2004. Information constraints and the precision of adaptation: sex ratio manipulation in wasps. *Proceedings of the National Academy of Sciences U.S.A.* 101:10363–10367.

Singer, T. L., K. E. Espelie, and D. S. Himmelsbach. 1992. Ultrastructural and chemical examination of paper and pedicel from laboratory and field nests of the social wasp *Polistes metricus* Say. *Journal of Chemical Ecology* 18:77–86.

Sinha, A., S. Premnath, K. Chandrashekara, and R. Gadagkar. 1993. *Ropalidia rufoplagiata*: a polistine wasp society probably lacking a permanent reproductive division of labor. *Insectes Sociaux* 40:69–86.

Sledge, M. F., A. Fortunato, S. Turillazzi, E. Francescato, R. Hashim, G. Moneti, and G. R. Jones. 2000. Use of Dufour's gland secretion in nest defense and brood nutrition by hover wasps (Hymenoptera, Stenogastrinae). *Journal of Insect Physiology* 46:753–761.

Smith, A. R., S. O'Donnell, and R. L. Jeanne. 2002. Evolution of swarm communication in eusocial wasps (Hymenoptera: Vespidae). *Journal of Insect Behavior* 15:751–764.

Smith, R. H., and M. R. Shaw. 1980. Haplodiploid sex ratios and the mutation rate. *Nature* 287:728–729.

Snell, G. D. 1932. The role of male parthenogenesis in the evolution of the social Hymenoptera. *The American Naturalist* 66:381–384.

Snelling, R. 1952. Notes on nesting and hibernation of *Polistes*. *Pan-Pacific Entomologist* 29:177.

Snyder, L. E. 1992. The genetics of social behavior in a polygynous ant. *Naturwissenschaften* 79:525–527.

Snyder, L. E. 1993. Non-random behavioural interactions among genetic sub-groups in a polygynous ant. *Animal Behaviour* 46:431–439.

Sober, E. 1984. *The nature of selection: Evolutionary theory and philosophical focus*. Cambridge: MIT Press.

Sober, E., and D. S. Wilson. 1998. *Unto others: The evolution and psychology of unselfish behavior*. Cambridge, MA: Harvard University Press.

Solís, C. R. and J. E. Strassmann. 1990. Presence of brood affects caste differentiation in the social wasp, *Polistes exclamans* Viereck (Hymenoptera, Vespidae). *Functional Ecology* 4:531–541.

Spradbery, J. P. 1971. Seasonal changes in the population structure of wasp colonies (Hymenoptera: Vespidae). *Journal of Animal Ecology* 40:501–523.

Spradbery, J. P. 1972. A biometric study of seasonal variation in worker wasps (Hymenoptera: Vespidae). *Journal of Entomology (A)* 47:61–69.

Spradbery, J. P. 1973. *Wasps: An account of the biology and natural history of solitary and social wasps*. Seattle: University of Washington Press.

Spradbery, J. P. 1975. The biology of *Stenogaster concinna* Van der Vecht with comments on the phylogeny of the Stenogastrinae (Hymenoptera Vespidae). *Journal of the Australian Entomological Society* 14:309–318.

Spradbery, J. P. 1993. Queen brood reared in worker cells by the social wasp, *Vespula germanica* (F.) (Hymenoptera: Vespidae). *Insectes Sociaux* 40:181–190.

Stahlhut, J. K., and D. P. Cowan. 2004. Single-locus complementary sex determination in the inbreeding wasp *Euodynerus foraminatus* Saussure (Hymenoptera : Vespidae). *Heredity* 92:189–196.

Starks, P. T. 1998. A novel 'sit and wait' reproductive strategy in social wasps. *Proceedings of the Royal Society of London B* 265:1407–1410.

Starr, C. K. 1976. Nest reutilization by *Polistes metricus* (Hymenoptera: Vespidae) and possible limitation of multiple foundress associations by parasitoids. *Journal of the Kansas Entomological Society* 49:142–144.

Starr, C. K. 1978. Nest reutilization in North American *Polistes* (Hymenoptera: Vespidae): two possible selective factors. *Journal of the Kansas Entomological Society* 51:394–397.

Starr, C. K. 1982. P. Marchal, 1897 (A translation of La castration nutriciale chez les Hyménoptères sociaux). *Sphecos* 5:26–27.

Starr, C. K. 1985. A simple pain scale for field comparison of hymenopteran stings. *Journal of Entomological Science* 20:225–232.

Stearns, S. C. 1992. *The evolution of life histories*. Oxford: Oxford University Press.

Stein, K. J., and R. D. Fell. 1992. Seasonal comparison of weight, energy reserve, and nitrogen changes in queens of the baldfaced hornet (Hymenoptera: Vespidae). *Environmental Entomology* 21:148–155.

Steiner, A. L. 1983. Predatory behavior of solitary wasps V. Stinging of caterpillars by *Euodynerus foraminatus* (Hymenoptera: Eumenidae). Weakening of the complete four-sting pattern. *Biology of Behaviour* 8:11–26.

Strambi, C., A. Strambi, and R. Augier. 1982. Protein level in the haemolymph of the wasp *Polistes gallicus* L. at the beginning of imaginal life and during overwintering. Action of the strepsiterian parasite *Xenos vesparum* Rossi. *Experientia* 38:1189–1191.

Strand, M. R., and M. Grbic. 1997. The development and evolution of polyembryonic insects. In *Current Topics in Developmental Biology, Vol 35*.

Strand, M. R., and M. Grbic. 1999. Life history shifts and alterations in the early development of parasitic wasps. *Invertebrate Reproduction and Development* 36:51–56.

Strassmann, J. E. 1979. Honey caches help female paper wasps (*Polistes annularis*) survive Texas winters. *Science* 204:207–209.

Strassmann, J. E. 1981. Evolutionary implications of early male and satellite nest production in *Polistes exclamans* colony cycles. *Behavioral Ecology and Sociobiology* 8:55–64.

Strassmann, J. E. 1983. Nest fidelity and group size among foundresses of *Polistes annularis* (Hymenoptera: Vespidae). *Journal of the Kansas Entomological Society* 56:621–634.

Strassmann, J. E. 1985. Worker mortality and the evolution of castes in the social wasp *Polistes exclamans*. *Insectes Sociaux* 32:275–285.

Strassmann, J. E. 1989a. Group colony foundation in *Polistes annularis* (Hymenoptera: Vespidae). *Psyche* 96:223–236.

Strassmann, J. E. 1989b. Early termination of brood rearing in the social wasp, *Polistes annularis* (Hymenoptera: Vespidae). *Journal of the Kansas Entomological Society* 62:353–362.

Strassmann, J. E. 1996. Selective altruism towards closer over more distant relatives in colonies of the primitively eusocial wasp, *Polistes*. In *Natural history and evolution of paper-wasps* (pp. 190–201), edited by S. Turillazzi and M. J. West-Eberhard. Oxford: Oxford University Press.

Strassmann, J. E., A. Fortunato, R. Cervo, S. Turillazzi, J. M. Damon, and D. C. Queller. 2004. The cost of queen loss in the social wasp *Polistes dominulus* (Hymenoptera: Vespidae). *Journal of the Kansas Entomological Society* 77:343–355.

Strassmann, J. E., C. R. Hughes, and D. C. Queller. 1990. Colony defense in the social wasp, *Parachartergus colobopterus*. *Biotropica* 22:324–327.

Strassmann, J. E., R. E. Lee, R. R. Rojas, and J. G. Baust. 1984a. Caste and sex differences in cold-hardiness in the social wasps, *Polistes annularis* and *Polistes exclamans* (Hymenoptera, Vespidae). *Insectes Sociaux* 31:291–301.

Strassmann, J. E., and D. C. Meyer. 1983. Gerontocracy in the social wasp, *Polistes exclamans*. *Animal Behaviour* 31:431–438.

Strassmann, J. E., D. C. Meyer, and R. L. Matlock. 1984b. Behavioral castes in the social wasp, *Polistes exclamans* (Hymenoptera: Vespidae). *Sociobiology* 8:211–224.

Strassmann, J. E., and D. C. Queller. 1989. Ecological determinants of social evolution. In *The genetics of social evolution* (pp. 81–101), edited by M. D. Breed and R. E. Page, Jr. Boulder, CO: Westview Press.

Strassmann, J. E., D. C. Queller, and C. R. Hughes. 1988. Predation and the evolution of sociality in the paper wasp *Polistes bellicosus*. *Ecology* 69:1497–1505.

Strassmann, J. E., C. R. Solís, C. R. Hughes, K. F. Goodnight, and D. C. Queller. 1997. Colony life history and demography of a swarm-founding social wasp. *Behavioral Ecology and Sociobiology* 40:71–77.

Strassmann, J. E., B. W. Sullender, and D. C. Queller. 2002. Caste totipotency and conflict in a large-colony social insect. *Proceedings of the Royal Society of London B* 269:263–270.

Strouthamer, R., R. F. Luck, and W. D. Hamilton. 1990. Antibiotics cause parthenogenetic *Trichogamma* (Hymenoptera/Trichogrammatidae) to revert to sex. *Proceedings of the National Academy of Sciences U.S.A.* 87:2424–2427.

Stuart, R. J., and R. E. Page. 1991. Genetic component to division of labor among workers of a leptothoracine ant. *Naurwissenschaften* 78:375–377.

Stubblefield, J. W., and E. L. Charnov. 1985. Some conceptual issues in the origin of eusociality. *Heredity* 57:181–187.

Šula, J., D. Kodřik and R. Socha. 1995. Hexameric haemolymph protein related to adult diapause in the red firebug, *Pyrrhocoris apterus*. *Journal of Insect Physiology* 41:793–800.

Sumner, S., M. Sasiraghi, W. Foster, and J. Field. 2002. High reproductive skew in tropical hover wasps. *Proceedings of the Royal Society of London B* 269:179–186.

Suzuki, T. 1985. Mating and laying of female-producing eggs by orphaned workers of a paper wasp, *Polistes snelleni* (Hymenoptera: Vespidae). *Annals of the Entomological Society of America* 78:736–739.

Suzuki, T. 1997. Worker mating in queen-right colonies of a temperate paper wasp. *Naturwissenschaften* 84:304–305.

Suzuki, T. 1998. Paradox of worker reproduction and worker mating in temperate paper wasps, *Polistes chinenesis* and *P. snelleni* (Hymenoptera Vespidae). *Ethology, Ecology & Evolution* 10:347–359.

Suzuki, T., and M. Ramesh. 1992. Colony founding in the social wasp, *Polistes stigma* (Hymenoptera Vespidae), in India. *Ethology, Ecology & Evolution* 4:333–341.

Taffe, C. A. 1978. Temporal distribution of mortality in a field population of *Zeta abdominale* (Hymenoptera) in Jamaica. *Oikos* 31:106–111.

Talbot, M. 1943. Population studies of the ant, *Prenolepis imparis* Say. *Ecology* 24:31–44.

Talbot, M. 1945. Population studies of the ant, *Myrmica schenckii* ssp. *emeryana* Forel. *Annals of the Entomological Society of America* 38:365–372.

Talbot, M. 1951. Populations and hibernating conditions of the ant *Aphaenogaster (Attomyrma) rudis* Emery (Hymenoptera: Formicidae). *Annals of the Entomological Society of America* 44:302–307.

Talbot, M. 1954. Populations of the ant *Aphaenogaster (Attomyrma) treatae* Forel on abandoned fields of the Edwin S. George Reserve. *Contributions from the Laboratory of Vertebrate Biology, University of Michigan* 69:1–9.

Tang-Martinez, Z. 2001. The mechanisms of kin discrimination and the evolution of kin recognition in vertebrates: a critical re-evaluation. *Behavioural Processes* 53:21–40.

Tarpy, D. R., and R. E. Page. 2002. Sex determination and the evolution of polyandry in honey bees (*Apis mellifera*). *Behavioral Ecology and Sociobiology* 52:143–150.

Tarpy, D. R., and R. E. Page, Jr. 2001. The curious promiscuity of queen honey bees (*Apis mellifera*): evolutionary and behavioral mechanisms. *Annales Zoologici Fennici* 38:255–265.

Taylor, L. H. 1922. Notes on the biology of certain wasps of the genus *Ancistrocerus* (Eumenidæ). *Psyche* 29:48–65.

Terra, W. R., C. D. Santos, and A. F. Ribeiro. 1990. Ultrastructural and biochemical basis of the digestion of nectar and other nutrients by the moth *Erinnyis ello*. *Entomologia Experimentalis et Applicata* 56:277–286.

Thomas, R. K. 2001. Lloyd Morgan's Canon: a history of misrepresentation. http://eprints.yorku.ca/archive/00000017/00/MCWeb.htm.

Trivers, R. 1985. *Social evolution*. Menlo Park, CA: Benjamin Cummings.

Trivers, R. L., and H. Hare. 1976. Haplodiploidy and the evolution of the social insects. *Science* 191:249–263.

Trostle, G. E., and P. F. Torchio. 1986. Notes on the nesting biology and immature development of *Euparagia scutellaris* Cresson (Hymenoptera, Masaridae). *Journal of the Kansas Entomological Society* 59:641–647.

Truman, J. W., and L. M. Riddiford. 1999. The origins of insect metamorphosis. *Nature* 401:447–452.

Tschinkel, W. R., and D. F. Howard. 1978. Queen replacement in orphaned colonies of the fire ant, *Solenopsis invicta*. *Behavioral Ecology and Sociobiology* 3:297–310.

Tsuchida, K., N. Nagata, and J. Kojima. 2002. Diploid males and sex determination in a paper wasp, *Polistes chinensis antennalis* (Hymenoptera, Vespidae). *Insectes Sociaux* 49:120–124.

Tsuchida, K., T. Saigo, S. Tsujita, and K. Takeuchi. 2004. Early male production is not linked to a reproductive strategy in the Japanese paper wasp, *Polistes chinensis antennalis* (Hymenoptera : Vespidae). *Journal of Ethology* 22:119–121.

Turillazzi, S. 1980. Seasonal variations in the size and anatomy of *Polistes gallicus* (L) (Hymenoptera, Vespidae). *Monitore zoologico italiano (Nuova serie)* 14:63–75.

Turillazzi, S. 1983a. Extranidal Behavior of *Parischnogaster nigricans serrei* (Du Buysson) (Hymenoptera, Stenogastrinae). *Zeitschrift für Tierpsychologie* 63:27–36.

Turillazzi, S. 1983b. Patrolling behavior in males of *Parischnogaster nigricans serrei* (Du Buysson) and *P. mellyi* (Saussure) (Hymenoptera, Stenogastrinae). *Accademia Nazionale dei Lincei, Rendiconti della Classe di Scienze fisiche, matematiche e naturali, Serie 8* 72:153–157.

Turillazzi, S. 1985a. Brood rearing behavior and larval development in *Parischnogaster nigricans serrei* (Du Buysson) (Hymenoptera, Stenogastrinae). *Insectes Sociaux* 32:117–127.

Turillazzi, S. 1985b. Function and characteristics of the abdominal substance secreted by wasps of the genus *Parischnogaster* (Hymenoptera Stenogastrinae). *Monitore zoologico italiano (Nuova serie)* 19:91–99.

Turillazzi, S. 1985c. Egg deposition in the genus *Parischnogaster* (Hymenoptera, Stenogastrinae). *Journal of the Kansas Entomological Society* 58:749–752.

Turillazzi, S. 1985d. Colonial cycle of *Parischnogaster nigricans serrei* (Du Buysson) in West Java (Hymenoptera, Stenogastrinae). *Insectes Sociaux* 32:43–60.

Turillazzi, S. 1986. Colony composition and social behavior of *Parischnogaster alternata* Sakagami (Hymenoptera Stenogastrinae). *Monitore zoologico italiano (Nuova serie)* 20:333–347.

Turillazzi, S. 1987. Distinguishing features of the social behaviour of Stenogastrinae wasps. In *Chemistry and biology of social insects* (pp. 492–495), edited by J. Eder and H. Rembold. München: Verlag J. Peperny.

Turillazzi, S. 1989. The origin and evolution of social life in the Stenogastrinae (Hymenoptera, Vespidae). *Journal of Insect Behavior* 2:649–661.

Turillazzi, S. 1990. Social biology of *Liostenogaster vechti* Turillazzi 1988 (Hymenoptera Stenogastrinae). *Tropical Zoology* 3:69–87.

Turillazzi, S. 1991. The Stenogastrinae. In *The social biology of wasps* (pp. 74–98), edited by K. G. Ross and R. W. Matthews. Ithaca, NY: Cornell University Press.

Turillazzi, S. 1994. Protection of brood by means of Dufours gland secretion in 2 species of stenogastrine wasps. *Ethology, Ecology & Evolution* (Special Issue) 3:37–41.

Turillazzi, S. 1999. New species of *Liostenogaster* van der Vecht 1969, with keys to adults and nests (Hymenoptera Vespidae Stenogastrinae). *Tropical Zoology* 12:335–358.

Turillazzi, S., and S. Carfi. 1996. Adults and nest of *Liostenogaster pardii* n. sp. (Hymenoptera Stenogastrinae). *Tropical Zoology* 9:19–30.

Turillazzi, S., R. Cervo, and F. R. Dani. 1997. Intra and inter-specific relationships in a cluster of stenogastrine wasp colonies (Hymenoptera Vespidae). *Ethology, Ecology & Evolution* 9:385–395.

Turillazzi, S., and E. Francescato. 1989. Observations on the behaviour of male stenogastrine wasps (Hymenoptera, Vespidae, Stenogastrinae). *Actes Colloques des Insectes Sociaux* 5:181–187.

Turillazzi, S., and E. Francescato. 1990. Patrolling behavior and related secretory structures in the males of some stenogastrine wasps (Hymenoptera, Vespidae). *Insectes Sociaux* 37:146–157.

Turillazzi, S., and E. Francescato. 1994. Notes on the nest architecture and brood rearing of *Polybioides raphigastra* (Vespidae, Polistinae). *Rendiconti Lincei, Accademia Nazionale dei Lincei, Classe di Scienze fisiche, mathematiche e naturali—Scienze fisiche e naturali* 9:367–375.

Turillazzi, S., E. Francescato, A. B. Tosi, and J. M. Carpenter. 1994. A distinct caste difference in *Polybioides tabidus* (Fabricius) (Hymenoptera, Vespidae). *Insectes Sociaux* 41:327–330.

Turillazzi, S., and M. H. Hansell. 1991. Biology and social behavior of 3 species of *Anischnogaster* (Vespidae, Stenogastrinae) in Papua New Guinea. *Insectes Sociaux* 38:423–437.

Turillazzi, S., and L. Pardi. 1981. Ant guards on nests of *Parischnogaster nigricans serrei* (Buysson) (Stenogastrinae). *Monitore zoologico italiano (Nuova serie)* 15:1–7.

Turillazzi, S., and L. Pardi. 1982. Social behavior of *Parischnogaster nigricans serrei* (Hymenoptera, Vespoidea) in Java. *Annals of the Entomological Society of America* 75:657–664.

Turillazzi, S., and A. Ugolini. 1979. Rubbing behavior in some European *Polistes* (Hymenoptera Vespidae). *Monitore zoologico italiano (Nuova serie)* 13:129–142.

Turillazzi, S., and M. J. West-Eberhard, editors. 1996. *Natural history and evolution of paper-wasps*. Oxford: Oxford University Press.

Ueno, T. 1997. Host age preference and sex allocation in the pupal parasitoid *Itoplectis naranyae* (Hymenoptera: Ichneumonidae). *Annals of the Entomological Society of America* 90:640–645.

Ueno, T. 1999. Host-size-dependent sex ratio in a parasitoid wasp. *Researches on Population Ecology* 41:47–57.

Uyenoyama, M. K. 1984. Inbreeding and the evolution of altruism under kin selection: effects on relatedness and group structure. *Evolution* 38:778–795.

Uyenoyama, M. K., and B. O. Bengtsson. 1982. Towards a genetic theory for the evolution of the sex ratio. III. Parental and sibling control of brood investment ratio under partial sib-mating. *Theoretical Population Biology* 22:43–68.

van Alphen, J. J. M., and L. E. M. Vet. 1986. An evolutionary approach to host finding and selection. In *Insect parasitoids* (pp. 23–61), edited by J. K. Waage and D. Greathead. London: Academic Press.

Van Hooser, C. A., G. J. Gamboa, and T. G. Fishwild. 2002. The function of abdominal stroking in the paper wasp, *Polistes fuscatus* (Hymenoptera Vespidae). *Ethology, Ecology & Evolution* 14:141–148.

Varman, A. R. 1978. Amino-acid and sugar composition of silk of weaver-ants, *Oecophylla*. *Current Science* 47:827–828.

Vecht, J. van der. 1977. Studies of oriental Stenogastrinae (Hymenoptera Vespoidea). *Tijdschrift voor Entomologie* 120:55–75.

Veenendaal, R. L., and T. Piek. 1988. Predatory behaviour of *Discoelius zonalis*. *Entomologische Bererichten* 48:8–12.

Vernier, R. 1997. Essai d'analyse cladistique des generes d'Eumeninae (Vespidae, Hymenoptera) représentés en Europe septentrionale, occidentale et centrale. *Bulletin de la Société Neuchâteloise des Sciences naturelles* 120:87–98.

Vilhelmsen, L. 2001. Phylogeny and classification of the extant basal lineages of the Hymenoptera (Insecta). *Zoological Journal of the Linnean Society* 131:393–442.

Vilhelmsen, L. 2003. Larval anatomy of Orussidae (Hymenoptera). *Journal of Hymenoptera Research* 12:346–354.

Vilhelmsen, L., N. Isidoro, R. Romani, H. H. Basibuyuk, and D. L. J. Quicke. 2001. Host location and oviposition in a basal group of parasitic wasps: the subgenual organ, ovipositor apparatus and associated structures in the Orussidae (Hymenoptera, Insecta). *Zoomorphology* 121:63–84.

Villesen, P., P. J. Gertsch, J. Frydenberg, U. G. Mueller, and J. J. Boomsma. 1999. Evolutionary transition from single to multiple mating in fungus growing ants. *Molecular Ecology* 8:1819–1825.

Vinson, S. B. 1976. Host selection by insect parasitoids. *Annual Review of Entomology* 21:109–133.

Vollrath, F. 1986. Eusociality and extraordinary sex ratios in the spider *Anelosimus eximius* (Araneae: Theridiidae). *Behavioral Ecology and Sociobiology* 18:283–287.

Waage, J. K. 1986. Family planning in parasitoids: adaptive patterns of progeny and sex allocation. In *Insect parasitoids* (pp. 63–95), edited by J. K. Waage and D. Greathead. London: Academic Press.

Wade, M. J. 1980. Kin selection: its components. *Science* 210:665–667.

Wade, M. J. 1982. The effect of multiple inseminations on the evolution of social behaviors in diploid and haplo-diploid organisms. *Journal of Theoretical Biology* 95:351–368.

Wade, M. J. 1985. The influence of multiple inseminations and multiple foundresses on social evolution. *Journal of Theoretical Biology* 112:109–121.

Wade, M. J. 2001. Maternal effect genes and the evolution of sociality in haplo-diploid organisms. *Evolution* 55:453–458.

Wade, M. J., and F. Breden. 1981. Effect of inbreeding on the evolution of altruistic behavior by kin selection. *Evolution* 35:844–858.

Wafa, A. K., and S. G. Sharkawi. 1972. Contribution to the biology of *Vespa orientalis* Fab. [Hymenoptera: Vespidae]. *Bulletin de la Societe Entomologique d'Egypte* 56:219–226.

Waldbauer, G. P., and D. P. Cowan. 1985. Defensive stinging and Müllerian mimicry among eumenid wasps (Hymenoptera: Vespoidea; Eumenidae). *The American Midland Naturalist* 113:198–199.

Wasbauer, M. S. 1995. Pompilidae. In: *The Hymenoptera of Costa Rica* (pp. 522–539), edited by P. E. Hanson and I. D. Gauld. London: The Natural History Museum.

Watson, J. D. 1968. *The double helix*. New York: Atheneum.

Way, M. J. 1963. Mutualism between ants and honeydew-producing Homoptera. *Annual Review of Entomology* 8:307–344.

Wcislo, W. T. 1997. Are behavioral classifications blinders to studying natural variation? In *The evolution of social behavior in insects and arachnids* (pp. 8–13), edited by J. C. Choe and B. J. Crespi. Cambridge: Cambridge University Press.

Weber, N. A. 1966. Fungus-growing ants. *Science* 153:587–604.

Weis-Fogh, T. 1964a. Functional design of the tracheal system of flying insects as compared with the avian lung. *Journal of Experimental Biology* 41:207–227.

Weis-Fogh, T. 1964b. Diffusion in insect wing muscle, the most active tissue known. *Journal of Experimental Biology* 41:229–256.

Wenseleers, T., and J. Billen. 2000. No evidence for *Wolbachia*-induced parthenogenesis in the social Hymenoptera. *Journal of Evolutionary Biology* 13:277–280.

Wenseleers, T., F. Ito, S. Van Borm, R. Huybrechts, F. Volcaert, and J. Billen. 1998. Widespread occurrence of the microorganism *Wolbachia* in ants. *Proceedings of the Royal Society of London B* 265:1447–1452.

Wenzel, J. W. 1987. *Ropalidia formosa*, a nearly solitary paper wasp from Madagascar (Hymenoptera: Vespidae). *Journal of the Kansas Entomological Society* 60:679–699.

Wenzel, J. W. 1989. Endogenous factors, external cues, and eccentric construction in *Polistes annularis* (Hymenoptera: Vespidae). *Journal of Insect Behavior* 2:679–699.

Wenzel, J. W. 1990. A social wasp's nest from the Cretaceous period, Utah, USA, and its biogeographic significance. *Psyche* 97:21–29.

Wenzel, J. W. 1991. Evolution of nest architecture. In *The social biology of wasps* (pp. 480–519), edited by K. G. Ross and R. W. Matthews. Ithaca, NY: Cornell University Press.

Wenzel, J. W. 1992. Extreme queen-worker dimorphism in *Ropalidia ignobilis*, a small-colony wasp (Hymenoptera: Vespidae). *Insectes Sociaux* 39:31–43.

Wenzel, J. W. 1993. Application of the biogenetic law to behavioral ontogeny: a test using nest architecture in paper wasps. *Journal of Evolutionary Biology* 6:229–247.

Wenzel, J. W., and J. M. Carpenter. 1994. Comparing methods: adaptive traits and tests of adaptation. In *Phylogenetics and ecology* (pp. 79–102), edited by P. Eggeleton and R. I. Vane-Wright. London: Academic Press.

Werren, J. H. 1980. Sex ratio adaptations to local mate competition in a parasitic wasp. *Science* 208:1157–1160.

Werren, J. H. 1983. Sex ratio evolution under local mate competition in a parasitic wasp. *Evolution* 37:116–124.

Werren, J. H., U. Nur, and C.-I. Wu. 1988. Selfish genetic elements. *Trends in Ecology and Evolution* 3:297–302.

West, M. J. 1967. Foundress associations in polistine wasps: hierarchies and the evolution of social behavior. *Science* 157:1584–1585.

West Eberhard, M. J. 1969. The social biology of polistine wasps. *Miscellaneous Publications, Museum of Zoology, University of Michigan* 140:1–101.

West-Eberhard, M. J. 1975. The evolution of social behavior by kin selection. *The Quarterly Review of Biology* 50:1–33.

West-Eberhard, M. J. 1977. Morphology and behavior in the taxonomy of *Microstigmus* wasps. In: *Proceedings of the eighth international congres* [sic] *of the International Union for the Study of Social Insects* (pp. 123–125), edited by J. de Wilde. Wageningen: Centre for Agricultural Publishing and Documentation.

West-Eberhard, M. J. 1978a. Polygyny and the evolution of social behavior in wasps. *Journal of the Kansas Entomological Society* 51:832–856.

West-Eberhard, M. J. 1978b. Temporary queens in *Metapolybia* wasps: nonreproductive helpers without altruism? *Science* 200:441–443.

West-Eberhard, M. J. 1979. Sexual selection, social competition and evolution. *Proceedings of the American Philosophical Society* 123:222–234.

West-Eberhard, M. J. 1982. The nature and evolution of swarming in tropical social wasps (Vespidae, Polistinae, Polybiini). In *Social insects in the tropics*, vol. 1 (pp. 97–128), edited by P. Jaisson. Paris: Université Paris-Nord.

West-Eberhard, M. J. 1987a. Observations on *Xenorhynchium nitidulum* (Fabricius) (Hymenoptera, Eumeninae), a primitively social wasp. *Psyche* 94:317–323.

West-Eberhard, M. J.1987b. The epigenetic origins of insect sociality. In *Chemistry and biology of social insects* (pp. 369–372), edited by J. Eder and H. Rembold. München: Verlag J. Peperny.

West-Eberhard, M. J. 1987c. Flexible strategy and social evolution. In *Animal societies: Theories and facts* (pp. 35–51), edited by Y. Itô, J. L. Brown and J. Kikkawa. Tokyo: Scientific Societies Press Ltd.

West-Eberhard, M. J. 1988. Phenotypic plasticity and "genetic" theories of insect sociality. In *Evolution of social behavior and integrative levels* (pp. 123–133), edited by G. Greenberg and E. Tobach. New Jersey: Lawrence Erlbaum Associates.

West-Eberhard, M. J. 1989. Phenotypic plasticity and the origins of diversity. *Annual Review of Ecology and Systematics* 20:249–278.

West-Eberhard, M. J. 1992a. Behavior and evolution. In *Molds, molecules, and Metazoa:*

Growing points in evolutionary biology (pp. 59–79), edited by P. R. Grant and H. S. Horn. Princeton, NJ: Princeton University Press.

West-Eberhard, M. J. 1992b. Genetics, epigenetics, and flexibility: a reply to Crozier. *The American Naturalist* 139:224–226.

West-Eberhard, M. J. 1996. Wasp societies as microcosms for the study of development and evolution. In *Natural history and evolution of paper-wasps* (pp. 290–317), edited by S. Turillazzi and M. J. West-Eberhard. Oxford: Oxford University Press.

West-Eberhard, M. J. 2003. *Developmental plasticity and evolution*. New York: Oxford University Press.

Weyrauch, W. 1935. *Dolichovespula* und *Vespa*. Vergleichende Übersicht über zwei wesentliche Lebenstypen bei sozialen Wespen. Mit Bezugnahme auf die Frage nach der fortschrittlichkeit tierischer Organisation. *Biologisches Zentralblatt* 55:484–524.

Wheeler, G. C., and J. Wheeler. 1979. Larvae of the social Hymenoptera. In *Social Insects*, vol. 1 (pp. 287–338), edited by H. R. Hermann. New York: Academic Press.

Wheeler, W. M. 1910. *Ants: Their structure, development and behavior*. New York: Columbia University Press.

Wheeler, W. M. 1918. Study of some ant larvae with a consideration of the origin and meaning of social habits among insects. *Proceedings of the American Philosophical Society* 57:293–343.

Wheeler, W. M. 1923. *Social life among the insects*. New York: Harcourt, Brace.

Wheeler, W. M. 1928. *The social insects*. London: Kegan Paul, Trench, Trubner & Co., Ltd.

Whitfield, J. B. 1992a. Phylogeny of the non-aculeate Apocrita and the evolution of parasitism in the Hymenoptera. *Journal of Hymenoptera Research* 1:3–14.

Whitfield, J. B. 1992b. The polyphyletic origin of endoparasitism in the cyclostome lineages of Braconidae (Hymenoptera). *Systematic Entomology* 17:273–286.

Whitfield, J. B. 1998. Phylogeny and evolution of host-parasitoid interactions in Hymenoptera. *Annual Review of Entomology* 43:129–151.

Wiernasz, D. C., C. L. Perroni, and B. J. Cole. 2004. Polyandry and fitness in the western harvester ant, *Pogonomyrmex occidentalis*. *Molecular Ecology* 13:1601–1606.

Wigglesworth, V. B. 1972. *The principles of insect physiology*, 7th ed. London: Methuen.

Williams, F. X. 1919. Philippine wasp studies: descriptions of new species and life history studies. *Experiment Station of the Hawaiian Sugar Planters Association, Entomology Series* 14:19–186.

Williams, F. X. 1927. *Euparagia scutellaris* Cresson, a masarid wasp that stores its cells with the young of a curculionid beetle. *The Pan-Pacific Entomologist* 4:28–29.

Williams, G. C. 1966. *Adaptation and natural selection*. Princeton, NJ: Princeton University Press.

Williams, G. C. and D. C. Williams. 1956. Natural selection of individually harmful social adaptations among sibs with special reference to social insects. *Evolution* 11:32–39.

Wilson, E. O. 1966. Behaviour of social insects. In *Insect Behaviour* (pp. 81–96), edited by P. T. Haskell. London: Symposia of the Royal Entomological Society of London.

Wilson, E. O. 1971. *The insect societies*. Cambridge, MA: Belknap Press of Harvard University Press.

Wilson, E. O. 1975. *Sociobiology: The new synthesis*. Cambridge, MA: Belknap Press of Harvard University Press.

Wilson, E. O. 1994. *Naturalist*. Washington, DC: Island Press/Shearwater Books.

Wilson, E. O. 2005. Kin selection as the key to altruism: its rise and fall. *Social Research* 72:159–166.

Wilson, E. O., and B. Hölldobler. 2005. Eusociality: origin and consequences. *Proceedings of the National Academy of Sciences U.S.A.* 102:13367–13371.

Winston, M. L. 1987. *The biology of the honey bee*. Cambridge, MA: Harvard University Press.

Wirtz, P. 1973. Differentiation in the honeybee larva: a histological, electron-microscopical and physiological study of caste induction in *Apis mellifera mellifera* L. *Mededelingen Landbouwhogeschool Wageningen* 73:1–155.

Woyke, J. 1963. What happens to diploid drone larvae in a honeybee colony. *Journal of Apicultural Research* 2:73–75.

Wyatt, G. R. 1961. The biochemistry of hemolymph. *Annual Review of Entomology* 6:75–102.

Yamane, S., S. F. Sakagami, and R. Ohgushi. 1983. Multiple behavioral options in a primitively social wasp, *Parischnogaster mellyi*. *Insectes Sociaux* 30:412–415.

Yamane, Sk. 1976. Morphological and taxonomic studies on vespine larvae, with reference to the phylogeny of the subfamily Vespinae (Hymenoptera: Vespidae). *Insecta Matsumarana (new series)* 8:1–45.

Yamane, Sô. 1969. Preliminary observations on the life history of two polistine wasps, *Polistes Snelleni* and *P. biglumis* in Sapporo, northern Japan. *Journal of the Faculty of Science, Hokkaido University* 17:78–105.

Yamane, Sô., and T. Kawamichi. 1975. Bionomic comparison of *Polistes biglumis* (Hymenoptera, Vespidae) at two different localities in Hokkaido, Northern Japan, with Reference to its probable adaptation to cold climate. *Kontyû* 43:213–232.

Yamane, Sô., J. Kojima, and Sk. Yamane. 1983. Queen/worker size dimorphism in an Oriental wasp, *Ropalidia montana* Carl (Hymenoptera: Vespidae). *Insectes Sociaux* 30:416–422.

Yamane, Sô., and T. Okazawa. 1977. Some biological observations on a paper wasp, *Polistes (Megapolistes) tepidus malayanus* Cameron (Hymenoptera, Vespidae) in New Guinea. *Kontyû* 45:283–299.

Yamashita, O. 1996. Diapause hormone of the silkworm, *Bombyx mori*: structure, gene expression and function. *Journal of Insect Physiology* 42:669–679.

Yanega, D. 1988. Social plasticity and early-diapausing females in a primitively social bee. *Proceedings of the National Academy of Sciences U.S.A.* 85:4374–4377.

Yanega, D. 1989. Caste determination and differential diapause within the first brood of *Halictus rubicundus* in New York (Hymenoptera: Halictidae). *Behavioral Ecology and Sociobiology* 24:97–107.

Yanega, D. 1996. Sex ratio and sex allocation in sweat bees (Hymenoptera: Halictidae). *Journal of the Kansas Entomological Society* 69 (supplement):98–115.

Yocum, G. D. 2001. Differential expression of two *HSP70* transcripts in response to cold shock, thermoperiod, and adult diapause in the Colorado potato beetle. *Journal of Insect Physiology* 47:1139–1145.

Yocum, G. D. 2003. Isolation and characterization of three diapause-associated transcripts from the Colorado potato beetle, *Leptinotarsa decemlineata*. *Journal of Insect Physiology* 49:161–169.

Yoshikawa, K. 1962. Introductory studies on the life economy of polistine wasps. I. Scope of problems and consideration on the solitary stage. *Bulletin of the Osaka Museum of Natural History* 15:3–27.

Yoshikawa, K., R. Ohgushi, and S. F. Sakagami. 1969. Preliminary report on entomology of the Osaka City University 5th scientific expedition to southeast Asia 1966 – With descriptions of two new genera of stenogasterine [sic] wasps by J. van der Vecht. *Nature and Life in Southeast Asia* 6:153–182.

Young, A. M. 1979. Attacks by the army ant *Eciton burchelli* on nests of the social paper wasp *Polistes erythrocephalus* in northeastern Costa Rica. *Journal of the Kansas Entomological Society* 52:759–768.

Young, A. M. 1986. Natural history notes on the social paper wasp *Polistes erythrocephalus* Latreille (Hymenoptera: Vespidae: Polistinae) in Costa Rica. *Journal of the Kansas Entomological Society* 59:712–722.

Zucchi, R., S. F. Sakagami, F. B. Noll, M. R. Mechi, S. Mateus, M. V. Baio, and S. N. Shima. 1995. *Agelaia vicina*, a swarm-founding polistine with the largest colony size among wasps and bees (Hymenoptera: Vespidae). *Journal of the New York Entomological Society* 103:129–137.

Zucchi, R., Sô. Yamane, and S. F. Sakagami. 1976. Preliminary notes on the habits of *Trimeria howardii*, a Neotropical communal masarid wasp, with description of the mature larva (Hymenoptera: Vespoidea). *Insecta Matsumarana, new series* 8:47–57.

Author Index

Subject Index